Henry Wood Elliott

An Arctic Province

Alaska and the Seal Islands

Henry Wood Elliott

An Arctic Province
Alaska and the Seal Islands

ISBN/EAN: 9783743318229

Manufactured in Europe, USA, Canada, Australia, Japa

Cover: Foto ©berggeist007 / pixelio.de

Manufactured and distributed by brebook publishing software
(www.brebook.com)

Henry Wood Elliott

An Arctic Province

CONTENTS.

CHAPTER I.

CHAPTER II.

CHAPTER III.

CHAPTER IV.

CHAPTER V.

CHAPTER VI.

CHAPTER IX.

CHAPTER X.

CHAPTER XI.

CHAPTER XII.

CHAPTER XIII.

CHAPTER XIV.

LIST OF ILLUSTRATIONS.

FULL-PAGE ILLUSTRATIONS.

ILLUSTRATIONS IN TEXT.

OUR ARCTIC PROVINCE.

CHAPTER I.

DISCOVERY, OCCUPATION, AND TRANSFER.

The Legend of Bering's Voyage.—The Discovery of Russian America, or Alaska, in July, 1741.—The Return Voyage and Shipwreck of the Discoverer.— The Escape of the Survivors.—They Tell of the Furs and Ivory of Alaska. —The Rush of Russian Traders.—Their Hardy Exploration of the Aleutian Chain, Kadiak, and the Mainland, 1760-80, inclusive.—Fierce Competition of the Promyshlineks finally Leads to the Organization and Domination of the Russian American Company over all Alaska, 1799.—Its Remarkable Success under Baranov's Administration, 1800-18, inclusive.—Its Rapid Decadence after Baranov's Removal.—Causes in 1862-64 which Led to the Refusal of the Russian Government to Renew the Charter of the Russian American Company.—Steps which Led to the Negotiations of Seward and Final Acquisition of Alaska by the U. S. Government, 1867.

THE stolid, calm intrepidity of the Russian is not even yet well understood or recognized by Americans. No better presentation of this character of those Slavic discoverers of Alaska can be made than is the one descriptive of Veit Bering's voyage of Russian-American fame, in which shipwreck and death robbed him of the glory of his expedition. No legend of the sea, however fanciful or horrid, surpasses the simple truth of the terror and privation which went hand-in-hand with Bering and his crew.

Flushed with the outspoken favor of his sovereign, Bering and his lieutenant, Tschericov, sailed east from Petropaulovsky, Kamchatka, June 4, 1741; the expedition consisted of two small sail-vessels, the *St. Peter* and the *St. Paul*. They set their course S. S. E., as low as the 50th degree of north latitude, then they decided to steer directly east for the reported American continent. A few days later a violent storm arose, it separated the rude ships, and the two commanders never met in life again.

While groping in fog and tempest on the high seas, Bering drifted one Sunday (July 18th) upon or about the Alaskan mainland coast ; he disembarked at the foot of some low, desolate bluffs that face the sea near the spot now known to us as Kayak Island, and in plain view of those towering peaks of the St. Elias Alps.   He passed full six weeks in this neighborhood, while the crew were busy getting fresh food-supplies, water, etc., when, on the 3d of September, a storm of unwonted vigor burst upon them, lasted seven days, and drove them out to sea and before it, down as far as 48° 8′ north latitude, and into the lonely wastes of the vast Pacific. Scurvy began to appear on board the *St. Peter ;* hardly a day passed without recording the death of some one of the ship's company, and soon men enough in health or strength sufficient to work the vessel could not be mustered.   A return to Kamchatka was resolved upon.

Bering became surly and morose, and seldom appeared on deck, and so the second in command, "Stoorman" Vachtel, directed the dreary cruise.   After regaining the land, and burying a sailor named Shoomagin on one of a group of Alaskan islets that bear his name to-day, and making several additional capes and landfalls, they saw two islands which, by a most unfortunate blunder, they took to be of the Kurile chain, and adjacent to Kamchatka.   Thus they erred sadly in their reckoning, and sailed out upon a false point of departure.

In vain they craned their necks for the land, and strained their feeble eyes ; the shore of Kamchatka refused to rise, and it finally dawned upon them that they were lost—that there was no hope of making a port in that goal so late in the year.   The wonderful discipline of the Russian sailors was strikingly exhibited at this stage of the luckless voyage : in spite of their debilitated and emaciated condition, they still obeyed orders, though suffering frightfully in the cold and wet ; the ravages of scurvy had made such progress that the steersman was conducted to the helm by two other invalids who happened to have the use of their legs, and who supported him under the arms !   When he could no longer steer from suffering, then he was succeeded by another no better able to execute the labor than himself.   Thus did the unhappy crew waste away into death and impotency.   They were obliged to carry few sails, for they were helpless to reef or hoist them, and such as they had were nearly worn out ; and even in this case they were unable to renew them by replacing from the stores, since there were no seamen strong enough on the ship to bend new ones to the yards and booms.

Soon rain was followed by snow, the nights grew longer and darker, and they now lived in dreadful anticipation of shipwreck ; the fresh water diminished, and the labor of working the vessel became too severe for the few who were able to be about. From the 1st to the 4th of November the ship had lain as a log on the ocean, helpless and drifting, at the sport of the wind and the waves. Then again, in desperation, they managed to control her, and set her course anew to the westward, without knowing absolutely anything as to where they were. In a few hours after, the joy of the distressed crew can be better imagined than described, for, looming up on the gray, gloomy horizon, they saw the snow-covered tops of high hills, still distant however, ahead. As they drew nearer, night came upon them, and they judged best, therefore, to keep out at sea " off and on " until daybreak, so as to avoid the risk of wrecking themselves in the deep darkness. When the gray light of early morning dawned, they found that the rigging on the starboard side of the vessel was giving way, and that their craft could not be much longer managed ; that the fresh water was very low, and that sickness was increasing frightfully. The raw humidity of the climate was now succeeded by dry, intense cold ; life was well-nigh insupportable on shipboard then, so, after a brief consultation, they determined to make for the land, save their lives, and, if possible, safely beach the *St. Peter.*

The small sails were alone set ; the wind was north ; thirty-six fathoms of water over a sand bottom ; two hours after they decreased it to twelve ; they now contrived to get over an anchor and run it out at three-quarters of a cable's length ; at six in the evening this' hawser parted ; tremendous waves bore the helpless boat on in toward the land through the darkness and the storm, where soon she struck twice upon a rocky reef. Yet, in a moment after, they had five fathoms of water ; a second anchor was thrown out, and again the tackle parted ; and while, in the energy of wild despair, prostrated by sheets of salty spray that swept over them in bursts of fury, they were preparing a third bower, a huge combing wave lifted that ark of misery—that band of superlative human suffering—safely and sheer over the reef, where in an instant the tempest-tossed ship rested in calm water ; the last anchor was dropped, and thus this luckless voyage of Alaskan discovery came to an end.

Bering died here, on one of the Commander Islands,* where he had been wrecked as above related ; the survivors, forty-five souls in number, lived through the winter on the flesh of sea-lions, the sea-cow,† or manatee, and thus saved their scanty stock of flour ; they managed to build a little shallop out of the remains of the *St. Peter*, in which they left Bering Island—departed from this scene of a most extraordinary shipwreck and deliverance—on August 16, 1742, and soon reached Petropaulovsky in safety the 27th following. In addition to an authentic knowledge of the location of a great land to the eastward, the survivors carried from their camp at Bering Island a large number of valuable sea-otter, blue-fox, and other peltries, which stimulated, as no other induce-ment could have done, the prompt fitting out and venture of many new expeditions for the freshly discovered land and islands of Alaska.

So, in 1745, Michael Novidiskov first, of all white men, pushed over in a rude open wooden shallop from Kamchatka, and landed on Attoo, that extreme western islet of the great Aleutain chain which forms upon the map a remarkable southern wall to the green waters of Bering Sea. No object of geographical search was in this hardy fur-hunter's mind as he perilled his life in that adventure—far from it ; he was after the precious pelage of the

---

* Bering's Island—he was wrecked on the east coast, at a point under steep bluffs now known as "Kommandor." Scarcely a vestige of this shipwreck now remains there.

† That curious creature is extinct. It formerly inhabited the sea-shores of these two small islands. The German naturalist Steller, who was the sur-geon of Bering's ship, has given us the only account we have of this animal's appearance and habits; it was the largest of all the Sirenians ; attained a length sometimes of thirty feet. When first discovered it was extremely abundant, and formed the main source of food-supply for the shipwrecked crew of Be-ring's vessel. Twenty-seven years afterward it became extinct, due to the merciless hunting and slaughter of it by the Russians, who, on their way over to Alaska from Kamchatka, always made it an object to stop at Bering or Cop-per Island and fill up large casks with the flesh of this sea-cow. Its large size, inactive habits, and clumsy progress in the water, together with its utter fear-lessness of man, made its extinction rapid and feasible.

I make the restoration from a careful study of the details of Steller's description.

THE RHYTINA, OR SEA-COW (Extinct)

The flesh of this animal constituted the chief food supply of Bering's shipwrecked crew, 1741-'42

sea-otter, and like unto him were all of the long list of Russian explorers of Alaskan coasts and waters. These rough, indomitable men ventured out from their headquarters at Kamchatka and the Okotsk Sea in rapid succession as years rolled on, until by the end of 1768–69 a large area of Russian America was well determined and rudely charted by them.*

The history of this early exploration of Russian America is the stereotyped story of wrongs inflicted upon simple natives by ruthless, fearless adventurers—year in and year out—the eager, persistent examination of the then unknown shores and interior of Alaska by tireless Cossacks and Muscovites, who were busy in robbing the aborigines and quarrelling among themselves. The success of the earliest fur-hunters had been so great, and heralded so loudly in the Russian possessions, that soon every Siberian merchant who had a few thousand rubles at his order managed to associate himself with some others, so that they might together fit out a slovenly craft or two and engage in the same remunerative business. The records show that, prior to the autocratic control of the old Russian American Company over all Alaska in 1799, more than sixty distinct Russian trading companies were organized and plying their vocation in these waters and landings of Alaska.

They all carried on their operations in essentially the same manner : the owner or owners of the shallop, or sloop, or schooner, as it might be, engaged a crew on shares; the cargo of furs brought back by this vessel was invariably divided into two equal subdivisions—one of these always claimed by the owners who had furnished the means, and the other half divided in such a manner as the navigator, the trader, and the crew could agree upon between themselves: Then, after this division had been made, each participant was to give one-tenth part of his portion, as received above, to the Government at St. Petersburg, which, stimulated by such generous swelling of its treasury, never failed to keep an affectionate eye upon its subjects over here, and encouraged them to the utmost limit of exertion.

---

* The order of this search and voyaging has been faithfully recorded by Ivan Petroff in his admirable compendium of the subject. (See Tenth Census U. S. A., Vol. VIII.) While this narrative may be interesting to a historian, yet I deem it best not to inflict it upon the general reader. Also in "Bancroft's History of Alaska," recently published at San Francisco, it is graphically and laboriously described.

This Imperial impetus undoubtedly was the spur which caused most of that cruel domination of the Russians over a simple people whom they found at first in possession of their new fur-bearing land ; the thrifty traders managed to do their business with an exceedingly small stock of goods, and, where no opposition was offered, these unscrupulous commercial travellers ordered the natives out to hunt and turn over all their booty, not even condescending to pay them, except a few beads or strips of tobacco, "in return for their good behavior and submission to the crown !" Naturally enough, the treacherous Koloshes of Sitka, the dogged Kadiakers, the vivacious Eskimo or Innuits, and even the docile Aleutes, would every now and then arise and slaughter in, their rage and despair a whole trading post or ship's crew of Russians ; but these outbreaks were not of preconcerted plan or strength, and never seriously interrupted the iron rule of Slavonian oppression.

The rapidly increasing number of competitors in the fur trade, however, soon began to create a scarcity of the raw material, and then the jealousies and rivalries of the trading companies began in turn to vent themselves in armed struggles against each other for possession and gain. This order of affairs quickly threw the whole region into a reign of anarchy which threatened to destroy the very existence of the Russians themselves. Facing this deplorable condition, one of the leading promoters of the fur-trading industry in Alaska saw that, unless a bold man was placed at the head of the conduct of his business, it would soon be ruined. This man he picked out at Kargopol, Siberia, and on August 18, 1790, he concluded a contract with Alexander Baranov, who sailed that day from the Okotsk, and who finally established that enduring basis of trade and Russian domination in Alaska which held till our purchase in 1867 of all its vested rights and title.

The wild savage life which the Russians led in these early days of their possession of this new land—their bitter personal antagonisms and their brutal orgies—actually beggar description, and seem well-nigh incredible to the trader or traveller who sojourns in Alaska to-day. It is commonly regarded as a rude order of existence up there among ourselves now ; and when we come to think back, and contrast the stormy past with the calm present, it is difficult to comprehend it ; yet it is not so strange if it be remembered that they were practically beyond all reach of authority, and lived for many consecutive years in absolute non-restraint.

It is easy to trace the several steps and understand the motives which led to our purchase of Alaska. There was no subtle statecraft involved, and no significance implied. The Russian Government simply grew weary of looking after the American territory, which was an element of annually increasing cost to the Imperial treasury, and was a source of anxiety and weakness in all European difficulties. It became apparent to the minds of the governing council at St. Petersburg that Russians could not, or at least, would not settle in Russian America to build up a state or province, or do anything else there which would redound to the national honor and strength. This view they were well grounded in, after the ripe experience of a century's control and ownership.

One period in that history of Russian rule afforded to the authorities much rosy anticipation. This interval was that season in the affairs of the Russian American Company which was known as Baranov's administration, in which time the revenues to the crown were rich, and annually increasing. But Baranov was a practical business man, while every one of his successors, although distinguished men in the naval and army circles of the home government, was not. Comment is unnecessary. The change became marked; the revenues rapidly declined, and the conduct of the operations of the company soon became a matter of loss and not of gain to the stockholders and to the Imperial treasury. The history, however, of the rise and fall of this great Russian trading association is a most interesting one; much more so even than that of its ancient though still surviving, but decrepit rival, the Hudson's Bay Company.

Those murderous factional quarrels of the competing Russian traders throughout Alaska in 1790–98 finally compelled the Emperor Paul to grant, in 1799, much against his will, a charter to a consolidation of the leading companies engaged in American fur-hunting, which was named the Russian American Company. It also embraced the Eastern Siberian and Kamchatkan colonies. That charter gave to this company the exclusive right to all the territory in Alaska, Kamchatka, and the Siberian Okotsk, and Kurile districts, and the privileges conferred by this charter were very great and of the most autocratic nature; but at the same time the company was shrewdly burdened with deftly framed obligations, being compelled to maintain, at its own expense, the new government of the country, a church establishment, a military force, and,

at various points in the territory, ample magazines of provisions and
stores to be used by the Imperial Government for its naval vessels
or land troops whenever ordered. At a time when all such stores
had to be transported on land·trails over the desolate wastes of Siberia from Russia to the Okotsk, this clause in the franchise was
most burdensome, and really fatal to the financial success of the
company.

The finesse of the Russian authorities is strikingly manifested in
that charter, which ostensibly granted to the Russian American Company all these rights of exclusive jurisdiction to a vast domain without selfishly exacting a single tax for the home treasury ; but in
fact it did pay an immense sum annually into the royal coffers in
this way. The entire fur trade in those days was with China, and
all the furs of Alaska were bartered by the Russians with the Mon-
gols for teas, which were sold in Russia and Europe. The records
of the Imperial treasury show that the duties paid into it by this
company upon these teas often exceeded two millions of silver
rubles annually.*

The company was also obliged, by the terms of their charter, to
make experiments in the establishment of agricultural settlements
wherever the soil and climate of Alaska would permit. The natives
of Alaska were freed from all taxes in skins or money, but were

---

* The Russian currency is always expressed in kopecks and in rubles.
Gold coinage there is seldom ever seen, and was never used in Alaska. The
following table explains itself :

1 copper kopeck = 1 silver kopeck.	15 silver kopecks = 1 petecaltin.
2 copper kopecks = 1 grösh.	20 silver kopecks = 1 dvoogreevenik.
3 copper kopecks = 1 alteen.	25 silver kopecks = 1 chetvertak.
5 copper kopecks = 1 peetak.	50 silver kopecks = 1 polteenah.
5 silver kopecks = 1 peetak.	100 silver kopecks = 1 ruble.
10 silver kopecks = 1 greevnah.	

The silver ruble is nearly equal to seventy-five cents in our coin. The
paper ruble fluctuates in Russia from forty to fifty cents, specie value ; in
Alaska it was rated at twenty cents, silver. Much of the "paper" currency
in Alaska during Russian rule was stamped on little squares of walrus hide.

A still smaller coin, called the "*polooshka*," worth ¼ kopeck, has been used
in Russia. It takes its name from a hare-skin, "*ooshka*," or "little ears,"
which, before the use of money by the Slavs, was one of the lowest articles
of exchange, *pol* signifying *half*, and *polooshka*, *half a hare's skin*. From another small coin, the "*deinga*" (equal to ½ kopeck in value), is derived the
Russian word for money, *deinguh* or *deingie*.

obliged to furnish to the company's order certain quotas of sea-otter hunters every season, all men between the ages of eighteen and fifty being liable to this draft, though not more than one-half of any number thus subject could be enlisted and called out at any one time.

The management of this great organization was vested in an administrative council, composed of its stockholders in St. Petersburg, with a head general office at Irkutsk, Siberia—a chief manager, who was to reside in Alaska, and was styled "The Governor," and whose selection was ordered from the officers of the Imperial navy not lower in rank than post-captain. That high official and Alaskan autocrat had an assistant, also a naval officer, and each received pay from the Russian Company, in addition to their regular governmental salaries, which were continued to them by the Crown.

In cases of mutiny or revolt the powers of the governor were absolute. He had also the fullest jurisdiction at all times over offenders and criminals, with the nominal exception of capital crimes. Such culprits were supposed to have a preliminary trial, then were to be forwarded to the nearest court of justice in Siberia. Something usually "happened" to save them the tedious journey, however. The Russian servants of the company—its numerous retinue of post-traders, factors, and traders, and laborers of every class around the posts—were engaged for a certain term of years, duly indentured. When the time expired the company was bound to furnish them free transportation back to their homes, unless the unfortunate individuals were indebted to it; then they could be retained by the employer until the debt was paid. It is needless to state in this connection that an incredibly small number of Russians were ever homeward bound from Alaska during these long years of Muscovitic control and operation. This provision of debtor vs. creditor was one which enabled the creditor company to retain in its service any and all men among the humbler classes whose services were desirable, because the scanty remuneration, the wretched pittance in lieu of wages, allowed them, made it a matter of utter impossibility to keep out of debt to the company's store. Even among the higher officials it is surprising to scan the long list of those who, after serving one period of seven years after another, never seemed to succeed in clearing themselves from the iron grasp of indebtedness to the great corporation which employed them.

As long as the Russian Company maintained a military or naval force in the Alaskan territory, at its own expense, these forces were entirely at the disposal of its governor, who passed most of his time in elegant leisure at Sitka, where the finest which the markets and the vineyards of the world afforded were regularly drawn upon to supply his table. No set of men ever lived in more epicurean comfort and abundance than did those courtly chief magistrates of Alaska who succeeded the plain Baranov in 1818, and who established and maintained the vice-regal comfort of their physical existence uninterruptedly until it was surrendered, with the cession of their calling, in 1867.

The charter of the Russian American Company was first granted for a period of twenty years, dating at the outset from January 1, 1799. It also had the right to hoist its own colors, to employ naval officers to command its vessels, and to subscribe itself, in its proclamation or petition, "Under the highest protection of his Imperial Majesty, the Russian American Company." It began at once to attract much attention in Russia, especially among moneyed men in St. Petersburg and Moscow. Nobles and high officials of the Government eagerly sought shares of its stock, and even the Emperor and members of his family invested in them, the latter making their advances in this direction under the pretext of donating their portions to schools and to charitable institutions. It was the first enterprise of the kind which had ever originated in the Russian Empire, and, favored in this manner by the Crown, it rose rapidly into public confidence. A future of the most glowing prosperity and stability was prophesied for it by its supporters—a prosperity and power as great as ever that of the British East India Company—while many indulged dreams of Japanese annexation and portions of China, together with the whole American coast, including California.

But that clause in the charter of the company, which ordered that the chief manager of its affairs in Alaska should be selected from the officers of the Imperial navy, had a most unfortunate effect upon the successful conduct of the business, as it was prosecuted throughout Russian America. After Baranov's suspension and departure, in the autumn of 1818, not a single practical merchant or business man succeeded him. The rigid personal scrutiny and keen trading instinct which were so characteristic of him, were followed immediately by the very reverse ; hence the dividends began to diminish every year, while the official writing, on the other hand, became suddenly more voluminous, graphic, and declared a steady increase of prosperity. Each succeeding chief manager, or governor, vied with the reports of his predecessors in making a record of great display in the line of continued explorations, erection of buildings, construction of ships of all sizes, and the establishment of divers new industries and manufactories, agriculture, etc.

The second term of the Russian American Company's charter expired in 1841, and the directors and shareholders labored most industriously for another renewal ; the Crown took much time in consideration, but in 1844 the new grant was confirmed, and rather increased the rights and privileges of the company, if anything ; still matters did not mend financially, the affairs of the large corporation were continued in the same reckless management by one governor after the other—with the same extravagant vice-regal display and costly living—with useless and abortive experiments in agriculture, in mining and in shipbuilding, so that by the approach of the lapsing of the third term of twenty years' control, in 1864, the company was deeply in debt, and though desirous of continuing the business, it now endeavored to transfer the cost of maintaining its authority in Alaska to the home Government ; to this the Imperial Cabinet was both unwilling and unable to accede, for Russia had just emerged from a disastrous and expensive war, and was in no state of mind to incur a single extra ruble of indebtedness which she could avoid. In the meantime, pending these domestic difficulties between the Crown and the company, the charter expired ; the Government refused to renew it, and sought, by sending out commissioners to Sitka, for a solution of the vexed problem.

Now, if the reader will mark it, right at this time and at this juncture, arose the opportunity which was quickly used by Seward, as Secretary of State, to the ultimate and speedy acquisition of Russian America by the American Union. Those difficulties which the situation revealed in respect to the affairs of the Russian Company conflicting with the desire of the Imperial Government, made much stir in all interested financial circles. A small number of San Francisco capitalists had been for many years passive stockholders in what was termed by courtesy the American Russian Ice Company—it being nothing more than a name really, inasmuch as very little ever was or has been done in the way of shipping ice to California from Alaska. Nevertheless these gentlemen quickly con-

ceived the idea of taking the charter of the Russian Company themselves, and offered a sum far in excess of what had accrued to the
Imperial treasury at any time during the last forty years' tenure of
the old contract. The negotiations were briskly proceeding, and
were in a fair way to a successful ending, when it informally became known to Secretary Seward, who at once had his interest excited in the subject, and speedily arrived at the conclusion that if it
was worth paying $5,000,000 by a handful of American merchants
for a twenty years' lease of Alaska, it was well worthy the cost of
buying it out and out in behalf of the United States; inasmuch as
leasing it, as the Russians intended to, was a virtual surrender of it
absolutely for the period named. In this spirit the politic Seward
approached the Russian Government, and the final consummation of
Alaska's purchase was easily effected,* May, 1867, and formally
transferred to our flag on the 18th of October following.

If the Russian Government had not been in an exceedingly
friendly state of mind with regard to the American Union, this somewhat abrupt determination on its part to make such a virtual gift
of its vast Alaskan domain would never have been thought of in St.
Petersburg for a moment. Still, it should be well understood from
the Muscovitic view, that in presenting Russian America to us, no
loss to the glory or the power of the Czar's Crown resulted; no surrender of smiling hamlets, towns or cities, no mines or mining, no
fish or fishing, no mills, factories or commerce—nothing but her
good will and title to a few thousand poor and simple natives, and
a large wilderness of mountain, tundra-moor and island-archipelago
wholly untouched, unreclaimed by the hand of civilized man. Russia then, as now, suffered and still suffers, from an embarrassment of
just such natural wealth as that which we so hopefully claim as
our own Alaska.

---

* $7,200,000 gold was paid by the United States into the Imperial treasury
of Russia for the Territory of Alaska; it is said that most of this was used in
St. Petersburg to satisfy old debts and obligations incurred by Alaskan enterprises, attorneys' fees, etc. So, in short, Russia really gave her American possessions to the American people, reaping no direct emolument or profit whatsoever from the transfer.

CHAPTER II.

FEATURES OF THE SITKAN REGION.

The Vast Area of Alaska.—Difficulty of Comparison, and Access to her Shores
save in the Small Area of the Sitkan Region.—Many Americans as Officers
of the Government, Merchants, Traders Miners, etc., who have Visited
Alaska during the last Eighteen Years.—Full Understanding of Alaskan
Life and Resources now on Record.—Beautiful and Extraordinary Features
of the Sitkan Archipelago.—The Decaying Town of Wrangel.—The
Wonderful Glaciers of this Region.—The Tides, Currents and Winds.—
The Forests and Vegetation Omnipresent in this Land-locked Archipelago.
—Indigenous Berries.—Gloomy Grandeur of the Cañons.—The Sitkan
Climate.—Neither Cold nor Warm.—Excessive Humidity.—Stickeen Gold
Excitement of 1862 and 1875.—The Decay of Cassiar.—The Picturesque
Bay of Sitka.—The Romance and Terror of Baranov's Establishment there
in 1800–1805.—The Russian Life and Industries at Sitka.—The Contrast
between Russian Sitka and American Sitka a Striking One.

> "For hot, cold, moist and dry, four champions fierce
> Strive here for mastery."—MILTON.

THE general contour of Alaska is correctly rendered on any and
all charts published to-day ; but it is usually drawn to a very much
reduced scale and tucked away into a corner of a large conventional map of the United States and Territories, so that it fails,
in this manner, to give an adequate idea of its real proportion—
and does not commonly impress the eye and mind, as it ought to,
at first sight. But a moment's thoughtful observation shows the
vast landed extent between that extreme western point of Attoo
Island in the occident, and the boundary near Fort Simpson in the
orient, to be over 2,000 miles ; while from this Alaskan initial post
at Simpson to Point Barrow, in the arctic, it covers the limit of
1,200 geographical miles.* The superficial magnitude of this region

---

* The superficial area of Alaska is 512,000 square miles ; or, in round
numbers, just one sixth of the entire extent of the United States and Territories. Population in 1880: Whites, 430; Creole, 1,756 ; Eskimo, 17,617;
Aleut, 2,145 ; Athabascan, 3,927 ; Thlinket, 6,763—total, 33,426.

is at once well appreciated when the largest States or Territories are each held up in contrast.

The bewildering indentation and endless length of the coast, the thousands of islands and islets, the numerous volcanoes and towering peaks, and the maze of large and small rivers, make a comparison of Alaska, in any other respect than that of mere superficial area, wholly futile when brought into contrast with the rest of the North American continent.   Barred out as she is from close communion with her new relationship and sisterhood in the American Union by her remote situation, and still more so by the unhappiness of her climate, she is not going to be inspected from the platforms of flying express trains ; and, save the little sheltered jaunt by steamer from Puget Sound to Sitka and immediate vicinity, no ocean-tourists are at all likely to pry into the lonely nooks and harbors of her extended coasts, surf-beaten and tempest-swept as they are every month in the year.

But, in the discharge of official duty, in the search for precious metals, coal and copper, in the desire to locate profitable fishing ventures, and in the interests of natural science, hundreds of energetic, quick-witted Americans have been giving Alaska a very keen examination during the last eighteen years.   The sum of their knowledge throws full understanding over the subject of Alaskan life and resources, as viewed and appreciated from the American basis ; there is no difficulty in now making a fair picture of any section, no matter how remote, or of conducting the reader into the very presence of Alaska's unique inhabitants, anywhere they may be sought, and just as they live between Point Barrow and Cape Fox, or Attoo and the Kinik mouth.

In going to Alaska to-day, the traveller is invariably taken into the Sitkan district, and no farther ; naturally he goes there and nowhere beyond, for the best of all reasons : he can find no means of transportation at all proper as regards his safety and comfort which will convey him outside of the Alexander archipelago.   To this southeastern region of Alaska, however, one may journey every month in the year from the waters of the Columbia River and Puget Sound, in positive pleasure, on a seaworthy steamer fitted with every marine adjunct conducive to the passenger's comfortable existence in transit ; it is a landlocked sea-trip of over eighteen hundred miles, made often to and from Sitka without tremor enough on the part of the vessel even to spill a brimming glass of

water upon the cabin table. If fortunate enough to make this trip
of eight or nine hundred miles up, and then down again, when the
fog is not omnipotent and rain not incessant, the tourist will record
a vision of earthly scenery grander than the most vivid imagination
can devise, and the recollection of its glories will never fade from
his delighted mind.

If, however, you desire to visit that great country to the westward and the northwest, no approach can be made via Sitka—no
communication between that region and this portion of Alaska ever
takes place, except accidentally ; the traveller starts from San Francisco either in a codfishing schooner, a fur-trader's sloop, or steamer,
and sails out into the vast Pacific on a bee-line for Kadiak or
Oonalashka ; and, from these two chief ports of arrival and departure, he laboriously works his way, if bent upon seeing the country,
constantly interrupted and continually beset with all manner of
hindrances to the progress of his journey by land and sea. These
physical obstructions in the path of travel to all points of interest ·
in Alaska, save those embraced in the Sitkan district, will bar out
and deprive thousands from ever beholding the striking natural
characteristics of a wonderful volcanic region in Cook's Inlet and
the Aleutian chain of islands. When that time shall arrive in the
dim future which will order and sustain the sailing of steamers in
regular rotation of transit throughout the waters of this most interesting section, then, indeed, will a source of infinite satisfaction
be afforded to those who love to contemplate the weird and the
sublime in nature ; meanwhile, visits to that region in small sailing-
craft are highly risky and unpleasant—boisterous winds are chronic
and howling gales are frequent.

The beautiful and extraordinary features of preliminary travel
up the British Columbia coast will have prepared the mind for a
full enjoyment and comprehension of your first sight of Alaska.
If you are alert, you will be on deck and on good terms with the
officer in charge when the line is crossed on Dixon Sound, and the
low wooded crowns of Zayas and Dundas Islands, now close at hand,
are speedily left in the wake as the last landmarks of foreign soil.
To the left, as the steamer enters the beautiful water of Clarence
Straits, the abrupt, irregular, densely wooded shores of Prince of
Wales Island rise as lofty walls of timber and of rock, mossy and
sphagnous, shutting out completely a hasty glimpse of the great
Pacific rollers afforded in the Sound ; while on the right hand you

turn to a delighted contemplation of those snowy crests of the
towering coast range which, though thirty and fifty miles distant,
seem to fairly be in reach, just over and back of the rugged tree-
clad elevations of mountainous islands that rise abruptly from
the sea-canal in every direction. Not a gentle slope to the water
can be seen on either side of the vessel as you glide rapidly ahead ;
the passage is often so narrow that the wavelets from the steamer's

Lodges in a Vast Wilderness.

wheel break and echo back loudly on your ear from the various
strips of ringing rocky shingle at the base of bluffy intersections.

If, by happy decree of fate, fog-banks do not shut suddenly
down upon your pleased vision, a rapid succession of islands and
myriads of islets, all springing out boldly from the cold blue-green
and whitish-gray waters which encircle their bases, will soon tend
to confuse and utterly destroy all sense of locality ; the steamer's
path seems to be in a circle, to lead right back to where she started
from, into another equally mysterious labyrinthine opening : then

the curious idiosyncrasy possesses you by which you seem to see in the scenery just ahead an exact resemblance to the bluffs, the summits and the cascades which you have just left behind. Your emphatic expression aloud of this belief will, most likely, arouse some fellow-passenger who is an old voyageur, and he will take a guiding oar : he will tell you that the numerous broad smooth tracks, cut through the densely wooded mountain slopes from the snow lines above abruptly down to the very sea below, are the paths of avalanches ; that if you will only crane your neck enough so as to look right aloft to a certain precipice now almost hanging 3,000 feet high and over the deck of the steamer, there you will see a few small white specks feebly outlined against the grayish-red background of the rocks—these are mountain goats ; he tells you that those stolid human beings who are squatting in a large dug-out canoe are " Siwashes," halibut-fishing—and as these savages stupidly stare at the big " Boston" vessel swiftly passing, with uplifted paddles or keeping slight headway, you return their gaze with interest, and the next turn of the ship's rudder most likely throws into full view a " rancherie," in which these Indians permanently reside ; your kindly guide then eloquently describes the village and descants with much vehemence upon the frailties and shortcomings of " Siwashes " in general—at least all old-stagers in this country agree in despising the aboriginal man. On the steamer forges through the still, unruffled waters of intricate passages, now almost scraping her yard-arms on the face of a precipitous headland—then rapidly shooting out into the heart of a lovely bay, broad and deep enough to float in room and safety a naval flotilla of the first class, until a long, unusually low, timbered point seems to run out ahead directly in the track, when your guide, giving a quick look of recognition, declares that Wrangel* town lies just

---

* When the Cassiar mines in British Columbia were prosperous, Wrangel was a very busy little transfer-station—the busiest spot in Alaska ; then between four and five thousand miners passed through every spring and fall as they went up to and came down from the diggings on the Stickeen tributaries above : they left a goodly share, if not most, of their earnings among the store and saloon keepers of Wrangel. The fort is now deserted—the town nearly so ; the whole place is rapidly reverting to the Siwashes. Government buildings erected here by the U. S. military authorities, which cost the public treasury $150,000, were sold in 1877, when the troops were withdrawn, for a few hundreds. The main street is choked with decaying logs and stumps. A recent visitor declares, upon looking at the condition of this place

2

around it, and you speedily make your inspection of an Alaskan hamlet.

Owing to the dense forest-covering of the country, sections of those clays and sands which rest in most of the hollows are seldom seen, only here and there where the banks of a brook are cut out, or where an avalanche has stripped a clear track through the jungle, do you get a chance to see the soil in southeastern Alaska. There are frequent low points to the islands, composed, where beaten upon by the sea, of fine rocky shingle, which form a flat of greater or less width under the bluffs or steep mountain or hill slopes, about three to six feet above present high-water mark; they become, in most cases, covered with a certain amount of good soil, upon which a rank growth of grass and shrubbery exists, and upon which the Indians love to build their houses, camp out, etc. These small flats, so welcome and so rare in this pelagic wilderness, have evidently been produced by the waves acting at different times in opposing directions.

In all of those channels penetrating the mainland and intervening between the numerous islands from the head of Glacier Bay and Lynn Canal down to the north end of Vancouver's Island, marks, or glacial scratchings, indicative of the sliding of a great ice-sheet, are to be found, generally in strict conformity with the trend of the passages, wherever the rocks were well suited for their preservation; and it is probable that the ice of the coast range, at one time, reached out as far west as the outer islands which fringe the entire Alaskan and British Columbian coast. Many of the boulders on the beaches are plainly glaciated; and, as they are often bunched in piles upon the places where found, they seem to have not been disturbed since they were dropped there. The shores are

---

in the summer of 1883: "Fort Wrangel is a fit introduction to Alaska. It is most weird and wild of aspect. It is the key-note to the sublime and lonely scenery of the north. It is situated at the foot of conical hills, at the head of a gloomy harbor, filled with gloomy islands. Frowning cliffs, beetling crags, stretch away on all sides surrounding it. Lofty promontories guard it, backed by range after range of sharp, volcanic peaks, which in turn are lost against lines of snowy mountains. It is the home of storms. You see that in the broken pines on the cliff-sides, in the fine wave-swept rocks, in the lowering mountains. There is not a bright touch in it—not in its straggling lines of native huts, each with a demon-like totem beside it, nor in the fort, for that is dilapidated and fast sinking into decay."

GRAND GLACIER: ICY BAY

everywhere abrupt and the water deep. The entire front of this
lofty coast-range chain, that forms the eastern Alaskan boundary
from the summit of Mt. St. Elias to the mouth of Portland Canal,
is glacier-bearing to-day, and you can scarcely push your way to the
head of any cañon, great or small, without finding an eternal ice-
sheet anchored there : careful estimation places the astonishing aggregate of over 5,000 living glaciers, of greater or less degree, that
are silently but forever travelling down to the sea, in this region.

Those congealed rivers which take their origin in the flanks of
Mt. Fairweather * and Mt. Crillon† are simply unrivalled in frigid
grandeur by anything that is lauded in Switzerland or the Hima-
layas, though the vast bulk of the Greenland ice-sheets is, of course,
not even feebly approximated by them ; the waters of the channels
which lead up from the ocean to the feet of these large glaciers of
Cross Sound and Lynn Canal, are full of bobbing icebergs that
have been detached from the main sheet, in every possible shape
and size—a detachment which is taking place at intervals of every
few moments, giving rise, in so doing, to a noise like parks of artillery ; but, of course, these bergs are very, very small compared
with those of Greenland, and only a few ever escape from the intricate labyrinth of fiörds which are so characteristic of this Sitkan
district. An ice-sheet comes down the cañon, and as it slides into
the water of the canal or bay, wherever it may be, the pressure exerted by the buoyancy of the partially submerged mass causes it to
crack off in the wildest lines of cleavage, and rise to the surface in
hundreds and thousands of glittering fragments ; or again, it may
slide out over the water on a rocky bed, and, as it advances, break
off and fall down in thundering salvos, that ring and echo in the
gloomy cañons with awe-inspiring repetition. At the head and
around the sides of a large indentation of Cross Sound there are
no less than five immense, complete glaciers, which take their origin
between Fairweather and Crillon Mountains, each one reaching and
discharging into tide-water : here is a vast, a colossal glacier in full
exhibition, and so easy of access that the most delicate woman
could travel to, and view it, since an ocean-steamer can push to its
very sea-walls, without a moment's serious interruption, where from
her decks may be scanned the singular spectacle of an icy river from
three to eight miles wide, fifty miles long, and varying in depth

---

* 14,708 feet.   † 13,400 feet.

from fifty to five hundred feet. Between the west side of this frozen bay and the water, all the ground, high and low, is covered by a mantle of ice from one thousand to three thousand feet thick !

Here is an absolute realism of what once took place over the entire northern continent—a vivid picture of the actual process of degradation which the earth and its life were subjected to during that long glacial epoch which bound up in its iron embrace of death just about half of the globe.* This startling exhibition of a mighty glacier with its cold, multitudinous surroundings in Cross Sound, is alone well worth the time and cost of the voyage to behold it, and it alone. There is not room in this narrative for further dwelling upon that fascinating topic, for a full description of such a gelid outpouring would in itself constitute a volume.

Throughout this archipelago of the Sitkan district, the strongest tidal currents prevail : they flow at places like mill-races, and again they scarcely interfere with the ship or canoe. The flood-tides usually run northward along the outer coasts, and eastward in Dixon's Entrance ; the weather, which is generally boisterous on the ocean side of the islands, and on which the swell of the Pacific never ceases to break with great fury, is very much subdued inside, and the best indication of these tidal currents is afforded by the streaming fronds of kelp that grow abundantly in all of these multitudinous fiörds, and which are anchored securely in all depths, from a few feet to that of seventy fathoms : when the tide is running through some of those narrow passages, especially at ebb, it forms, with the whip-like stems of seaweed, a true rapid with much white water, boiling and seething in its wild rushing ; these alternations between high and low water here are exceedingly variable—the spring-tides at some places are as great as eighteen feet of rise, and a few miles beyond, where the coast-expansion is great, it will not be more than three or four feet.

Those baffling tides and the currents they create, together with gusty squalls of rain or sleet, and irregular winds, render the navigation of this inside passage wholly impracticable for sailing-vessels—they gladly seek the open ocean where they can haul and fill away to advantage even if it does blow "great guns ;" the high mural walls of the Alexander fiörds on both sides, usually, of the

---

* I am aware that geologists do not all subscribe to this view, which was the doctrine of Agassiz.

channels, cause the wind to either blow up them, or down : it literally funnels through with terrific velocity when the "southeasters" prevail, and nothing, not even the steamer, braves the fury of such a storm. —

The great growth of trees everywhere here, and the practical impenetrability of these forests on foot, owing to brush and bushes, all green and growing in tangled jungle, is caused by the comparative immunity of this country from the scourge of forest fires : this is due to a phenomenal dampness of the climate—it rains, rains, and drizzles here two-thirds of the time. The heaviest rains are local, usually occurring on the western or ocean slopes of the islands where the sea-winds, surcharged with moisture, first meet a barrier to their flow and are thrown up into the cooler regions of the atmosphere. It will be often noticed, from the steamer, that while heavy rain is falling on the lofty hills and mountains of Prince of Wales Island, it is clear and bright directly over the Strait of Clarence to the eastward, and not far distant. June and July are the most agreeable seasons of the year in which to visit the Sitkan district, as a rule.

Many thoughtful observers have questioned the truth of the exuberant growth of forestry peculiar to this region, as being due to that incessant rainfall mentioned above ; no doubt, it is not wholly so ; but yet, if the ravages of fire ran through the islets of the archipelago, as it does in the interior slightly to the eastward, the same order of vegetation here would be soon noted as we note it there to-day ; everywhere that you ascend the inlets of the mainland, the shores become steep and rocky, with no beach, or very little ; the trees become scrubby in appearance, and are mingled with much dead wood (brulé). Scarcely any soil clothes the slopes, and extensive patches of bare rock crop out frequently everywhere.

Although the forest is omnipresent up to snow-line in this great land-locked Sitkan district, yet it differs much in rankness of growth and consequent value; it nowhere clothes the ridges or the summits, which are 1,500 to 2,000 feet above tide-level; these peaks and rocky elevations are usually bare, and show a characteristically green-gray tint due to the sphagnous mosses and dwarfed brier and bushes peculiar to this altitude, making an agreeable and sharp contrast to that sombre and monotonous line of the conifers below. The variety is limited, being substantially confined to three evergreens, the spruce (*Abies sitkensis* and *menziesii*), the hemlock (*Abies merten-siana*), and the cedar (*Thuja gigantea*). The last is the most valu-able, is found usually growing near the shores, and never in great quantities at any one place; wherever a sheltered flat place is found, there these trees seem to grow in the greatest luxuriance. In the narrower passages, where no seas can enter, the forest seems almost to root in the beach, and its branches hang pendent to the tides, and dip therein at high water. Where a narrow beach, capped with warm sands and soil, occurs in sheltered nooks, vividly green grass spreads down until it reaches the yellow seaweed " tangle " that grows everywhere in such places reached by high tide, for, owing to the dampness of the climate, a few days exposure at neap-tides fails to injure this fucoid growth. Ferns, oh! how beautiful they are!—also grow most luxuriantly and even abundantly upon the fallen, rotting tree-trunks, and even into the living arboreal boughs, and green mosses form great club-like masses on the branches.

Large trunks of this timber, overthrown and dead, become here at once perfect gardens of young trees, moss, and bushes, even though lying high above the ground and supported on piles of yet earlier windfall. Similar features characterize the littoral forests of the entire landlocked region of the northwest coast, from Puget Sound to the mouth of Lynn Canal.

In addition to these overwhelmingly dominant conifers already specified, a few cottonwoods and swamp-maples and alders are scat-tered in the jungle which borders the many little streams and the large rivers like the Stickeen, Tahko, and Chilkat. Crab-apples (*Pyrus rivularis*) form small groves on Prince of Wales Island, where the beach is low and capped with good soil. Then on the exposed, almost bare rocks of the western hilltops of the islands of the archi-pelago, a scrub pine (*Pinus contorta*) is found; it also grows in small clumps here and there just below the snow-line on the moun-tains generally. Berries abound; the most important being the sal-lal (*Gaultheria shallow*)—they are eaten fresh in great quantities, and are also dried for use in winter—and another small raspberry (*Rubus sp.*), a currant (*Ribes sp.*), and a large juicy whortleberry. Of course these berries do not have the flavor or body which we prize at home in our small fruits of similar character—but up here they, in the absence of anything better or as good, are eaten with avidity and relish, even by the white travellers who happen to be

around when the fruit is ripe ; wild strawberries appear in sheltered nooks ; a wild gooseberry too is found, but it, like the crab-apple of Prince of Wales Island, is not a favorite—it is drastic.

We find in many places throughout this district highland moors, which constitute the level plateau-summits of ridges and mountain foothills ; these areas are always sparsely timbered, covered by a thick carpet of sphagnous heather, and literally brilliant in June and July with the spangled radiance of an extensive variety of flowering annuals and biennials. In these moorland mantles, which are usually soaked full of moisture so as to be fairly spongy under foot, cranberries flourish, of excellent flavor, and quite abundant, though, compared with our choice Jersey and Cape Cod samples, they are very small.

Certainly the scenery of this Venetian wilderness of Lower Alaska is wonderful and unrivalled—the sounds, the gulfs, bays, fiörds, and river-estuaries are magnificent sheets of water, and the snow-capped peaks, which spring abruptly from their mirrored depths, give the scene an ever-changing aspect. At places the ship seems to really be at sea, then she enters a canal whose lofty walls of sye-nite, slate, and granite shut out the light of day, and against which her rigging scrapes, and the passenger's hand may almost touch—a hundred thousand sparkling streams fall in feathery cascades, adown their mural heights, and impetuous streams beat themselves into white foam as they leap either into the eternal depths of the Pacific or its deep arms.

Probably no one point in the Sitkan archipelago is invested by nature with a grander, gloomier aspect than is that region known as the eastern shore of Prince Frederick's Sound, where the moun-tains of the mainland drop down abruptly to the seaside ; here a spur of the coast range, opposite Mitgon Islet, presents an unusu-ally dreadful appearance, for it rises to a vast height with an inclin-ation toward and over the water : the serrated, jagged summits are loaded with an immense quantity of ice and snow, which, together with the overhanging masses of rock, seems to cause its sea-laved base to fairly totter under that stupendous weight overhead ; the passage beneath it, in the canoe of a traveller, is simply awful in its dread suggestion, and few can refrain from involuntary shuddering as they sail by and gaze upward.

A word about the Sitkan climate : you are not going to be very cold here even in the most severe of winters, nor will you complain of heat in the most favorable of summers; it may be best epitomized by saying in brief that the weather is such that you seldom ever find a clean cake of ice frozen in the small fresh-water ponds six inches thick; and you never will experience a summer warm enough to ripen a head of oats. The first impression usually made upon the visitor is that it is raining, raining all the time, not a pouring rain or shower, then clearing up quickly, but a steady "driz–driz–drizzle"; it rained upon the author in this manner seventeen consecutive days in October, 1866, accompanied by winds from all points of the compass. Therefore, by contrast, the relatively clear and dry months of June and July in the archipelago are really delightful— clear and pleasant in the sun, and cool enough for fires indoors—then you have about eighteen hours of sunshine and six hours of twilight.

It is very seldom that the zero-point is ever recorded at tide-level during winter here, though in January, 1874, it fell to $-7°$ Fah.; the thermometer at no time in the winter preceding registered lower than $11°$ above. A late blustering spring and an early, vigorous winter often join hands over a very backward summer—about once or twice every five years; these are the backward seasons; then the first frost in the villages and tidal bottoms occurs about the 28th to 31st of October, soon followed by the rain turning to snow, being as much as three feet deep on the level at times. Severe thunderstorms, with lightning, often take place during these violent snowfalls in the winter—strange to say they are not heard or seen in the summer! Snow and rain and sleet continue till the end of April— sometimes as late as the 10th of May, before giving way to the enjoyable season of June and July. Then again the mild winters are marked by no frost to speak of—perhaps the coldest period will have been in November, little or no snow, six or seven inches at the most, and much clear and bracing weather.

The average rainfall in the Sitkan district is between eighty-four and eighty-six inches annually—it is a very steady average, and makes no heavier showing than that presented by the record kept on the coast of Oregon and Vancouver's Island. A pleasant season in the archipelago will give the observer about one hundred fair days; the rest of the year will be given over to rain, snow, and foggy-shrouds, which wet like rain itself.* A most careful search during

---

* The chief signal officer of the U. S. Army has had a number of meteoro-logical observers stationed at half a dozen different posts in Alaska, and has

the last hundred years has failed to disclose in all the extent of this Sitkan region an arable or bottom-land piece large enough to represent a hundred-acre farm, save in the valley of the Tahkoo River, where for forty or fifty miles a low, level plateau extends, varying in width from a few rods to half a mile, between the steep mountain walls that compass it about. Red-top and wild timothy grasses grow here in the most luxuriant style, as they do for that matter everywhere else in the archipelago on little patches of open land along the streams and sea-beaches ; the humidity of the climate makes the cost of curing hay, however, very great, and prevents the profitable ranging of cattle.

We have strayed from the landing which we made at Wrangel, and, returning to the contemplation of that town, candor compels an exclamation of disappointment—it is not inviting, for we see nothing but a straggling group of hastily erected shanties and frame store-houses, which face a rickety wharf and a dirty trackway just above the beach-level ; a dense forest and tangled jungle spring up like a forbidding wall at the very rear of the houses, which are supplemented by a number of Indian rancheries that skirt the beach just beyond, and hug the point ; this place, however, though now in sad decline, was a place of much life and importance during 1875–79, when the Cassiar gold-excitement in British Columbia, via the Stickeen River, drew many hundreds of venturesome miners up here, and through Wrangel en route. This forlorn spot was still earlier a centre of even greater stir and activity, for, in 1831, the Russians, fearing that they would be forced into war with the Hudson's Bay people, made a quick movement, came down here from Sitka, and built a bastioned log fortress right where the present Siwash rancheries stand. Lieutenant Zarenbo, who engineered the construction, called his work " Rédoute Saint Dionys," and had scarcely got under cover when he was attacked by several large bateaux, manned

---

had this service fully organized up there during the last ten years ; the inquirer can easily gain access to a large amount of published data touching this subject.

The mean temperature of the year will run throughout the months in the Sitkan region about as follows—an average, for the time, of 44° 7' Fah.

January,	29° 2'	May,	45° 5'	September,	51° 9'
February,	36° 4'	June,	55° 3'	October,	49° 2'
March,	37° 8'	July,	55° 6'	November,	36° 6'
April,	44° 7'	August,	56° 4'	December,	30° 2'

by employés of the great English company ; he fired upon them,
beat them off, and held his own so well that the grateful Baron Von
Wrangel, who then was governor-in-chief, bestowed the name of
the plucky officer upon the large, rugged island which overshadows
the scene of the conflict, and which it bears upon every chart to-
day.*

Again, in 1862, the solitude of Wrangel was broken by the
sudden eruption of over two thousand British Columbia and Cali-
fornian miners, who rushed up the Stickeen River on a gold
"excitement." Quite a fleet of sail and steam-vessels hung about
the place for a brief season, when the flurry died out, and the restless gold-hunters fled in search of other diggings, taking all their
belongings with them.

The steamer does not tarry long at Wrangel ; a few packages
fall upon the shaky wharf, the captain never leaves the bridge, and
in obedience to his tinkling bell, the screw scarce has paused ere it
starts anew, and the vessel soon heads right about and west, out to
the open swell of the great Pacific ; but it takes six or seven hours
of swift travel over the glassy surface of Clarence Strait to pass the
rough heads of Kuprianov Island on the right, flanked by the
sombre, densely wooded elevations of Prince of Wales on the left.
The lower, yet sharper spurs of the straggling Kou forests force
our course here directly to the south. It is said that more than
fifteen hundred islands, big and little, stud this archipelago from
Cape Disappointment to Cross Sound. You will not attempt to
count them, but readily prefer to believe it is so. From the great
bulk of Vancouver's Land to the tiny islet just peeping above
water, they are all covered to the snow-line from the sea-level with
an olive-green coniferous forest—islands right ahead, islands on
every side, islands all behind. You stand on deck and wonder
where the egress from the unruffled inland lake is to be as you
enter it ; no possible chance to go ahead much faster, is your constant thought, which keeps following every sharp turn of the vessel
as she rapidly swings right about here, there, and everywhere, in
following the devious path of this weird course to her destination.

Unless the fog shuts down very thick, the darkness of night
does not impede the steamer's steady progress, for the pilot sees

---

* Zarenbo Island—it blocks the northern end of Clarence Strait, and affords many varied vistas of rare scenic beauty.

the mountain tops loom up darker against the blue-black sky, and with unerring certainty he guides the helm. When the ship is running through tide-rapids in the night, the boiling phosphorescence of the foaming waters, as they rush noisily under our keel, gives a fresh zest to the novelty of the cruise, and the pilot's cries of command ring out in hoarse echoes over the surging tumult below; meanwhile, the passengers anxiously and nervously watch the unquiet turns of the trembling vessel—then suddenly the helm is put up, and the steamer fairly bounds out of still water and the leeward of Coronation Island, into the rhythmic roll of the vast billows of the Pacific, which toss her in strange contrast to the even keel that has characterized our long, land-locked sea-voyage up to this moment. The wrinkled, rugged nose of Cape Ommaney looms right ahead in the north, and soon we are well abreast of the mountainous front to the west coast of Baranov Island, running swiftly into Sitka Sound.*

Cape Ommaney is a very remarkable promontory; it is a steep, bluffy cliff, with a round, high rocky islet, lying close by and under it. The eastern shore of that cape takes a very sharp northerly direction, and thus makes this southern extremity of Baranov Island an exceedingly narrow point of land. An unlucky sailor, Isaac Wooden, fell overboard from Vancouver's ship the *Discovery*, when abreast of it and homeward bound, Sunday, August 24, 1794, and —was drowned, after having safely passed through all the perils of that most remarkable voyage, extended as it was over a period of four consecutive years' absence from home. The rock bears the odd name "Wooden" in consequence.

The location of New Archangel, or Sitka village, is now conceded to be the one of the greatest natural beauty and scenic effect that can be found in all Alaska. The story of its occupation by the Russians is a recitation of violent deeds and unflinching courage on the part of the iron-willed Baranov and his obedient servants: he led the way down here from Kadiak first, of all white men, in 1799, after hearing the preliminary report of exploration made two years previously by his lieutenant, Captain James Shields, an English adventurer and shipbuilder, who entered the service of the Russian Company in the Okotsk. Baranov, though small in stature,

---

* Sitka port is on the west coast of Baranov Island; north latitude 57° 02' 52"; west longitude 135° 17' 45".

was possessed of unusual physical endurance and muscular strength. He was absolutely fearless ; he never allowed any obstacle, no matter how serious, which the elements or savage men were perpetually raising, to check his advances.   He loved to travel and explore, and possessed rare executive or governing power over his rude and boisterous followers.   He soon realized that the establishment of the headquarters of the company at St. Paul, Kadiak Island, was disadvantageous, and quickly resolved to settle himself permanently in the Bay of Sitka, or Norfolk Sound,· where he could communicate with the vessels of other nations and purchase supplies of them.   Late in the autumn of 1799 he sailed to this port in the brig *Catherine*, accompanied by a large fleet of Aleutian and Kadiak sea-otter hunters with their bidarkas, or skin-canoes.   So abundant were sea-otters then, now so rare, that, with the assistance of these native hunters, he secured over fifteen hundred prime otter-skins in less than a month ; then satisfied with the trading resources of the locality, Baranov began the construction of a stockaded post, the site selected for which was on the main island, about six miles to the northward of the Sitkan town-site of to-day.   During the winter of 1799–1800 he and his whole force were busily engaged in the erection of substantial log houses and the surrounding stockade at this location.   In the spring, two American fur-trading vessels made their appearance here, and the owners began to carry on a brisk traffic with the native Sitkans, right under the eyes of Baranov.   Knowing that this must be stopped, the energetic Russian hastened back to Kadiak and set the machinery in motion to that end.   But his absence in the meantime from Sitka was improved upon by the Koloshians, who, acting in preconcerted plan, utterly destroyed the post.   These savages on a certain day, when most of the garrison was far outside of the stockade, hunting and fishing, rushed in, several thousand of them, upon a few armed men, surrounded the block-house, assailed it from all sides at once, and soon forced an entrance.   They massacred the defenders to a man, including the commander, Medvaidniekov, and carried off more than three thousand sea-otter pelts from the warehouses.

During this wild and bloody fight an English ship was lying at anchor far down the harbor, some ten miles from the scene ; three Russians and five Aleutes only, out of the hunting parties absent at the time of the attack, managed to secrete themselves in the woods, and hide until they could gain the decks and protection of this vessel, and thus acquaint her captain, Barber, of the outrage ; he contrived to entice two of the leading Sitkan chiefs on board of his ship, plied them with drink, and soon had them securely ironed, and then, having quite a battery of guns, he was able to make his own terms for their release; this was done after the surrender of eighteen women (captured outside of the stockade) and 2,000 sea-otter skins was made to Barber, who at once sailed for Kadiak. Here the British seaman demanded from Baranov the salvage of 50,000 rubles for rescuing his men and women and property ; with this demand the Russian could not or would not comply; but, after many days in amicable argument, Captain Barber received and accepted 10,000 rubles in full settlement.

While the lurid light of the burning wreck of this first Sitkan post was flashing over the sound, and the Koloshes were howling and dancing around it in their fiendish exultation, nearly two hundred Aleutian hunters were surprised and slaughtered at various points in the vicinity, and a party of over one hundred of these simple natives perished almost to a man, on the same day, from eating poisonous mussels which they detached from the rocks in the strait that separates Baranov Island from Chichagov ; that canal still bears the name commemorative of this dreadful accident—it is called "Pogeebshie"* or "Destruction" Strait.

The enraged Russian manager was unable, by reason of a complicated flood of troubles with his subordinates elsewhere, to revisit Sitka until the spring of 1804 ; he then came down from Kadiak in a squadron consisting of three small sloops, in all considerably less than 100 tons burden ; these craft he had built and fitted out in Prince William Sound and Yakootat Bay during the preceding winter. He had with him forty Russians and three hundred Aleutian sea-otter hunters. With this small force the indomitable man resolved to attack and subjugate a body of not less than five or six thousand fierce, untamed savages, who were flushed with their cruel successes, and eager to shed more blood. He was unexpectedly strengthened by the sudden appearance in the bay of the *Neva*, 400 tons, which had sailed from London to Kadiak, and arrived just after Baranov's departure, but Captain Lissiansky, learning of the object of his trip, determined to assist in rebuilding the

---

* Not "Peril," as it is translated by American geographers and printed on all of our Alaskan maps.

Sitkan post and to punish the Indians, so he sailed at once for the place.

Baranov found the Sitkans all entrenched behind a huge stockade that was thrown up on the same lofty rocky site of the governor's castle in the town to-day.   They reviled him and defied him, taunted him with his misfortunes, and easily succeeded in exciting him to a ferocious attack, in which, despite his demoniacal bravery, he was beaten off at first with the loss of eleven white sailors and hunters, he himself badly wounded, together with Lieutenants Arbuzov and Povalishin.   The darkness of a violent rain

The Castle of Baranov: 1809–1827.
[*Wholly remodelled and rebuilt by his successors.*]

and sleet storm, with night close at hand, caused a cessation, for the time, of further hostilities, but in the morning the ship and the little sloops approached the beach and opened upon the startled savages a hot bombardment—the splintering of their log bastions and the terrible, unwonted noise accompanying, was too much for their self-control, and though, during the whole day they refused to fly, yet when night again came round they abandoned their fortification, and retreated silently and quickly in canoes to Chatham Strait.

The Russians then took possession of the present town-site of Sitka.   The rocky eminence which the savages had so bravely held was cleared of their rude barricades, and the foundations were laid then to the castle that still stands so conspicuous. Around this nucleus the Russian settlement soon sprang up in a few months, a high stockade was then erected between the village and the Indian rancheries, which still stands in part to-day; it was bastioned and fortified with an armament of three-pounder brass guns. From this time on the supremacy of the Russians was never questioned by the Indians of the Sitkan archipelago. The reckless daring of Baranov, evinced by his personal bearing at the head of a handful of men in repeated attacks upon the castle-rock encampment was exaggerated by the savages in repetition among themselves, until his name to them became synonymous with a charmed life and supreme authority. Baranov himself called this spot the final headquarters of the Russian American Company, and henceforth it became so, and it was officially known as New Archangel ; but the tribal name of the savages who lived just outside the stockade fence was "Seet-kah," and soon the present designation was used by all visitors and Russians alike, brevity and euphony making it "Sitka."

It is not probable that the beautiful vistas of this sound influenced Baranov in the slightest when he selected it for his base of operations ; but there must have been mornings and evenings when this hardy man looked at them with some responsive pleasure, for certainly the human being who could remain insensible to their scenic glories must be one without a drop of warm blood in his veins. Those high-peaked summits of the Baranov Mountains, which overshadow the town on the east, destroy, in a great measure, the effects of sunrise ; but the transcendent glow of sundown colors is the glow that floods the crown and base of Mt. Edgecumb on the western horizon of the bay, and repeats its radiance in tipping with golden gild the host of tiny islets which stud the flashing waters, to burn in lingering brightness on the peaks of Verstova and her sister hills, when all else is in darkness or its shades around about.

The most characteristic and expressive single view of Sitka is that one afforded from Japan Island, which is close by and right opposite the town : the place was in its greatest architectural grandeur prior to the departure of the Russians, in 1866. The lofty peak which rises abruptly back of the village is Verstova, to the bald summit of which a champagne picnic by the Russians was religiously made every summer. Although the mountain is slightly under three thousand feet in altitude and seemingly right at hand,

yet the journey to its crest is one that taxes the best physical energies of strong men. The forest is so dense, so damp, the underbrush so thick and so tangled, that the walk requires a supreme
bodily effort, if the trip be made up there and back in the same day.

This view from "Yahponskie" gives an exceedingly good idea of
the ultra-mountainous character of Baranov Island, much better
than any power of verbal description can. It also illustrates the
futility of land travel in the Sitkan archipelago, and affords ample
reason for the utter absence of all roads, even footpaths, in that entire region; it also preserves the somewhat imposing front which
the extensive warehouses and official quarters of the Russian American Company presented in 1866, before their transfer to us, and
the ravages of fire and that decay which has since well-nigh destroyed them; it recalls the shipyards and the brass and iron foun-
dries and machine-shops that have not even a vestige of their existence on that ground to-day, and it outlines a larger Indian village
than the one we find there now.

For the objects of self-protection and comfort the Russians
built large apartment-houses or flats, and lived in them at Sitka.
Several of these dwellings were 150 feet in length by 50 to 80
feet in depth, three stories high, with huge roof-attics. They
were constructed of big spruce logs, smoothly trimmed down
to $12' \times 12'$ timbers. These were snugly dovetailed at the corners,
and the expansive roof covered with sheet-iron. The exteriors
were painted a faint lemon-yellow, while the iron roof everywhere
glistened with red-ochre. The windows were uniformly small, but
fitted very neatly in tasteful casemates, and usually with double
sashes. Within, the floors were laid of whipsawed planks, tongued
and grooved by hand and highly polished. The inner walls were
"ceiled" up on all sides and overhead by light boards, and usually
papered showily. The heavy, unique Russian furniture was moved
in upon rugs of fur and tapestry, and then these people bade defiance to the elements, no matter how unruly, and led therein the
most enjoyable of physical lives. The united testimony of all travellers, who were many, and who shared the hospitality of the Russians at Sitka, is one invariable tribute to the excellence and the
comfort of their indoor living at New Archangel.

The shipyard of Sitka was as complete as any similar establishment in the Russian Empire. It was actively employed in boat and
sail-vessel building, being provided with all sorts of workshops and

A GLIMPSE OF SITKA

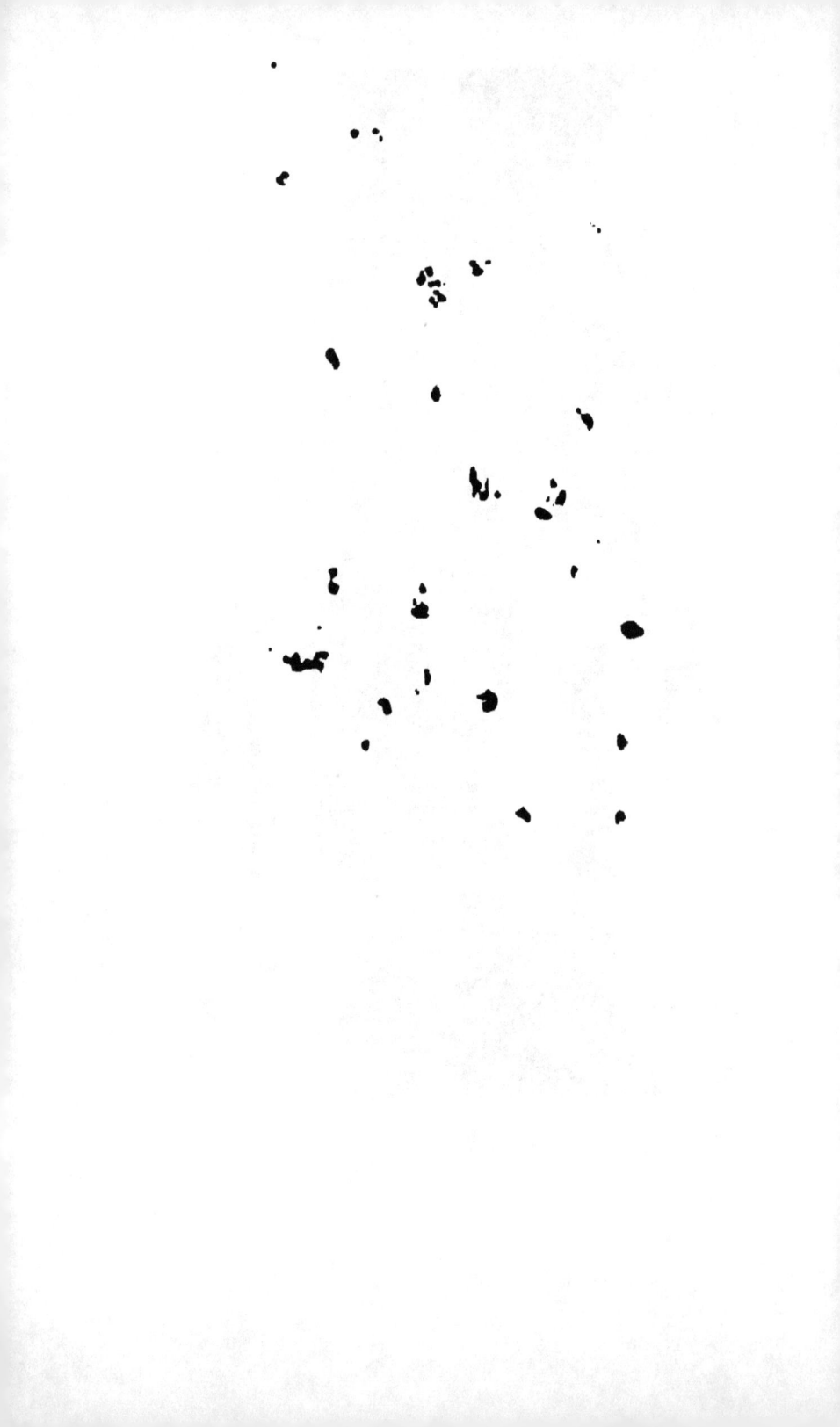

materials. Experiments were also instituted and prosecuted, to some extent, in making bricks, so much prized in the construction of the big conventional Russian "stoves," the turning of wooden-ware, the manufacture of woollen stuffs from the crude material brought up from California; but the great cost of importing skilled labor from far-distant Russia, and the relative expense of maintaining it here, caused the financial failure of all these undertakings. Much money was also wasted in attempting to make iron out of the different grades of ore found in many sections of the country. · The only real advantage that the company ever reaped from the workshops at Sitka was that which accrued to it from the manufacture of agricultural implements, which it sold to the indolent rancheros of California and Mexico. Thousands of the primitive ploughshares and rude hoes and rakes used in those countries then were made here; also axes, hatchets, and knives were turned out by industrious Muscovites for Alaskan post-trading. The foundry was engaged most of the time in making the large iron and brazen bells which every church and mission from Bering's Straits to Mexico called for. Most of these bells are still in use or existence, and give ample evidence of skilful workmanship, and of this early development of a unique industry on our northern coast.

Naturally enough the contrast of what the Russian Sitka was, with what the American Sitka is to-day, is a striking one : then a force of six or eight hundred white men, with wives and families, busily engaged as above sketched, directed by a retinue of fifty or sixty subalterns of the governor, lived right under the windows of his castle and within the stockade ; then the Greek-Catholic Bishop of all Alaska also resided there, with a staff of fifteen ordained priests and scores of deacons all around him, maintained regardless of expense, at this time, by the Imperial Government in that ecclesiastical pomp so peculiar to this Oriental Church—then a fleet of twelve to fifteen sailing-vessels, from ships in size to mere sloops, with two ocean-going steamers, made the waters of the bay their regular rendezvous, their hardy crews assisting to give life and stir to the town, shore, and streets—all this ordered by the concentration of the entire trade and commerce of Alaska at New Archangel.

Now, how different ! As you step ashore you scarcely pause to notice the handful of whites who have assembled on the wharf, but at once the impression of general decay is made upon your mind ; the houses, mostly the original Russian buildings, are settling here,

3

there, and everywhere, rotting on their foundations, and scarcely
more than half of them even occupied, while the combined population of some three hundred souls in number peers at you from every
corner.   The great majority of these people are the half-breeds, or
"Creoles," or the descendants of Indians and Russians ; some of
them are tall and well-formed, and a few of them good-looking, but
they are nearly all short-statured, abject, and apathetic.   Yet in
one respect Sitka has vastly improved under American supremacy
—she has become clean ; for although the Russian officers kept the
immediate surroundings of their residences in good order, still
they never looked after the conduct of the rest of the town.   There
were, in their time, no defined streets or sidewalks, and mud and
filth were knee-deep and most noisome.   Our military authorities,
however, who first took charge immediately after the transfer, and
who are proverbial for cleanliness and neatness in garrison life,
made the sanitary reformation of Sitka an instant and imperative
duty ; the slimy walks were soon planked, the muddy streets were
gravelled and curbed, the main street especially widened, the oldest
houses were repainted, and where dilapidated, repaired, and things
put into shape most thoroughly ; they also graded and sauntered
over the first wagon-road ever opened in Alaska, which they constructed, from the steamers' landing under the castle, back bordering the bay to Indian River, over a mile in length.

But the pomp and circumstance of the old castle—still the
most striking artificial feature now in all Alaska—will never wake
to the echoes of that proud and lavish 'hospitality which once
reigned within its walls, and when the flashing light in its lofty
cupola carried joy out over the dark waters of the sound to the
hearts of inbound mariners, who came safely into anchor by its
gleaming—the elegant breakfasts and farewell dinners given to
favored guests, where the glass, the plate, viands, wines and ap-
pointments were fit for regal entertainment itself—all these have
vanished, and naught but the uneven, slowly settling floors, warped
doors, and general mouldiness of the present hour greets the in-
quiring eye.   So heavy are its timbers, and so faithfully were they
keyed together, that in spite of neglect, the ravages of decay and
frequent vandalism, yet, in all likelihood, an age will elapse ere the
structure is removed by these destroying agencies now so actively
at work upon it.   Moved by the desire to preserve the salient
features of this historical structure, the author made, during one

clear June day, a pre-Raphaelitic drawing of it,* as his vessel swung
at anchor under its shadows ; in it the reader will observe that the
rocky eminence which it crowns is covered to the very foundations
and to the promenade cribbing that surrounds them, with a thick
growth of alders, stunted spruces, and other indigenous vegetation. That walk around the castle, which was artificially reared
thereon, gives a most commanding view of everything, over all
objects in the town and Indian village, and sweeps the landscape
and the sound. Another picture from the promenade walk under
the flagstaff is also given, in order that a faint effect may be conveyed to the reader of the exceeding beauty of the island-studded
Bay of Sitka. Descending and standing immediately under the
castle on the beach, to the right you have a perfect Alpine scene
as you look east along the pebbly shore to the living green flanks
of Mount Verstova, which carry your gaze up quickly over rolling
purplish curtains of fog to the snowy crest of it, and other lofty
crests *ad infinitum*, over far beyond. The little trading stores on
the left in this view hide the track so well known in Sitka as the
" Governor's Walk," for this is the only direction out to the saw-
mill in the middle distance, in which the earth lies smooth and
dry enough in all this archipelago for a clean mile-jaunt. These
still blue and green waters are alive with food-fishes, while the
dense coverts on the mountains harbor grouse and venison in
lavish supply ; the oyster and the lobster you have not, but the
clam and the crab are here in overwhelming abundance and excellence. "Ah!" you exclaim, "if it were not for this eternal rain,
this everlasting damp precipitation, how delightful this place would
be to live in !"

* This building, as it stands upon its foundations, is 140 feet in length by
70 feet in width—two stories with lofts, capped with the light-house cupola ;
these foundations rest upon the summit of the rock, 60 feet above tide-water.

CHAPTER III.

ABORIGINAL LIFE OF THE SITKANS.

The White Man and the Indian Trading.—The Shrewdness and Avarice of the
Savage.—Small Value of the entire Land Fur Trade of Alaska.—The
Futile Effort of the Greek Catholic Church to Influence the Sitkan In-
dians.—The Reason why Missionary Work in Alaska has been and is
Impotent.—The Difference between the Fish-eating Indian of Alaska and
the Meat-eating Savage of the Plains.—Simply One of Physique.—The
Haidahs the Best Indians of Alaska.—Deep Chests and Bandy Legs from
Canoe-travel.—Living in Fixed Settlements because Obliged To.—Large
"Rancheries" or Houses Built by the Haidahs.—Communistic Families.
—Great Gamblers.—Indian "House-Raising Bees."—Grotesque Totem
Posts.—Indian Doctors "Kill or Cure."—Dismal Interior of an Indian
"Rancherie."—The Toilet and Dress of Alaskan Siwashes.—The Unwrit-
ten Law of the Indian Village.—What Constitutes a Chief.—The Tribal
Boundaries and their Scrupulous Regard.—Fish the Main Support of
Sitkan Indians.—The Running of the Salmon.—Indians Eat Everything.
—Their Salads and Sauces.—Their Wooden Dishes and Cups, and Spoons
of Horn.—The Family Chests.—The Indian Woman a Household Drudge.
—She has no Washing to Do, However.—Sitkan Indians not Great
Hunters.—They are Unrivalled Canoe-builders.—Small-pox and Measles
have Reduced the Indians of the Sitkan Archipelago to a Scanty Number.
—Abandoned Settlements of these Savages Common.—The Debauchery of
Rum among these People.—The White Man to Blame for This.

> "Think you that yon church steeple
> Will e'er work a change in these wild people?"

Our people living now in the Sitkan district are engaged either in
general trading with the Indians, in prospecting for "mineral," or
actively mining; and, also, in a small fashion, in canning salmon
and rendering dog-fish and herring oil. Perhaps we can give a fair
idea of the traders by introducing the reader to one of them and
his establishment just as we find him at Sitka. In a small
frame one-story house, not usually touched by paint, the trader
shelters a general assortment of notions and groceries, but princi-

pally tobacco, molasses, blankets of all sizes and colors, cotton prints and cheap rings, beads, looking-glasses, etc.; he stands behind a rude counter, with these wares displayed to best advantage on the rough shelves at his back; a wood-burning stove diffuses a genial glow, but no chairs or benches are convenient. A "Siwash" * and his squaw deliberately and gravely enter. The Indian slowly looks up and down the room, and then proceeds to price every object within his vision, no matter whether he has the least idea of purchasing or not; this is the prelude and invariable habit of a Sitkan Indian, and it arouses an immense amount of suppressed profanity on the part of the outwardly courteous trader. But our savage has come in this time bent upon buying, and selling also; his female partner has a bundle carefully done up under her blanket, and which she wholly concealed when she squatted down on her haunches the moment after entering the door; she also has a number of small silver coins in her mouth, for, funny as it may seem, this worthy pair have carefully agreed upon what they shall spend in the store before coming in; so the woman has taken out from the leathern purse which hangs on her breast and under her chemise, the exact amount, and, returning the pouch to the privacy of her bosom, she places the available coin in her mouth for safe keeping *ad interim.*

Finally the Indian, in the course of half an hour, or perhaps a whole half-day in preliminary skirmishing, boldly reaches down for his bundle in the squaw's charge; then having, by so doing, given the trader to fully understand that he has something to sell, as well as desiring to buy, he reaches out for the groceries, the cloth, the tobacco, or whatever he may have fully decided to purchase; a long argument at once ensues as to the bottom cash price, and in every case of doubt the squaw decides; all the articles are done up in brown paper and neatly tied with attractive parti-colored twine. Then the dusky woman arises, with an indescribably vacant stare, bends over the counter and lets the jingling silver drop upon it, pausing just a moment until the tired but triumphant trader counts and sweeps it, still moist, into his till.

Now the Siwash, having bought, proceeds to sell, and he does it in his own peculiar way. He unrolls his package of furs; he

---

* All savages are called by this name up here—the sex being indicated by "buck" and "squaw." Children are called "pappooses."

eloquently discourses as he strokes each pelt out on the counter, in
turn praising its size and its quality ; the trader in the meanwhile
sharply keeps one eye on the savage and one eye on the furs, and,
after the story of their capture and quality has been told over the
third or fourth time, he asks, "How much?" The crafty hunter
promptly demands more than they would retail at in London ; the
trader answers with great emphasis and a most disgusted head-
shake, "no ;" he then offers just half or one-third the sum named,
whereupon the Indians, affecting great contempt, both shout out
"klaik !" which sounds like Poe's "Raven"—roll up their furs and
hustle out in a huff, still repeating, in sonorous unison, "klaik,
klaik"—(no, no).   Then they go to the rival trader's establishment,
and to all of them in turn, even if there are half a dozen, not
leaving one of them unvisited ; they finally finish the rounds in the
course of a week or two, and then quietly march back to that
trader who offered the most, and laying their peltries down in
perfect silence on his counter, hold out a grimy hand for the exact
sum he had previously proffered.

   In this shrewd and aggravating manner does the simple untu-
tored savage of the northwest coast deal with white traders—are
they swindled, do you think?   From the beginning to the end of
any transaction you may have with an Alaskan Indian you will be
met with the keenest understanding on his part of the full value in
dollars and cents of whatsoever he may do for you or sell.   When,
however, the Hudson Bay or the Russian Company held an exclusive franchise in this district, then the Indian had no alterna-
tive but the single post-trader's terms ; and then the white man's
profits were enormous.   But now, with the keen rivalry of competing stores, the trader barely makes a living anywhere in Alaska
to-day, while the Indian gets the best of every bargain—vastly
better compared with his former experience.

   The fur trade, however, in the whole Sitkan district is now of
small commercial importance ; thirty or forty thousand dollars annually will more than express its gross value.   This great shrinkage
is due to the practical extermination of the sea-otter in these
waters, while the brown and black bears, the mink and marten, the
beaver and the land-otter skins secured in this archipelago and its ·
mainland coast are not highly valued by furriers, inasmuch as the
climate here is never cold enough to give them that depth and
gloss of fur desired and so characteristic of those animals which

are taken, away back in the interior, where the temperature ranges from 20° to 40° below zero for months at a time. In early days, the Sitkan savages acted as middlemen, receiving these choice pelt-ries from the back-country Indians, who were never permitted by the coast tribes to come down to the sea—and then trading the

The Sitkan Chimes.

stock anew in their own right over to the Russian and English posts, they reaped a large advance. Now, however, the indepen-dent white trader penetrates to the interior himself, and the Alas-kan Siwashes mourn the loss of those rich commissions which once accrued to their emolument and consequence. The irruption, also, of the restless, tireless, wandering miners throughout Alaska and

British Columbia, who, prospecting in every ravine and cañon, never let an opportunity pass to trade and trap for good furs, has also contributed to this total stagnation of the business in the Sitkan region.

The finest structure in Sitka to-day is the Greek church, which alone did not pass from the custody of its original owners at the time of the transfer. This building has been kept in repair, so that its trim and unique architecture never fails to arrest the visitor's attention and challenge inspection, especially of the interior. We find the service of the church rich and profuse in silverware, candelabra, ornately framed pictures—oil-paintings of the saints—and rich vestments; two priests officiate, a reader chants rapid automatism, and a choir of small boys respond in shrill but pleasing orisons; instrumental music is banished from the services of the Greek Church, and so are pews, chairs, and hassocks; the Creole congregation, men, women, and children, stand and kneel and cross themselves, erect and bowed, for hours and hours at a time during certain festivals, never moving a step from their positions. The men stand on the right side of the vestibule, facing the altar, while the women all stand by themselves, on the left, the children at option as they enter. No one looks to the one side or the other, but every face is riveted upon the priest, who says little, and is busily engaged in symbolic worship.

The Indians do not enter here, nor did they ever; for them the Russians erected a small chapel, which still stands on the site of its first location; it is built against the inner side of the stockade, and, like the old Lutheran church lower down in the town, it is fast going to ruin; the door is secured by one of those remarkable Muscovitic padlocks—it is eight or ten inches long, five or six wide, and three deep; these singular locks must be seen to be appreciated in all of their clumsy strength. This little faded place of savage worship was the scene in 1855 of the second and last stand ever made by the Sitkan Indians in revolt against the Russians. Those savages, brooding over some petty indignities received from the whites, became suddenly inflamed with passion, and a swarm of armed warriors from the adjacent rancheries rushed, one dusky evening, upon the fortified palisade surrounding the village, and began to cut and tear it down. The Russians opened their brass batteries of grape and round-shot upon the infuriated, yelling natives from the several block-houses which commanded the stockade, but the Siwashes returned the fire fearlessly with their smooth-bore muskets, and succeeded in getting possession of this chapel, behind the stout logs of which they were sheltered and able to do deadly execution with their rifles in picking off the Russian officers and men, as they hurried to and from the bastions and through the streets of the town. When, however, one of the company's vessels hauled off the beach opposite the Indian village, and trained her guns upon it and its people, the savages humbly sued for mercy, and have remained in abasement ever since.

Contemplating this Indian church at Sitka, which has stood here for nearly three quarters of a century, and then glancing over it and into the savage settlement that nestles in its shadow, it is im-

Old Indian Chapel at Sitka.
[*Greek Catholic Church, June 9, 1874.*]

possible to refrain from expressing a few thoughts which arise to my mind over the subject of the Indian in regard to his conversion, to the faith and practices of our higher civilization. Nearly a whole century has been expended, here, of unflagging endeavor to better and to change the inherent nature of these Indians—its full result is before our eyes. Go down with me through the smoky, reeking, filthy rancheries and note carefully the attitude and occupation of these savages, and contrast your observation with that so vividly recorded of them by Cook, Vancouver, Portlock, and Dixon, and many other early travellers, and tell me in what manner have they advanced one step higher than when first seen by white men full a hundred years ago. You cannot escape the conclusion with

this tangible evidence in your grasp, that in attempting to civilize
the Alaskan Indian the result is much more like extermination, or
lingering, deeper degradation to him than that which you so earnestly desire. The cause of this failure of the missionary and the
priest is easy to analyze : it is due to the demoralizing precept and
example of those depraved whites who always appear on the field
of the Indian mission, sooner or later ; if they could be shut out,
and the savage wholly uninfluenced by their vicious lives, then the
story of Alaskan Indian salvage might be very different. Still, the
thought will always come unbidden and promptly—these savages
were created for the wild surrounding of their existence ; expressly
for it, and they live happily in it : change this order of their life,
and at once they disappear, as do the indigenous herbs and game
before the cultivation of the soil and the domestication of animals.

The Indians of Alaska, however, will never call upon the Government for food and reservations—there is a great abundance on
the earth and in the waters thereof for them ; living as they do all
down at tide-water, at the sole source of their subsistence, they are
within the quick reach of a gunboat ; the overpowering significance
of that they fully understand and fear. There is a huge wilderness
here for them which the white man is not at all likely to occupy,
even in part, for generations of his kind to come, yet unborn.

Sitka is the seat of that Alaskan civil government* which Congress, after much deliberation, ordered in 1884 ; but the governor
lives here in much humbler circumstances than did his Slavonian
predecessors. As it would require a small fortune to rehabilitate the
"castle," the present chief-magistrate resides in one of those neatly
built houses which the military authorities erected shortly after they
took charge in 1867–68 ; it is not at all commanding, but has a
pleasant vista from its windows over the parade ground, and the
steamers' landing.

While the most impressive feature of the Sitkan archipelago is
unquestionably that of the awe-inspiring solemnity and grand

---

* This Act wisely does not establish a full-fledged form of territorial government in Alaska, because the lack of a suitable population to maintain
it reputably was conclusively shown by the census returns of 1880 : it creates
an executive and a judiciary ; it extends certain laws of the United States
relating to crimes, customs, and mining, over Alaska, and provides for their
enforcement. The land laws of the United States should also be made opera-
tive in Alaska, they are expressly omitted in the present act.

beauty of its strange wilderness, yet the most interesting single idea is the Indian and the life he leads therein ; with the single exception of the substitution of a woollen blanket and a cotton shirt for his primitive skin garments, he is living here to-day just as he has lived away back to the time when his legends fail to recite, and centuries before the bold voyages of Cook and Vancouver, and the savage sea-otter fleet of Baranov, first discovered him and then made his existence known to the civilized world. True, some of the young fellows who have labored upon vessels and in the fish-canneries wear an every-day workingman's shirt and trousers, and speak a few words of English, understanding much more, yet the primeval simplicity of all Indian life in this district is substantially preserved.

These savages are fish-eaters, and as such they have a common bond of abrupt contrast in physique with their meat-eating brethren of the Rocky Mountains and the great plains ; but the traits of natural disposition are the same, the heart and impulse of the Haidah or the Tongass, are the heart and the impulse of the Sioux or the Cheyenne—the former moves nowhere except squatted in his shapely canoe, the other always bestrides a pony or mustang. This wide divergence in every-day action gives alone to these savages their strongly marked bodily separation ; the fish-eater is stooping as he stands, and though he has a deep chest and sinewy arms, yet his lower limbs are bowed, sprung at the knees, and imperfectly muscled ; while the meat-eater is erect and symmetrical, in fine physical outline from the crown of his head to his heels.

The various divisions or bands of the Indian population of the Sitkan archipelago and mainland * differ but little in their manner of life and customs, and speak closely related dialects of the same

---

* I. *Chillkahts:* Lynn Canal and Glacier Bay.
II. *Hooniahs :* Chichagov Island and islets.
III. *Auks :* North end Admiralty Island.
IV. *Tahkoos:* Mainland, Stephen's Passage and Juneau City.
V. *Khootznahoos :* South end Admiralty Island.
VI. *Sitkas :* Baranov Island.
VII. *Kakho :* Kou and Kuprianov Islands, Prince Frederick Sound, mainland coast.
VIII. *Stickeens :* Wrangel, Zarenbo and Etholin Islands, Stickeen River mouth.
IX. *Haidah :* Prince of Wales Island.
X. *Tongass :* Mainland, Cape Fox to Cape Warde, and contiguous islands.

language.   The Haidahs are the best dispositioned and behaved.
They have been from the earliest times constantly in the habit of
making long and incessant canoe voyages ; and, taking into account
the ease with which all parts of this region can be reached on water,
it is rather surprising that any marked difference in language should
be found at all ; still, when we recall the knowledge which we have
of their fierce inter-tribal wars, it is not so strange ; this warfare,
however, was of the same barbarous character as that recognized in
all other American savages—it was the surprise and massacre of
helpless parties, never sparing old women, children or decrepit men.
These internecine family wars have undoubtedly been the sole cause
of the present subdivisions of the savages as we note them to-day.

In drawing the picture, faithfully, of any one Alaskan Indian, I
may say candidly that in so doing I give a truthfully defined image
of them all throughout the archipelago.   Physically the several
tribes of this region differ to some extent, but not near so much as
our colored people do among themselves ; the margin of distinction
up here between the ten or eleven clans, which ethnologists enume-
rate, is so slight that only a practised eye can declare them.   The
Haidahs possess the fairest skins, the best temper, and the best
physique ; while the ugly Sitkans and Khootznahoos are the darkest
and the worst.   But the coarse mouth, the width and prominence
of the cheek bones, and the relatively large size of the head for the
body, are the salient main departures from our ideal symmetry.

The body is also long and large, compared with the legs,
brought about by centuries of constant occupation in canoes and
the consequent infrequent land travel ; their hair is black and
coarse, unkempt, and never allowed, by the males, to fall below
their shoulders except in the case of their "shamans," or doctors.
A scattered, straggling mustache and beard is sometimes allowed
to grow upon the upper lip and chin, generally in the case of the
old men only, who finally grow weary of plucking it out by the
roots, which in youth they always did in sheer vanity.

Once in a while a face is turned upon you from a canoe, or in a
rancheric, which arrests your attention, and commands comment
as good-looking ; these instances are, however, rare—very, very
rare.   I think the Haidahs give more evidence in average physiognomy of possessing greater intelligence than that presented in the
countenances of their brethren ; while I deem the Sitkas and
Khootznahoos to be the most insensible—if they are as bright they

conceal the fact with astonishing success. Again, the ferocity and exceptionally savage expression of their faces, which Captain Cook and Vancouver saw and so graphically recorded, has faded out completely ; but in all other respects they agree to-day perfectly with those descriptions of these early voyagers. In those days firearms had not destroyed their faith in elaborate armaments of spear and bow and body armor-shields of wood and leather, so that they then appeared in much more elaborate costumes and varied pigments than they do now.

Each tribe has one or more large " rancheries," or villages, in which it lives, and which are always located at the level of the sea, just above tide and surf, at river-mouths, or on sheltered bays of the islands, or the mainland ; these rancheries, or houses, are built of solid, heavy timbers in the permanent villages, or thrown loosely together of lighter material in their temporary or camping stations. The general type of construction is the same throughout the archipelago, the most substantial houses being those of the Haidahs, who give more care to the accurate fitting together and ornamentation of their edifices than is shown elsewhere. They certainly show a greater constructive facility and mechanical dexterity, not only in the better style of house-building but in the greater number of, greater size of, and excessively elaborate carved totem posts. These peculiar adjuncts to Alaskan Indian architecture are small and shabby everywhere else when compared with the Prince of Wales exhibition.

All permanent villages are generally situated with regard to one great idea—easy access to halibut-fishing banks and such coast fisheries, which occupy the greater proportion of the natives' time in going to and coming from them when not actually engaged in fishing upon these chosen grounds ; therefore it happens that, occasionally, a village will be located on a rocky coast, bleak and exposed, though carefully placed at the same time so as to permit of the safe landing of canoes in rough water. These houses always face seaward, and stand upon some flat of soil, elevated a few feet above the high-tide mark, where below there is usually a sandy or gravelly beach upon which the fleet of canoes is drawn out, or launched from, as the owners come and go at all hours of the day and night. The houses are arranged side by side, either in close contact, or else a space of greater or less width between. A promenade or track is always left between the fronts of the houses and the edge of the bank, from

ten to thirty feet in width; it constitutes a street, and in which
the carved posts and temporary fish-drying frames, etc., are usually
planted. Also those canoes that are not in daily use, or will not
be used for some time, are invariably hauled up on this street, and
carefully covered by rush-mats or spruce-boughs, so as to protect
them from the weather, by which they might be warped or cracked.

The rancheries are themselves never painted by their rude archi-
tects and builders; they, however, soon assume a uniform, inconspicuous, gray color, and become yellowish-green in spots, or over-

A Haidah Rancherie.

grown with moss and weeds owing to the dampness of the climate.
If it were not for the cloud of bluish smoke that hovers over these
villages in calm weather, they would never be noticed from any considerable distance.

In localities where the encroachment of mountain and water
make the village area very scant, two rows of houses are occasionally formed, but in no instance whatever is any evidence given in
these Koloshian settlements of special arrangement of dwellings,
or of any set position for the house of the chief man of the village:
he may live either in the centre or at the extreme end of the row.

Each house usually shelters several families, in one sense of the term ; these are related to each other and under the tacitly acknowledged control of some elder, to whom the building is reputed to belong, and who is a person of greater or less importance in the tribe or village according to the amount of his property or cunning of his intellect.

Before some of these Siwash mansions a rude porch or platform is erected, upon which, in fair weather, a miscellaneous group of natives will squat in assembly, conversing, if squaws, or gambling, if men. The houses themselves are usually square upon their foundations, and vary much in size, some of them being a hundred feet square, while most of them are between fifty and sixty feet, the smaller rancheries being less than twenty. The gable end, and the entrance right under its plumb, always faces the street and beach-view ; the roof slopes down at a low pitch or angle on each side, with a projecting shelter erected right over the hole left in the roof-centre, intended for the escape of smoke—no chimneys were ever built. This shelter, or shutter, is movable, and is shifted by the Indian just as the wind and rain may drive ; the floor is oblong or nearly square, and, in the older and better constructed examples, is partly sunk in the earth, i.e., the ground has been excavated to a depth of six or eight feet in a square area, directly in the centre, with one or two large earthen steps or terraces left running around the sides of the cellar. A small square of bare dirt is left in the exact centre, again, of this hole, while the rest of the floor is covered with split planks of cedar ; the earthen steps which environ the lower floor are in turn faced and covered with cedar-slabs, and these serve not only for sleeping and lounging places, but also for the stowage, in part, of all sorts of boxes and packages of property and food belonging to the family ; the balance of these treasures usually hangs suspended, in all manner of ingenious contrivances, from the heavy beams and roof-poles overhead. The rancheries which are built to-day by Alaskan Indians nearly all stand on the surface of the ground without any excavation—a decided degeneracy.

The pattern of the Koloshian house is maintained with little variation throughout the archipelago, and has been handed down from remote antiquity. When, after extended confabulation, a number of Indians agree to build a house, several months are passed first in the forest by them, where they are engaged in fell-ing the trees and dressing the timbers necessary; when these logs and planks are finally hewn into shape (everything in this line is done with axes and the little adze-like hatchets so often described), they are tumbled into the water and towed around to the contemplated site of the new edifice. The great size of the beams and planks used in a big Indian rancherie make it imperative that a large number of hands co-operate in the work. The erection, therefore, of such a structure in all its stages, the cutting and hewing in the woods, the launching and towing of the timbers to the foundations, and their subsequent elevation and fitting, forms the occasion of a regular gathering, or "bee," that generally calls in whole detachments from neighboring villages, which is always the

Section Showing Arrangement of Interior of a Rancherie.

precursor to a grand "potlatch," or giving away of the portable property of the savage for whom the labor is undertaken.

Some of the larger houses have required the repeated assembling of a whole tribe, and the lapse of two or three years of time ere completion in all details, because the Siwash for whom the work has been done has regularly exhausted his available resources on each occasion, and has needed this interval, longer or shorter as it may have been, in which to accumulate a fresh stock of suitable property, especially blankets, with which to reward a renewed and continued effort. Dancing and gambling relieve the monotony of the labor, which, however, seldom ever is suffered to occupy more than two or three hours of each day, and is conducted in a perfect babel of guttural talk and noise, and the exultant shouting of the entire combination of men, women, and children, as the great beams are placed in position.* In the construction of these dwellings the savage uses no iron or wooden spikes, he "mortices" and "tenons" rudely but solidly everything that requires binding firmly ; in the lighter and temporary summer rancheries much use is made of cedar-root and bark-rope lashings to the same end. Within the last fifteen or twenty years the common use of small windows has been employed, the glazed sashes being purchased from the whites either at Victoria or else brought up to order by the traders ; these are inserted in the most irregular manner, usually on the sides under the eaves.

The oddly-carved totem posts, which appear in every village, sometimes like a forest of dead trees at distant sight, are, broadly speaking, divisible into two classes : that is to say, the clan or family pillars, and those erected as memorials of the dead. There has been too much written in regard to these grotesque features seeking to endow them with idolatry, superstitions, and other fancies of the savage mind. Nothing of the kind, in my opinion, belongs to the subject ; the image posts of the totem order are generally from 30 to 50 feet in height, with a diameter of 3 to 5 feet at the base, tapering slightly upward. They are often hollowed at the back, after the fashion of a trough, so that they can be the easier handled and put into position. Those grotesque figures which cover these posts from top to bottom, closely grouped together, have little or no serious significance whatever : they always display the totem of the owner, and a very marked similarity runs through the carvings of this character in each village, though they have a wide range of variation when one settlement is contrasted with another. I am unable to give any definite explanation, that is worthy of attention, of the real meaning of all those strange designs —perhaps, in truth, there is none ; they are simply ornamental doorways.

The smaller memorial posts are also generally standing in the

* The exact measurements of such a rancherie, and of which the author submits a careful drawing, were : Breadth in front of house, 54 feet 6 inches ; depth from front to back, "in the clear," 47 feet 8 inches ; height of ridge of roof, 16 feet 6 inches ; height of eaves, 10 feet 8 inches : girth of main vertical posts and horizontal beams, 9 feet 9 inches ; width of outer upright beams, 2 feet 6 inches, thickness, about 6 inches ; width of carved totem post in front of house, 3 feet 10 inches, height, (?) 50 feet.

village, upon the narrow border of land running between the houses
and the beach, but in no determinate relation to the buildings.
When a man falls before prostrating illness, his relatives call in the
medicine man, or "shaman," and also invite the friends of the family to the house of sickness, usually providing them with tobacco ;
soon the rancherie is full of curious friends, of smoke, and of the
abominable noise of the shaman. If the patient dies, the body is
not burned now, as it used to be prior to the advent of the whites,
but is bent double into a sitting posture, and enclosed in a square
cedar box, which has been made for this purpose by the joint labor
of the assembled Indians, or else they have subscribed and purchased it from some one of their number. This coffin is exactly
the same in shape and size as the box commonly used by every
Siwash family here for the reception of spare food, oil, etc., so that
there never is any delay or difficulty in getting one.

If the dead Koloshian is a man of only ordinary calibre, his
body is put, while still warm, into the wooden crib, and this is at
once carried out and stored away in a little tomb-house, which is
generally a small covered shed right behind the rancherie, or in the
immediate vicinity. This vault is also made by the united labor of
the men of the village, and paid for in the same manner as that indicated for the purchase of the coffin-box. In it may be placed but
a single body, then again it will contain several—all relatives, however. But should the deceased savage have been one of great importance, then the whole rancherie itself is given up to the reception
of the body, which is boxed and placed therein, sitting thus, in state,
perhaps for a year or more, no one removing any of the things, the
members of the family all vacating the premises, and seeking quarters elsewhere in the village. Now it becomes necessary, sooner or
later, to erect a carved post to the memory of this man. Again the
Indians collect for the purpose, and are repaid by a distribution of
property made by the deceased man's brother, or that relative to
whom the estate has come down, in order of descent. This
inheriting relative takes possession the moment the body of the
dead has been enclosed in its cedar casket, and not before.*

The doorway to the Alaskan house is usually a circular hole

* Whole volumes have been written upon this subject of the totem and
consanguinity among these savages of the northwest coast. Further description or discussion, in this instance, is superfluous.

through which the Indian must stoop to half his stature when he enters. It is generally from four to six feet from the ground, and is gained by a rude flight of stairs or a notched log leading up to it on the outside, and in the same manner down to the floor on the inside. As you enter, the whole interior seems dark—everything, at first, indistinct, and the only light being directly above and below the smoke-hole in the roof, for a blanket is dropped as a portière over the doorway the moment you pass within. In the centre of this gloomy interior, directly beneath a hole in the roof, is the fireplace, upon which logs are smouldering or fitfully blazing ; kettles of stewing fish, and oil and berries simmering under the care of some squatty, grimy squaws who surround it. If this house be a large one you will find within fifty or sixty Koloshes of both sexes, all ages, and in all conceivable attitudes, as they stand, sit, or lounge or sleep around the four sides of the deep terraced room, some cleaning firearms, others repairing fishing-tackle, or carving in wood or slate ; while others are idly staring into the fire, or, wrapped in their blankets, are sleeping with reiterated snoring. Against the walls, pendent from the black, sooty beams overhead, hang an infinite variety of personal effects peculiar to this life, such as fish-spears and hooks, canoe paddles, bundles of furs, cedar-bark lines and ropes, immense wooden skewers of dried salmon and halibut, while the boxes which contain the real wealth of such people —blankets,* tobacco, and cloths of cotton, and handkerchiefs of silk, are stowed away in the corners.

But odors that the civilized nose never before scented now rise thick and fast as you contemplate this interior, and the essential oils of rancid oolachan grease, decaying fish, and others, in rotation swift,

---

* The blanket is now, however, the general recognized currency among these people. It is the substitute among them of that unit of value, the beaver skin, which has been for so long the currency of the great Hudson Bay region. The blankets used in Alaskan trade are of all colors—green, blue, yellow, red, and white—of the very best woollen texture, none others will do. They are rated in value by the "points" or line-marks woven into the edge, the best and largest being a "four-point," the smallest and poorest being "one-point." The unit of value is a single "two and a half point" blanket, worth a little over $1.50. Everything is referred to this unit, even a large four-point blanket is said to be worth so many blankets. Traders not infrequently buy in blankets, taking them, when in good order, from the Indians as money, and selling them out again as trade demands.

of many shades of startling disgust, cause you to speedily turn and gladly seek, with no delay, the outer stairway, even though a tempest of rain and wind is beating down (with that fury which seems to be most pronounced in violence here as compared with the rest of the world, when it does storm in earnest). Here again it is not pleasant for us to tarry even in fair weather, inasmuch as the Koloshian has no idea of sewerage or of its need, the refuse—slops, bones, shells, fish-débris, and a medley of similar and worse nuisances are lazily thrown out of this doorway on either side and straight ahead, as they are from the entrance to every other rancherie in the village. A merciful growth of rank grass and mighty weeds charitably covers and assimilates much, but yet the atmosphere hangs heavy around our heads—we move away.

On ordinary occasions a head-covering is usually dispensed with, unless it be some old hat of our style. The squaws, however, fashion and often wear grass hats, made as they weave their fine basket-ware; they have the form of an obtuse cone, generally ornamented by conventional designs painted in black, blue, or red. The feet are almost invariably bare—too wet for moccasins. Painting the face is a very common practice; vermilion is the favorite pigment, and is usually rubbed in without the least regard to pattern or effect; blue and black colors also are used in the same manner, but I have never seen their limbs or bodies so treated, which is the common method of meat-eating savages, who always paint themselves with great care as to exact and symmetrical design. Here the faces of Alaskan Siwashes are thus daubed for the dance or for mourning; especially hideous are the mixtures of spruce-gum grease, and charcoal which you observe smeared over the countenances of the Sitkans, who do so chiefly to prevent unpleasant effects of the sun when it happens to shine out upon them as they are fishing or paddling extended journeys in their canoes, and who also give you an ugly reminder of their being in mourning by the same application.

Bracelets are beaten-out pieces of copper or brass wire and silver coins, highly polished, and worn chiefly by the women, who often carry several upon each arm. When worn upon the ankles they are forged in round sections, while for the wrist they are made quite flat. Tattooing once was universal, but is now going out of style; and, until quite lately, the females all wore labrets in the lower lips—this disgusting distortion is also being abated. Only among the very old women can this monstrosity now be found in its original form. Most of the middle-aged squaws still have a small aperture in the lower lip, through which a little silver, beaten tube, of the size of a quill, is thrust, and projects from the face, just above the chin, about a quarter of an inch. The younger women have not even this remnant of a most atrocious old custom. The ears are often pierced, and tiny shell ornaments, backed with thin sheet-silver or copper, are inserted ; and also the septum of the nose is perforated, of both sexes very generally, for the insertion of a silver ring, or a pendant of haliotis shell.

Each village has its *lex non scripta*, and is a law unto itself everywhere within the confines of the Alexander archipelago ; or, in different words, it conducts its affairs wholly without reference to any other village or savages—it is the largest unit in the Indian system of government. Living as they do in these settlements, where they know each other just as well and as familiarly as we know the individual members of our own private home circles, no matter whether the village contain a thousand souls or but half a dozen—there are no strangers in it. Every little daily incident of each other's simple life, every move that they make, what they capture in the forest or hook out from the sea, is regularly recounted in the rancheries over night. All engaged in precisely the same calling of fishing and hunting, naturally there is no room among them for the eager rivalries and passionate enterprises which our living stimulates and sustains. Therefore the routine of government is almost nothing in its detail—no laws appear to be necessary, and they are not acknowledged ; but any action tending to the injury of another, in person or property, lays the offender open to reprisals by the sufferer—usually atoned for—and the village feud, thus aroused, is soon satisfied by a payment in blankets, or other valuable property, to a full settlement. Injuries, thefts and murder, however, which, inflicted by the people of one village upon another, either close at hand or remote, have not always been adjusted in this amicable manner ; hence, from time immemorial, the disputants have been at war with each other in this region, and the result of these wars has been to divide them into the existing clans as we find them now. Their internecine warfare was carried on in true savage style. If the cause was one which concerned the whole village, then the chief of that settlement could implicitly count upon the services of every male Indian able to bear arms ; and although these savages

are fearless and brave, yet they know no open, fair fight—taught to
get his living by stratagem when fishing or hunting, so the Kolosh
advances in capturing his human enemies, just as all other Indians
have done and do.

Each village has a well-recognized head man, or chief, who, though
possessing much influence, still never has had, and does not now
enjoy, that absolute rule which is attributed to such Indians. He
is really a presiding elder over the several families in the hamlet,
and, without their consent, his decisions are futile or carry no
weight. He has no power to compel other members of the tribe
to work, hunt, or fish for him, and if he builds a house, or a canoe,
he has to hire them to labor by making the customary "potlatch,"
just as any other man of the tribe would do—only he must give a
little more. The social rules which exist among these savages show
many strange features, for though every rancheric has its freely-
acknowledged chief, yet they are divided into as many or more
families than there are houses, each one of which has its own regulations, and a subordinate authority of its own governing it, and it
alone.*

The Sitkan Indians trouble themselves very little about the interior country; but the coast line, and especially the margins of
rivers and streams, are duly divided up among the different families. These tracts are regarded as strictly private property, just as
we would regard them if fenced in as farms and cattle ranches—
and they are passed from one generation to the other in the line of
savage inheritance; they may be sold, or even rented by one family
desiring to fish, to gather berries, to cut timber, or to hunt on the
domain of another. So settled and so strict are these ideas of pro-
prietary and vested rights in the soil, that, on some parts of the
coast, corner-stones and stakes may be seen to-day set up there to
define the limits of such properties between savages, by savages;

---

* There are naturally in every clan certain individuals of hereditary Indian
wealth and a long pedigree, who speak in better language, who have a fine
physical presence, a more dignified bearing, and the self-possession and pride
of incarnate egotism. From these men the chiefs are selected, and although
the chieftainship is not necessarily hereditary, yet it is often retained in this
manner for many generations in one family. The covers of this volume,
however, cannot be expanded wide enough to permit the further discussion
and enumeration of a thousand and one singular points in this connection
which rise in the author's mind.

and furthermore, woe to the disreputable trespassing Siwash who steps over these boundaries and appropriates anything of value, such, for instance, as a stranded whale, shark, seal, or otter—berries, wreckage, or shell-fish.

The woods and the waters are teeming with animal life; the lofty semi-naked peaks harbor mountain goats in large flocks; the beautiful grouse of Sabine hides in the forest thickets; the land otter and the mule-eared deer haunt the countless ravines, valleys, and rivulet bottoms; salmon in fabulous numbers run up those streams, and big, brown and glossy black bears come down to fatten upon these spawning fish. But the Sitkan savage is indolent, and, though all this dietary abundance and variety is before him, he lives quite exclusively upon halibut and salmon, the former mostly fresh and the latter air-dried and smoked in the soot of his rancherie. Halibut he finds all the year round; salmon briefly run only at widely separated periods.

The halibut fishery is the one systematic regular occupation of the natives. These fish may be taken in all waters of the archipelago at almost any season, though on certain banks, well known to the Indians, they are more numerous at times. When the halibut are most active and abundant, the Koloshians take them in large quantities, fishing with a hook and line from their canoes, which are anchored over the favored spots by stones attached to cedar-bark ropes or cables. They still employ their own primitive, clumsy-looking hook in decided preference to using our own make. When the canoe is loaded to the gunwale by an alert fisherman, these halibut are brought in to some convenient adjacent point on the shore, where they are handed over to the women, who are there to take care of them, usually living in a temporary rancherie. They squat around the pile, rapidly clean the fish, removing the larger bones, head, fins, and tail, and cut it into broad, thin flakes. These are then hung on the poles of a wooden frame trellis, where, without salt, and by the wind and sun alone, sometimes aided by a slow fire underneath the suspended fish-meat, the flakes are sufficiently cured and dried; then they are packed away in those characteristic cedar boxes for future use.

A group of old and young squaws, half-nude, flecked with shining scales and splashed with blood, as they always are when at work upon a fine run of halibut or salmon—such a group is to be vividly remembered ever afterward, if you see it even but once. The little pappooses, entirely naked, with big heads and bellies, slender necks and legs, are running hither and thither in infantile glee and sport, always with a mouthful of raw ova or a handful of stewed fish from the kettle near by, while the babies, propped up in their stiff-backed lashings, croon and sleep away the time.

There are no rivers of any size flowing on the islands of the Sitkan archipelago ; but there are rapid rivulets and broad brooks in great numbers. Many of these are large enough to be known as "salmon rivers." The first run of those attractive fish usually takes place up some of the longest island-streams and the mainland rivers about July 10th to 20th. A month later a larger species begins to arrive from the depths of the ocean outside, and this run sometimes lasts, in a desultory manner, until January. These salmon, when they first appear, are fat and in superb condition and color ; but as they leave the salt water and take up their persistent, tireless ascent of fresh-water channels they become hook-jawed, lean, and pale-fleshed. They ascend very small streams in especially great numbers when these rivulets are swollen by the heavy rains of October, and, being easily caught and very large, they constitute the chief harvest of the Alaskan Indian—his meat and bread, in fact. They are either speared in the shallow estuaries or trapped in brush and split-stick weirs, which are planted in the streams. Everyone of the little salmon brooks has its owner in the Indian law. They are the private property of the several families or subdivisions of the clans. Those people always come out of their permanent village houses during the fishing period, and camp upon the banks of their respective water claims.

It is quite unnecessary to itemize all the species of food-fishes in the Alexander archipelago, for anything and everything that is at all abundant in the vicinity of an Indian rancherie is sure to be eaten ; trout, herring, flounders, rock-cod, and the rosy, glittering *sebastines* constitute minor details of the savage dietary. Codfish are taken in these waters, but not in great numbers, nor are they especially sought for. The spawn of the herring* is collected on spruce boughs, which the Indians carefully place at low-water on the spawning grounds ; then, when taken up, it is smoke-dried and stored away.

But the "loudest" feast of these savages consists of a box, just

---

* *Clupea mirabilis.*

opened, of semi-rotten salmon-roe. Many of the Siwashes have a custom of collecting the ova, putting it into wooden boxes, and then burying it below high-water mark on the earthen flats above. When decomposition has taken place to a great extent, and the mass has a most penetrating and far-reaching "funk," then it is ready to be eaten and made merry over. The box is usually un-covered without removing it from its buried position; the eager savages all squat around it, and eat the contents with every indica-tion on their hard faces of keen gastronomic delight—faugh!

The same ill-favored and heartily-hated "dog-fish"* of our Cape Cod fishermen is also very abundant in these far-away waters.

Indians Raking Oolochans and Herring.—Stickeen River.

Recently, the demand created for its oil by the tanneries of Oregon and California has made its capture by the Indians an important source of revenue to them; the oil rendered from its liver is readily sold by them to the white traders, who also have established a fishery for the purpose on Prince of Wales Island. These traders also are making good use of herring-oil, which is to be secured here in unfailing, abundant supply, to any quantity required.

The most grateful condiment to the Sitkan palate is rancid fish-oil, or oolachan "butter"—a semi-solid grease, with a fetid smell and taste; into this they always dip or rub their flakes of dried fish,

---

* *Squalus acanthias.*

their berries, in fact everything that they eat. A little wooden trencher or tub, grotesquely carved, always is to be seen (and smelled), placed alongside of the monotonous kettle of stewed fish, or pile of dried fish, which constitutes the regular spread for a full meal. And again, a very curious, soap-like use of this oil is made by the younger and more comely savages. An Indian never washes in water up in this wet and watery wilderness. I never have seen an attempt made to wash the face or hands with water, but they do rub oil vigorously over, and scrub it off bright and dry with a towel, or mop, of cedar-bark shreds or dry sedge-grass. The constant presence of this strong-flavored oil renders it a physical impossibility for a white man, not long-accustomed to its odor, to enter a rancherie and eat with the inmates, unless the pangs of starvation make him ravenous.

Whether from taste itself, or sheer indolence, the culinary art of these people is confined to the incessant simmering and boiling of everything which is not eaten raw, or ripe ; copper, sheet-iron, and brass kettles being now universally used, are the only decided innovation made by contact with ourselves in their aboriginal cooking outfit, though the introduction of tin and cheap earthenware dishes is growing more general every day. Most of the Indian household utensils are made of wood; they are fashioned in several forms or types, which appear to have been faithfully copied from early time. The berry and the food-trays are cut out of solid pieces of wood, the length being about one and one-third times as great as the width, while the depth is relatively small. In some of the large rancheries these trays, or troughs, are six to ten feet long ; the outer ends of those receptacles are generally carved richly in all sorts of fancy relief ; and, sometimes, the sides are grotesquely painted. A common form, and smaller in size, and a great favorite with the family, is boat-shaped, the hollow of the dish being oval ; the ends are provided with odd prow-shaped projections that serve as handles—one of these ends being usually carved into the head and fore-feet of some animal or bird, the other to represent its hind-feet and tail. These dishes are seldom more than eight or ten inches in length, and curve upward from the middle each way, like the "sheer," or the gunwale, of a clipper ship.

Water-dippers, pot-spoons and ladles are made from horns of the mountain sheep. They are steamed, bent, and pared down thin, carved and shaped so as to be exceedingly symmetrical, and well finished. The stew and berry-spoons in ordinary use are made

A STICKEEN SQUAW

Boiling Berries and Oil, Toasting Herrings, etc.

from the stiff, short black horns of the mountain goat, the handles often carved to represent a human form, animal, or bird. Knives of all sorts are now in use. Much ingenuity is often exhibited by the adaptation of old blades to new handles—in converting the large, flat blacksmithing files into keen weapons, and making fish-cleaning knives out of pieces of iron; thin, square or oblong sheets of this metal are so fitted into oblong wooden handles as to resemble the small hash-knives used in our kitchens.

But the Sitkan housekeeper glories in her boxes—great chests and little ones—in which she stores everything of value belonging to the family, except the dogs and the canoes. The big boxes, corded up with bark ropes, are her blanket and fur treasuries ; the smaller ones contain her oolachan "butter" and dried fish and meat. The larger chests are from two and a half to four feet square ; the lesser are between a foot to two feet. The sides of such a box are made of a single piece of thinly shaven cedar board, which by steaming is bent three times at a right angle, and pegged tightly and very neatly up to the fourth corner. The bottom is a separate solid plank, keyed in with little pegs very solidly, and water-tight ; the cover is cut out of a thick slab, and fits over and sets down heavily on the upper edge of the chest. Those boxes are all decorated in designs of the peculiar type so common among these savages, painted in black and white. The next desideratum of the squaw is a full supply of cedar-bark mats, which she plaits from strips of this material, and which are always spread out on the ground or rude plank floor when the Indian prepares to roll up in his blanket for slumber. Such mats are the pride of all Thlinket squaws, and vary much in texture and in pattern.

But the daily routine of the dusky housekeeper is a very different one indeed from that characteristic of woman's labor in caring for our homes. No sweeping or dusting in the Indian rancherie ; no bed-chambers to change the linen in and tidy up ; no kitchen or servants to look after ; nothing whatever of the kind. Yet the Indian matron is always busy. She has to hew the firewood and drag it in ; she has to carry water and attend to all of the rude cooking and filling of the trenchers ; she looks after the mats and the sewing of the children's fur and other garments—not much to be sure in the way of dressmaking—she has to make all of the tedious berry-trips, picking and drying of the fruit, as well as attending to the preservation, in the same manner, of the fish and game which

the man brings in.   She has an infinite amount of drudgery to do
in the line of gathering certain herbs, bark, and shell-fish.   Many
small roots indigenous to the country, containing more or less
starch, are eagerly sought after, dried, and stored away by the
women.   The inner sap-layer of the spruce and also that of the
hemlock—the cambium layer—is collected by cutting the trees
down and then barking the trunks for that object.   It is shaved off
in ribbons and eaten in great quantities, both fresh and dried, and
is considered very wholesome.   It is sweet, mucilaginous, but distinctly resinous in flavor.   The rank-growing seeds, shoots, and
leaf-stalks of the *Epilobium heracleum*, and many others, are
plucked and carried by the squaws in huge bundles to the family
fire, and there eaten by all hands, the stalks being dipped, mouthful after mouthful, in oil.

She has, however, no washing whatever of clothes to do for anybody, except what little she may see fit to do for herself; she never
treats the dishes even to that ordeal.   With all this, however, it
seems rather strange that the clothes of the Indians, consisting of
dresses, shirts and blankets for the men ; and for women, petticoats,
chemises, dresses (sometimes), and blankets also—that these articles
usually appear neat and tolerably clean—the children excepted, as
they are always dirty beyond all adequate description.   Every individual attends to his or her own washing—if the husband wants a
clean shirt, he washes it himself.

Before the introduction of the potato through early white fur-
traders, the only plant cultivated by the Alaskan savages was a potent weed which they grew as a substitute for tobacco—the importation of the latter, however, has taken its place entirely to-day, because the Virginian weed is far more pleasant.   But the old stone
mortars and pestles that are still to be found knocking around
the most venerable town-sites, bear evidence to the industry of
making native tobacco here ages ago.   This plant was prepared
for use by drying over a fire on a little frame stretcher, then bruised
to a powder in the stone mortars, then moistened and pressed into
cakes.   It was not smoked in a pipe, but, mixed with a little clam-
shell lime (burnt for the purpose), it was chewed or held in the
cheek, just as the Peruvian Indians use coca.*   Everybody knows

---

* This accounts for the puzzling appearance of ancient stone mortars and
pestles in Alaska, throughout the Sitkan region.   Ethnologists have endeav-

how fond Indians are of tobacco—there is no exception to the rule
in Alaska, and no excuse for attempting to recite in these pages the
well-worn story anew.

No domesticated animals, except dogs, are to be found with the
Alaskan Indians—no cats or fowls. The original breed of curs has
been very much disguised by imported strains ; the present natives
are gray and black, shaggy, wolfish beasts, about the size of a large
spitz dog. These cowardly, treacherous animals alone make a white
man's stay in an Indian village a burden to his existence.

The work bestowed by several of the Sitkan clans upon their so-
called potato gardens is hardly to be designated as the "cultivation" of that tuber. It forms to-day, this vegetable does, a very
important part of the food-supply, and where a white man takes·
hold of such a garden the result, in a small way, is very satisfactory ; but the Siwash finds that the greater part of the low, flat,
rich soil in this country is so thickly wooded that the task of clearing the ground is altogether too much for him to even consider,
much less undertake. But when he can find a place where an old
settlement once existed, though long abandoned—there the sites
of decayed rancheries are sure to be of rich, warm soil—such are
the spots which the Siwash calls his garden, and where his potatoes
are rudely planted, little or no attention being paid to the hoeing
and drilling which we deem essential, therefore the variety in use
has been run down so that the size and yield is very small, and the
quality watery and poor.

While we observe the very general possession of firearms in
every rancherie, and we hardly ever see a canoe-load of savages
unless the barrels of several muskets or rifles project over the
gunwale, yet these Sitkan Indians are not great hunters ; but the
potent fact that there is no place in all this region where foot travel
is practicable into the interior, or even along the coast margin itself, affords an excellent reason ; they do, however, kill a very considerable number of black bears every year, at two special seasons
therein, i.e., when these brutes are found prowling upon the sea-
beach. But they never follow bruin into the mountainous recesses, where he invariably retreats.

ored to reason that certain extinct tribes must have cultivated grain up here
of some kind and used it as food. I am indebted to the venerable Dr. W. F.
Tolmie for this fact, he showing me the mortars and giving the reason of their
use in December, 1866, at Victoria, B. C.

In the early spring, during that brief period when the weeds and grasses first grow green along the outskirts of the timber in warm sheltered nooks down by the tide-level, black bears come below from the cold, gloomy cañons above and feed upon the sprouting skunk-cabbage * and other succulent shoots, browsing here and rooting up there, these tender growths, just as hogs do in our orchards and clover-fields. Again, late in autumn, when the salmon rush up into the estuaries and through the shallows by countless myriads, bruin is once more tempted down to the sea-beaches, and again gets into trouble. In the same manner the Indians secure the beautiful little mule-deer,† which also loves tender vegetation, and in this love falls an easy prey to the silent approach of a canoe with its skulking crew. Geese and ducks, during winter months, spend much time on the quiet fiörds in large flocks, and constitute the chief gunning of the Siwashes, who shoot them from their canoes with the same old flint-lock trade-muskets first used by the whites a hundred years ago. The Indian admires this pattern still above all other patterns—despises the percussion-cap, which in this damp region often fails him, and the trader, knowing this weakness of the savage, always has a stock of these flint-lock muskets, newly made, on hand. A supply is steadily furnished by the Hudson's Bay Company at Victoria.

But the one thing of joy, of delight, and of infinite use to the native of the Sitkan archipelago is his canoe. Life, indeed, would be a sad problem for him were it not for this adjunct of his own creation. Upon its construction he lavishes the best of his thought, the height of his manual skill, and his infinite patience. The result of this attention is to fashion from a single cedar log a little vessel which challenges our admiration invariably, for its fine outline and its seaworthiness and strength.

All the canoes of this region have a common model, and are similar in type, though they differ much in details of shape and size. They are all made from the indigenous pine ‡ and giant cedar,§ the wood of which is light, durable, and worked very readily ; but it is apt to split parallel to its grain. This constitutes the only solici-

---

* *Lysichiton sp.*

† *Cervus columbianus*—a well-grown specimen weighs about one hundred and fifty pounds. Great numbers are taken in the Tahkoo region, though it is found everywhere.

‡ *Abies sitkensis.*   § *Thuja gigantea.*

tude of the Indian's mind. He keeps the canoe covered with mats and brush whenever it is hauled out, even for a few days, to avoid this danger, for whenever a canoe is heavily laden, and working, as it will do, in a rough channel, it is in constant danger of splitting at the cleavage lines of its grain, and thus jeopardizing its living as well as dead freight.

With an exception of the bow and stern-pieces, each canoe, no matter how large or how small, is made in the same manner and from a single log, which is roughed out in the forest, then towed around to the permanent village, where it is hauled up in front of the architect's house. Here he works upon it during winter months, usually in odd hours, employing nothing but his little adze-like hatchet and fire to assist in giving it shape and fine lines. The requisite expansion amidships, to afford that beam required, is effected by steaming with water and hot stones and the insertion of several thwart sticks. Canoes are smoothed outside and painted black, with a red or white streak under the gunwale in most cases; inside they bear the regular fine tooth-marks of the excavating adze, and are smeared with red-ochre. The paddles are usually made of yellow cypress, and a great variety of small wooden baling dippers are also provided, one or two for each canoe, because the water often slops over the gunwales in bad weather. The canoe itself is never suffered to leak. The average size is one of fifteen to twenty feet in length, which will carry from eight to ten savages, with baggage. One having a length of from thirty to thirty-five feet carries as many men. The smaller canoes of from twelve to thirteen feet are usually used by one or two savages in their quick, irregular trips to and from the village, and are easily launched and hauled out by one man.

It is very doubtful, indeed, whether the Sitkan ever took or takes any real enjoyment in hunting or fishing. If he does, it is never exhibited on his countenance or evidenced by his language. It is, in fact, the serious business of his life, and the steady routine of its prosecution has robbed him of every enjoyable sporting sensation which we love to experience when after fish or game. Perhaps, however, he may recall the thrill of that feeling which he felt when, as a boy, he was first taken out in his father's canoe to the halibut banks, and there permitted to bait a huge wooden hook and haul away upon the taut kelp-line when the "kambala" had swallowed it; but the necessity of going out to this shoal in all sorts of disagreeable weather every summer and winter of his sub-sequent existence, at very frequent intervals, soon destroyed pleas-urable emotions. Therefore, he fashions his acute-angled wooden hooks, his iron-tipped fish and seal spears, and polishes up his mus-ket with none of those enjoyable anticipations which possess the soul of a white sportsman.

In 1841–42 the best understanding of the Russian and English traders agreed in reporting a population of over twenty thousand Indians within the limits of the Alexander archipelago; to-day the same country can show no more than a scant seven thousand. The inroads that small-pox and measles have made, by which these savages were destroyed even as fire sweeps through and burns drought-withered thickets, leave little doubt as to the great numer-ical superiority of earlier days as compared with the present.. This decay and abandonment is everywhere exhibited now even in the per-manent villages, where houses have been deserted completely: some are shut up, mouldering, and rotting away upon their foundations; others, large and fit for the shelter of fifty or sixty natives, will be found tenanted by only two or three Siwashes. All the standing carved posts in this entire region, with rare exceptions, are, as a rule, more or less advanced into decay. A rank growth of weeds, dark and undisturbed in some cases, presses up close to inhabited houses, the traffic not being sufficient to keep them down. The original features of these settlements, in a few years more of this unchecked neglect and decay, will have entirely disappeared as they have already at Sitka. At the present hour, however, we can go among them, and readily call up to our minds what they once were when they were swarming with occupants who were dressed in tanned-leather shirts and sea-otter cloaks, as they thronged about the ships of Cook and Vancouver.

Slavery, which was originally firmly interwoven with the social fabric of these people, has been about abolished—slaves themselves to-day are very scarce, and are not much more so than in name. They were the captives taken in savage warfare between opposing clans, and were most horribly tortured and cruelly treated by their masters.

As a rule the young people marry young, after the stolid fashion of Indians. They approve of polygamy, but seldom do you find a man with more than one squaw, simply because the women do not contribute materially and primarily to the support of the family, and attend only to the accessory duties of it ; thus it becomes an increased tax upon the dull energies of the savage whenever he adds an extra woman to his household. The squaws are all well treated everywhere up here ; they have just as much to say as their lords and masters whenever the occasion of buying, selling, or hiring arises ; as to the children (we will not see many of them to-day), they are always kindly cared for by both parents, and the whole tribe is as indulgent, since they are constantly roaming about the village, after the custom of youngsters universally.

A candid verdict will result, in view of the surroundings of the Koloshian, that the only vice which can be legitimately charged up against him, or his kind, is the sin of gambling. To this dissipation the Alaskan savage is desperately prone ; the monotonous chant of the stick-shuffling players is ever on the air in the villages. These worthies sit on the ground, in a circle usually, in the centre of which a mat is spread ; six or seven small wooden pins about as large as the little finger of your hand, upon which various values are marked or carved, are taken into the hands of the first gambler, who thrusts them into a ball of soft teased cedar bark, or holds them under his blanket, then shuffles them rapidly, meanwhile shouting a deep guttural *hah-hah-ee-nah-hah!* the others watch him with lynx-like eyes for a few moments, when one of the players suddenly orders the shuffler to show his hands, in which the sticks are firmly clinched, and at the same time endeavors to guess the value of these sticks in either one hand or the other, which have been held up—he pauses a moment, then makes his decision, the clinched hand designated is opened, the little sticks fall to the mat, and the caller wins or loses just as he happens to hit the value expressed by the markings on these pins : if he guesses correctly he wins everything in the pot or pool, and takes up the wooden dice in turn, to shuffle, shout, and repeat for the rest of the circle. This game is usually sustained night and day, until some one of the party remains the winner of everything that the others started in with.

That wretched debauchery which an introduction of rum into the rancheries of these natives has caused, cannot be justly laid at the Indian's door ; this intense morbid craving for liquor among the Alaskan savages of this region is most likely due to the climate —it is not near so strong in the appetite of the natives who live east of the coast range. Although Congress has legislated, and

5

our officials have endeavored to carry out the prohibition statutes, yet the matter thus far is wholly beyond control—the savage cannot only smuggle successfully within these intricate watery channels, but he now thoroughly understands the distillation of rum itself from sugar and molasses.

There is something in this atmosphere which enables a white man to drink a great deal more with impunity than he can in any other section of the United States or Territories—the quantities of strong tea, the nips of brandy, wine, and cordials which he will swallow with perfect physical indifference, in the course of every day of his life, at Sitka for instance, would drive him to delirium in an exceedingly short time if repeated at San Francisco. Naturally enough, we find that the same craving for stimulants is reflected by Indian stomachs; and now that they have fully grasped the understanding of how to successfully satisfy that aching, no valid reason can be presented why the Thliuket will not continue to gratify a burning desire in this fatal direction to the ultimate extinction of his race. This fault of our civilization is far more potent to effect his worldly degeneration, than any one or all of our combined virtues are to regenerate his earthly existence.

CHAPTER IV.

THE ALPINE ZONE OF MOUNT ST. ELIAS.

The Hot Spring Oasis and the Humming-bird near Sitka.—The Value and Pleasure of Warm Springs in Alaska.—The Old "Redoubt" or Russian Jail.—The Treadwell Mine.—Futility of Predicting what may, or what will not Happen in Mining Discovery.—Coal of Alaska not fit for Steaming Purposes.—Salmon Canneries.—The Great "Whaling Ground" of Fairweather.—Superb and Lofty Peaks seen at Sea One Hundred and Thirty-five Miles Distant.—Mount Fairweather so named as the Whalemen's Barometer.—The Storm here in 1741 which Separated Bering and his Lieutenant.—The Grandeur of Mount St. Elias, Nineteen Thousand Five Hundred Feet.—A Tempestuous and Forbidding Coast to the Mariner. —The Brawling Copper River.—Mount Wrangel, Twenty Thousand Feet, the Loftiest Peak on the North American Continent.—In the Forks of this Stream.—Exaggerated Fables of the Number and Ferocity of the Natives.—Frigid, Gloomy Grandeur of the Scenery in Prince William Sound.—The First Vessel ever built by White Men on the Northwest Coast, Constructed here in 1794.—The Brig *Phœnix*, One Hundred and Eighty Tons, No Paint or Tar.—Covered with a Coat of Spruce-Gum, Ochre, and Whale-oil, Wrecked in 1799 with Twenty Priests and Deacons of the Greek Church on Board.—Every Soul Lost.—Love of the Natives for their Rugged, Storm-beaten Homes.

A BRONZED humming-bird* lies upon the author's table, that once hovered and darted over the waters of Sitka Sound. Its torn and rudely stuffed skin was given to him at Fort Simpson with the remark that it came from the hot springs just below New Archangel ; and that nowhere else in all of a vast wilderness, outside of the immediate vicinity of these springs, ever did or could a humming-bird be found. Should, therefore, a visitor to this Alaskan solitude chance to travel within it during the months of April and May, if

---

* *Selasphorus rufus*—it is common in California, Oregon, and parts of Washington Territory, and Southern British Columbia—never found north of Victoria on the coast, except as above stated ; it winters in Central America.

he will but follow the path of that wee brave bird, he will be led into a veritable green and fragrant oasis, encircled all round about with savage icy mountains and snowy forests.

Twenty miles south of Sitka, on the same island, in a pretty little bay sheltered by a score of tiny islets, there—from the sloping face of a verdant bank, the finest hot springs known to Alaska flow up and out to the sea. Fleecy clouds of steamy moisture rise over all to betray from a distance this delightful retreat ; the luxuriant vegetation, the variety of shrubs in full blossom here, when all botanical life about them is as dead as cold can make it, create thereon a spot in the early spring where all the senses of a traveller can rest with exquisite pleasure—the waters of the bay in front are covered with geese and ducks, while the rugged mountains that rise as a wall behind are teeming with deer and bear and grouse, secluded in the jungle.

The Indians, from time immemorial, have resorted to these hot waters of Baranov Island ; four distinct and freely flowing springs take their origin in those crevices and fissures of the feldspathic granite foundation of the earth hereabouts ; the temperature of the largest spring, at its source, is 150° to 160° Fah.; the waters are charged with sulphur to a very great extent. So jealous were the savages of any attempt among themselves which might savor of a monopoly of the use of these healing, beneficent warm streams, that no one tribe ever dared to build a village upon the site ; but, by tacit consent, all were allowed to camp thereon. Some Indians often came from a distance of three hundred miles away to enjoy the sanitary result of bathing here, a few days or a few weeks, as their troubles might warrant.

Naturally the Russians, burdened at Sitka with all diseases which flesh is heir to, turned their attention very promptly to this sanitarium ; they erected a small hospital and two spacious bathhouses over the springs, keeping everything in the strictest order and cleanliness, without and within doors. A sad change confronts us to-day—in so far as care of human hands ; but the savage Sitkan is here, exulting in his renewed supremacy.

The occurrence, however, of hot springs is quite frequent everywhere in this archipelago ; yet their extent and volume of outflow is not so great as evidenced by those we have just noticed of Baranov Island. Indians love to immerse their entire bodies in pools and eddies of these hot rivulets, which are cooled sufficiently by flowing a dozen or fifty yards from their origin over pebbly bottoms; Siwashes will soak themselves in this manner for hours at a time, with nothing but their heads visible. Though the Koloshian, like all others of his kind, never verbally complains, yet he is subject to acute rheumatism, to fevers, and to divers malignant cutaneous diseases; these springs, wherever known to him, are always well regarded as his happy relief and hope. Certain it is that when you behold the parboiled skin of a native, after bathing here, the fair almost white complexion really startles, for, prior to the immersion, he was a coppery brown or black.

Midway between these thermal fountains and Sitka is the site of an old Russian jail or prison; in a deep inlet, with no land in sight, but lofty mountains rising abruptly from the water's edge, is the "Redoubt." Here a small alpine lake empties itself in a foaming cascade channel of a few yards in width, that quickly plunges into a cañon, the perpendicular walls of which are a full thousand feet in mural height. The Russians erected mills of various kinds along the rapids to avail themselves of such abundant water-power; the buildings stood upon a bare rocky portion of the channel, and were kept in order by an old veteran in command; a squad of soldiers aided him; the fish, dried and salted salmon, which were required for the use of the company, were annually caught here as they swarmed up the cascade from the sea, into Gloobaukie Lake.

The great facility of travel afforded by these sheltered canals of the Alexander archipelago, has enabled and facilitated a most energetic and persistent search for gold and silver by our miners, but the rugged features of the country and its dense timber and jungle have rendered the progress of such investigation slow, and one of great physical difficulty. In the sands of every stream flowing between California and Cook's Inlet the "color" of gold can be found, but the paying quantities therein seldom warrant a mining camp or settlement. To-day the only mining rendezvous which we find in Alaska is a little village of rough cabins called "Juneau City," located on the north side of Gastineaux Channel, at a point near the upper end of that passage; near by, and adjacent, is established a large gold-quartz stamp-mill* on Douglas Island,

---

* The Treadwell Mine—free-milling gold ore; 120 stamps; employs 150 to 250 men—situated right at the tide-level.

where the mining experts feel justified in predicting a steady and inexhaustible yield of paying ore—it is paying handsomely at present.

This subject of what is, or what is not, a good mining region or investment is one to which no rational man can well afford to commit himself. Those who have had extended experience in these matters know that it is a topic which baffles the best investigator, and returns no safe answer to the most intelligent cross-examination. The true advice which can be honestly given is that which prompts every man interested to look and resolve wholly for himself, for he, in fact, knows just as much as anybody else. At the most, the finding of a rich or desirable lead of gold or silver in a new country is an accident or sheer opportunity of chance. Whether it will hold out, or end in a "pocket," is also only to be determined by working it for all it is worth. Once in a while a man makes a rich find, and is rewarded ; but an overwhelming majority of prospectors are ever wandering in fruitless, restless, tireless search for those golden ingots which are still hidden in the recesses of mountain ledges, or buried in the alluvium of river bottoms. The miners in Alaska embrace various nationalities—Australians and Canadians, Cornishmen and Californians, Oregonians and British Columbians predominate—but the number aggregated is not large.*

If gold or silver-quartz mines of free-milling ore (no matter how low the grade) can be located anywhere on the shores of these mountainous fiörds of the Alexander archipelago, their wealth will be great, because the transportation to them and from them is practically without cost. The expense of working such valuable quartz mines up a hundred or more miles from the sea, will result in abandonment, where reaching them involves frequent transfers of supplies, and the working season is cut by the rigor of winter to less than half or one-third of every year. The same mines, down within the dockage of an ocean-steamer in the Sitkan district would be a steady source of wealth and industry all the year round.·

The coal which is found here is not satisfactory for steamers' use—too heavily charged with sulphur. Copper ore is well-known, but not worked in competition with the Lake Superior and Arizona cheap outputs. At the present writing there are no active indus-

---

* Eight hundred, or a thousand, perhaps. They come and go suddenly, alternating in travel as the rumors relative to their occupation circulate.

tries whatsoever in the Sitkan archipelago beyond the energetic stamp-mill of the Treadwell Mine on Douglas Island, and the limited placer diggings of Juneau City. Until a market is created' for its large natural resources of food-fishes, the little canneries which our people have started here will not develop ; nor will the timber be of much commercial importance until the great reservoirs of the lower coast are exhausted. Statisticians and political economists can easily figure out the time when a population of twenty-five or thirty millions of our own people will be living upon the Pacific coast alone : then the real value of those latent resources * of the Sitkan watery wilderness must be patent to a most indifferent calculator.

With this survey of the Alexander archipelago fixed on our minds, we pass from it through the bold Cross Sound headlands that loom above those storm-churned swells of an open ocean, which break here in unceasing turmoil, and we sail out into an area that charts tell us is the "Fairweather ground," over which that superb peak itself and sister, Crillon, stand like vast sentinel-towers, rearing their immense bulk into many successive strata of clouds, until the elevation of thirteen thousand and fourteen thousand feet is reached, sheer and bold above the sea. This great expanse of the Pacific Ocean between us and Kadiak Island, five hundred and sixty miles to the west, and again down to Victoria, nine hundred miles farther to the south, was the rendezvous of the most successful and numerous whaling fleet that the history of the business records. In these waters the large "right" whale did most congregate, and the capture of it between 1846 and 1851 drew not less than three and four hundred ships with their hardy crews to this area backed by the Alaskan coast. They never landed, however, unless shipwrecked, which was a rare occurrence, but cruised "off and on" with the majestic head of Mount Fairweather as their point of arrival and departure.

---

* A few small saw-mills have been erected at several points in this Sitkan district to supply the local demand of trading-posts and mining-camps. With reference to quality or economic worth, the timber found herein may be classified as follows, in the order of its value : 1. Yellow cedar (*Cupressus nutkaensis*) and *Thuja gigantea*, the red variety. 2. Sitkan spruce (*Abies sitkensis*). This is the most abundant. 3. Hemlock (*Abies mertensiana*). 4. Balsam fir (*Abies canadensis*). The finest growth of this timber is found upon Prince of Wales Island, Admiralty, and Kou Islands, within the Alaskan lines.

When the whalemen saw the summit of that snow-clad peak un-veiled by clouds they were sure of fair weather for several consec-utive days afterward, hence the name. Early one June morning Captain Baker, of the *Reliance*, called the author up to see a moun-tain which was sharply defined in the warm, hazy glow of the dawn-ing sunrise on the horizon—there, bearing N.N.E.,* was the image of Mount Fairweather, just as clear cut as a cameo, and lofty as the ship's spars, though one hundred and thirty-five miles distant ! Closely associated and fully as impressive and quite as high, was the heavier form of the snowy Crillon.

That long stretch of more than four hundred miles of bare Alas-kan coast, between Prince William's Sound and Cape Spencer, which stands at the northern entrance to the Sitkan waters, is one that sustains very little human or animal life, and is so rough and is so bleak, that from September until May it is feared and avoided by the hardiest navigator. The flanks of Mounts Fairweather and Cril-lon rise boldly from the ocean at their western feet, and this sheer-ness of elevation undoubtedly gives them that effect of cloud-com-pelling, which does not lose its awe-inspiring power even when a hundred miles away. To the northward and westward of Fair-weather, however, the alpine range which it dominates abruptly sets back from the coast some forty or fifty miles, then turns about and faces the sea in an irregular, lofty half-moon of more than three hundred miles in length. A low table-land, or rolling shelf, is ex-tended at its base, intervening between the mountains and the wash of the Pacific. It is timbered with spruce quite thickly, and re-ported by the Indians to be the best berrying ground in all Alaska.

The Fairweather shore is a steep, woody one, much indented with roadstead coves or bays ; the coast line is hilly and uneven, with some rocks and rocky islets scattered along not far out from the surf. The sand-beaches which extend from Fairweather toward the feet of those under St. Elias are remarkably broad and exten-sive ; so much so that, from the ship's mast-head, large lagoons within the outer swell of the open ocean are frequently seen. These beach-locked estuaries communicate with the ocean by shal-

---

* Tuesday, June 13, 1874. It did not seem possible at first that the officer's observations were accurate, but the captain verified the ship's position anew, and confirmed the correctness of Lieutenant Glover's entry and sights : " bear-ing N.N.E., 135 m."

low breaks in the outer beach-wall of sand and gravel, across all of which the sea rolls with great violence.

Right under the towering slopes of Fairweather, as at St. Elias, is a large area of upland entirely destitute of verdure of any kind, except the brown and russet mosses and lichens ; huge, rugged masses of naked rocks are strewn about in every direction—an old prehistoric lava-flood, perhaps.

The coast, from the head of Cross Sound to Fairweather, is not sandy, but may be well described as the surf-beaten base of a frozen range of magnificent Alpine peaks.

In the centre of the arc of this grand crescent-range is the superb body and hoary crest of Mount St. Elias, which is, save Mount Wrangel, now known to be the loftiest peak on the North American coast ; the latter is slightly higher.   Triangulated from a base line in Yakootat, in 1874, by one * of the most accomplished mathematicians of the U. S. Coast Survey, the summit of that royal mountain was determined to be more than nineteen thousand feet above the level of the tide at the observer's feet.   It was under the shadow of this " bolshoi sopka " that Bering first saw the Continent of North America on the 18th of July, 1741, and undoubtedly he discerned it from a long distance, ere his boat landed.   Two days before anchoring, he records the fact that " the country had terrible high mountains, which were covered with snow."

When he finally landed (it was St. Elias' day), near a point that he named as he named the lofty central peak, Cape St. Elias, he found the temporary summer-houses of a band of natives ; those people themselves had fled in terror from an unwonted invasion, but the Russians soon had reason to regret their subsequent better understanding.

After the storm which parted Bering, early in June, from the company of the second vessel of his expedition, he had hoped to fall in with her ever afterward, and while eagerly scanning the coast and horizon about him for some sign of his lost comrades, the hand of fate caused him to turn to the northward, when, had his helm been set south, he would have met the object of his search. For the other vessel, the *St. Paul*, had proceeded on its solitary

---

* Marcus Baker.   Unfortunately no one connected with this Coast Survey Party was able to make an adequate drawing of the mountains, and it was so enveloped in clouds as to be partially invisible when the author cruised under its lee.

course, and anxiously sought the commander, until it, too, had sighted this same coast, three days earlier than had its storm-separated consort. Tschericov came to anchor off some distance from "steep and rocky cliffs"* in "lat. 56°," July 15. Weary and expectant, the captain sent his mate with the long-boat and a crew of ten or twelve of his best men away to the shore for the purpose of inquiry and for a fresh supply of water. The ship's boat disappeared behind the point sheltering a small wooded inlet; it and its men were never seen again by their shipmates. Troubled in mind, but thinking that the surf, perhaps, had stove the boat in landing, the captain sent his boatswain in the dingy with five men and two carpenters, all well armed, to furnish the necessary assistance. The small boat disappeared also, and it, too, was never seen again. At the same time a great smoke was constantly ascending from the shore. Shortly afterward two huge canoes, filled with painted, yelling savages, paddled out from the recesses of the bay, and lying at some distance from the ship, all howled, in standing chorus, "Agai—agai!" then, flourishing their rude arms, they rapidly returned to the shore. Sorrowfully the disturbed and distressed Tschericov turned his ship's course about and hurried home,† not knowing the fate of his men, unable to help them, and, to this day, no authentic inkling of what became of these Slavonian seamen has ever been produced. Unquestionably, they were tortured and destroyed.

The rains caught in the ship's sails filled the casks of the *Saint Paul*, since Tschericov, deprived of his boats and thoroughly alarmed, made no further attempts to land; but he had not the faintest idea of the presence, at that moment, of his superior officer in the same waters, and only a few leagues to the northward, who also, like himself was eagerly looking for his storm-parted consort. What a most remarkable voyage, this voyage of the discovery of

---

* That point, most likely, was Kruzov Island, and the bay into which the unhappy Russians were decoyed was Klokachev Gulf. This island forms the western shore of Sitka Sound.

† He reached Kamchatka on the 9th October following, with only forty-nine survivors out of his original crew of seventy. Bering never did ; he was shipwrecked and died on a bleak island, of the Commander group, December 8, 1741. They seem to have really sailed over this course of six thousand miles almost together, anxiously searching for each other, yet unconscious of their proximity.

the Alaskan region—what a chapter of disappointment, of hardship, and of death!

That bluffy sea-wall which forms a face to the low coast plateau at the feet of the St. Elias Alps is cut by no great river, nor indented by any noteworthy gulf or inlet, except at Yakootat Bay. Here a succession of precipitous glaciers sweep down from the lofty cradles of their birth to the waters of the sea, making an icy cliff of more than fifteen miles in breadth, where it breaks in constant reverberation and repetition. At the mouth of Copper River all silt carried down from old eroded glacial paths has been deposited for thousands and thousands of years, until a big deltoid chart of sea-water channels in muddy relief of bank and shoal has been formed, and through which the flood of an ice-chilled river takes its rapid course.

The gloomy, savage wildness of this region of supreme mountainous elevation, with its vast gelid sheets and precipitous cañons, its sombre forests and eternal snows, all as yet wholly unexplored, and only faintly appreciated as we can from the remote distance of shipboard observation—this region cannot remain much longer untrodden by the geologist and the naturalist, while the artist must accompany them if an adequate presentation is ever to be given of its weird, titanic realities.

The Mount St. Elias shore-line is made up of small projecting points, awash. These alternate with low cliffy or else white sandy beaches, which border a flat, rolling woodland country that extends back from the sea ten to thirty miles, where it suddenly laps and rises upon the lofty flanks of the Elias Alps. Into the ocean many rocky shoals and long sandy bars stretch for miles, and streams of white muddy glacial or snow waters rush into the surf at frequent intervals—hundreds of them.

There are sand-beaches and silt-shoals which extend from Cape Suckling, up seventy-five miles to Hinchinbrook Island, that stands as a gate-post to the entrance of Prince William's Sound: here is a long sand-ridge which is more than sixty miles in length and from three to seven miles broad, lying between the ocean and the mainland, which in turn is composed of low wooded uplands and of steep abrupt cliffs and hills that are quickly lost in the lofty snowy range of the Choogatch Alps. Through a section of this dreary sand-wall the impetuous flood of the Copper or Atna River forces its way, carrying its heavy load of glacial mud and silt far

into the ocean. How the winds do blow here! How the trader
dreads to tarry "off and on" this coast!

There are a few lonely places in this world, and the wastes of the
great Alaskan interior are the loneliest of them all. Those of Siberia are traversed occasionally by wandering bands, but those of
Alaska, never. The severe exigencies of climate there are such as
to substantially eliminate savage life, and to rear an impregnable
barrier to that of civilization.

When Alaska was first transferred, an estimate of many thousands of Indians inhabiting its vast interior was gravely made and
as gravely accepted by us; but a thorough investigation made by
our traders and officers of our Government during the last fifteen
years has exposed that error. Hundreds only live where thousands
were declared to exist. The Indians who live on the banks of the
Copper River are, perhaps, the most poverty-stricken of all their
kind in Alaska. Their shiftless spruce-bark rancheries and rude belongings are certainly the most primitive of their race, and render
that weird Russian legend of the massacre of Seribnickov in 1848,
which declared them so numerous and savage, absolutely grotesque.
They are perfectly safe as they live in their wild habitat. The cupidity of savage or civilized man never has and never will molest
them. But if half is true as to what they relate of huge glaciers
which empty into their river, then those that have been described
in Cross Sound have formidable rivals, which may yet prove to be
superiors, perhaps, although it seems incredible.

The Suchnito or Copper River has long been a bugbear, for the
Russians * years ago have returned from several unsuccessful at-

---

* When the surveying parties of the War Department were ascending Copper River last summer, certain Indians, who had been instrumental in slaying
the Russian party of Seribnickov in 1848, were very much alarmed. They
were sure that the fates had come for them at last. One of these natives, an
aged man, now wholly blind, was reported as saying that he was ready to die,
and knew what the white men wanted. This old fellow, Lieutenant Allen
says, was one of the finest-looking savages that he ever saw. The face of the
blind man was one of remarkable character—a large, massive head, high
aquiline nose, with a full, thin-lipped mouth and broad forehead. He was
totally blind and his hair white as snow.

The Russian party were sleeping in their sledges, which they compelled
the natives to draw while ascending the river. At a preconcerted signal the
unwilling Indians turned and brained their taskmasters with hatchets. These
natives had welcomed the Russians; but when they were made to perform

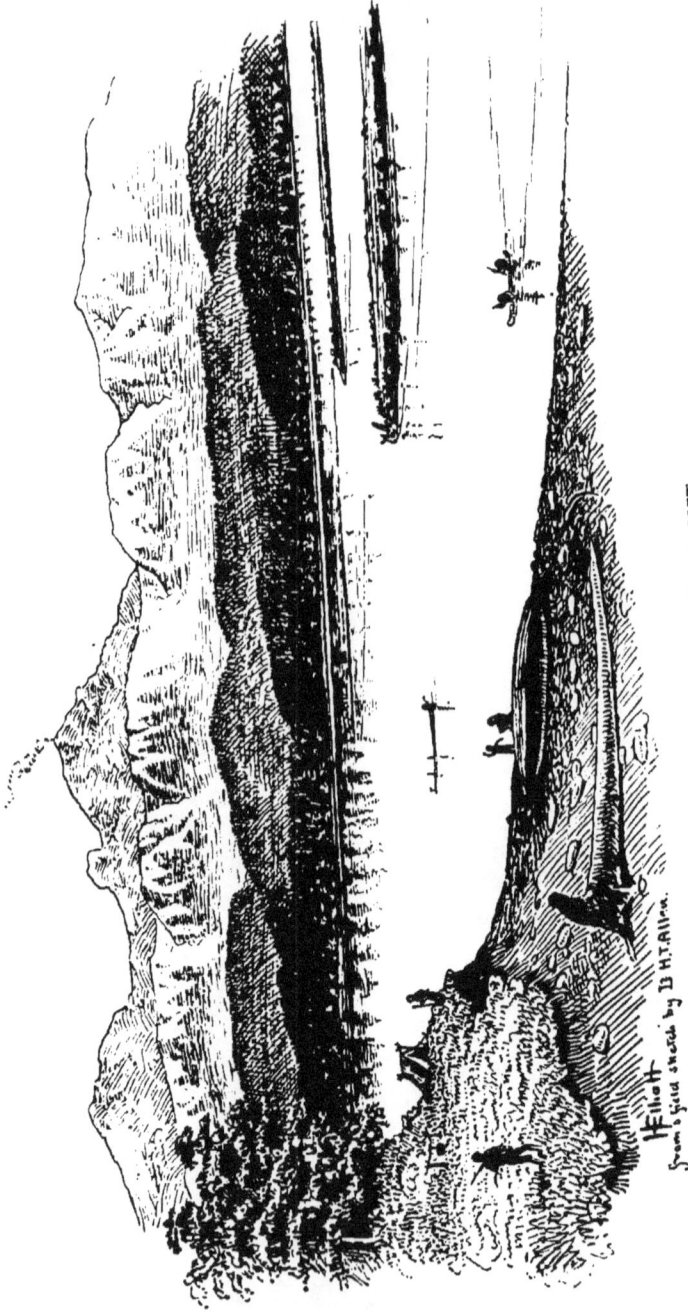

MT. WRANGEL : 20,000 FEET

In the Forks of Copper River : it is the loftiest Mountain on the North American Continent

tempts to ascend it, and gave the excuse of being driven out of the valley by savage and warlike natives. Recently it has been thoroughly explored, and the "savages" are found to be less than two hundred inoffensive natives, who constitute the whole population of this mysterious Atna or Maidnevskie region. But navigating the river is terrific labor, inasmuch as it is a continuous, swift rapid throughout its entire course.

This river is a short, turbulent, brawling stream, less than two hundred and fifty miles in length, but rising in the heart of a lofty and mighty mass of volcanic mountains. It receives a score of imposing glaciers, which almost rival those of Icy Bay in Cross Sound. The silt that these gelid rivers pour into its channel has given it a deltoid mouth of extended and most intricate area.

Triangulations made by an officer* of the Army last year declare that Mount Wrangel is the loftiest peak on the North American continent. The feet of this magnificent volcanic dome are washed by the forks of Copper River, which is eighteen thousand six hundred and forty feet below the apex of its smoking cap. Then the river at this point is more than two thousand feet above sea-level, so the vast altitude of more than twenty thousand feet for Mount Wrangel seems to be truthfully claimed.

The soil which borders the abrupt banks of the Copper River is entirely composed of glacial silt and gravel. It is moist and boggy in the driest seasons, covered with rank growing grasses and dense thickets of poplars, birches, and willows, that line the margins of the stream. The higher lands, as they rise from the narrow valley, are in turn clothed with a dense growth of spruce-forest, which gradually fades out into russet-colored areas of rock-sphagnum as the altitude increases to that point where nothing but the cold and frost-defying lichen can cling alive to the weather-splintered summits of alpine heights above.

Fish (salmon) are the chief reliance of these natives of Copper River; they depend almost wholly upon the annual running of those creatures. The difficulty of hunting is so great that the

---

the labor of dogs they turned upon their white oppressors, naturally. The massacre of Scribniekov and his party in this manner made the Indians very restless and determined in their opposition to further intercourse with the Russians. The memory of hostility has, however, died out, and nothing of the kind was shown to our people last year as they charted the valley and river.   Lieutenant H. T. Allen.

savage is content with shooting a few mountain sheep, a wandering moose or two, and, perhaps, a stray bear in the course of the year. Also, huckleberries and salmon berries are abundant on the sunshiny slopes of the high glacial river-terraces during August and September.

West of the Copper River mighty masses of the Choogatch Mountains rise directly from the sea without any intervening lowland, save at three tiny points upon which savage man has hastened to fix his abode. Many crests to this range on the north side of Prince William's Sound must have a mean elevation of over ten thousand feet, densely wooded with semper-virent coniferous forests up to a height of one thousand feet above sea-level, and covered with everlasting snowy blankets to within three or four thousand feet of the ocean at their bases. The body of Prince William's Sound is so forbidding in its dark grandeur that even the stolid Russians never tired of narrating its stirring impression upon their senses. Although the interior of this gulf is completely landlocked, being sheltered from the south by the islands of Noochek and Montague, yet it is by no means a safe or pleasant sheet of water to navigate, inasmuch as furious gales and "woollies" sweep down upon it from the steep mountain sides and cañons, so that, without even a moment's warning, the traveller's craft is suddenly stricken, and compelled to instantly run for shelter under the lee of some one of the hundreds of islands and capes which stud its waters or point its coast. Immense glaciers are descending from the cavernous inlets of the northern and eastern shores, and shedded fragments of ice; large and small, are cemented by the tide into large sheets, which are finally swept out and lost in the ocean.

The shores of these canals are formed of high, stupendous mountains that rise abruptly from the water's edge perpendicularly, and often overhanging. The dissolving snow upon their summits gives rise to thousands upon thousands of little cataracts, which fall with great impetuosity down their seamed sides and over sheer and rugged precipices. This fresh water, clear as crystal and cold as winter, thus descending into the green and blue salt sea, changes that tone to one of a strange whitish hue in its vicinity, as it also does in many fiörds of the Sitkan region. This peculiar flood always arrests attention and excites the liveliest curiosity in the mind of him who beholds it for the first time. Everywhere, save to the southward, mountains can be seen looming up in the background

VALDES GLACIER

View at the head of Valdes Inlet, Prince William's Sound: typical study of hundreds of such gelid rivers which discharge into the waters of this gloomy sound. A September sketch, made at low-tide

with snowy peaks and guttered ridges, and they attest the wild legends of their sullen grandeur which the first white men related who ever beheld them. These hardy sailors, when sent out in the ship *Three Saints* from Kadiak, in 1788, arrived in the Gulf of Choogatch, or this Sound of Prince William, during the month of May. They anchored in a little bay of Noochek Island, and there established a trading-station. This is the only post, Fort Constantine, or "Noochek," that has ever been located by our people in all this section of a vast wilderness ; to-day it is but little changed—a couple of trading-stores standing on the foundations of Ismailov's * erection, in which the only three white men now known to reside in all that region of alpine wonder are living, surrounded by a small village of sixty natives.

The large size of those spruce-trees on the southern slopes of Kenai Peninsula, Montague, and Noochek Islands of Prince William's Sound, so impressed the Russians that they established a shipyard at Resurrection Bay as early as 1794 ; by the close of that year they actually built and launched a double-decker, 73 feet long by 23 feet beam, of 180 tons burden—the first three-masted, full-rigged ship ever constructed on the west coast of the North American continent ; she was named the *Phœnix*, and as she slid from her ways into the unruffled waters of this far-away place the exultation and delighted plaudits † of her builders echoed in strange discord with the wild surrounding. Baranov had no paint or even tar, so that this pioneer ship was covered with a coat of spruce-gum, ochre, and whale-oil. A few small vessels only were built after this, inasmuch as the company found it much more economical to purchase in European yards the sailing-craft and steamers which it was obliged to employ : but, to-day the traces of the Russian ship-carpenter's

---

* Ivan Ismailov and Gayorgi Bochorov ; they went in the dual capacity of explorers and traders, lured into the undertaking by rumors which had prevailed at Kadiak respecting great numbers of sea-otters in this bay.

† Had these enthusiastic builders then been able to have foreseen the tragedy which this vessel precipitated, five years later, they would have scarcely thus expressed themselves, but rather have stood in silence, with bowed heads, as the work of their hands swept into the flood that embraced her. In 1799 she sailed from the Okotsk, bound for Sitka, with the newly-ordained Bishop Joasaph and twenty priests and deacons of the Greek Church; she was never seen or heard of afterward, nor was anything seen or heard of her passengers and crew—she took them with her to the bottom of the sea.

axe can be still plainly recognized at many points of the western
coast of the sound, and on Montague Island huge logs, as roughed
out nearly a full century ago, are lying now, as they lay then,
slightly decayed in many instances ; the anticipation which felled
them was never realized, and they have never been disturbed consequently.

In these early colonial Alaskan days, Fort St. Constantine, or
Noochek Island, was a very important trading-centre ; it was visited
by all the tribes living on the Mount St. Elias sea-wall to the
eastward as far as Yakootat, and also by the Copper Indians.
Then the sea-otter was abundant, and in its ardent chase those
Choogatch savages captured, incidentally, large numbers of black
and brown bears, marten, and mink.   Now, with the practical extermination of the sea-otter, we find a very poor lot of natives at
this once flourishing post ; but, for the means of a simple physical existence, they have no lack of an abundant supply of salmon,
seal-blubber and flesh—meat of the marmot, porcupine, and bear,
varied by the frequent killing of mountain sheep, which are found
all over this alpine range ; fine foxes are plentiful too.

These Indians live in houses partly underground, which we shall
describe as we visit Kadiak, and in purely race-characteristics those
people also closely resemble the Kadiak Eskimo. From the north of
the Copper River, however, toward the Sitkan archipelago, the
Koloshian or Thlinket is dominant in the form and features of
those savages which we find in a few small and widely separated
villages that exist on the narrow table-land between the high
mountains and the unbroken swell of the ocean.   These natives all,
however, agree in describing their country as an excellent hunting-
ground, well timbered, and traversed by numerous small streams
which take their rise in the glaciers and eternal snows of the St.
Elias Alps.

By some happy dispensation of the Creator every savage is so
constituted that here in Alaska, at least, he believes in his own particular area of existence as the very best realm of the earth—he
becomes homesick and refuses to be comforted if taken to California or Oregon, enters into a slow decline, and soon dies if not
returned to the dreary spot of his birth—a sad illustration of fatal
nostalgia.

An Alaskan Indian or Innuit has very little of what may be
styled true slavish superstition ; certainly he is credulous, but he

rather encourages it for the sake of the romance. He gives slight attention to augurs or omens ; he ventures out in search of food alike under all sorts of varying conditions of health and weather ; he has a few charms or amulets, but does not surrender to them by any means. Shamans, or sorcerers, never have had the influence with him that they have exerted in the barbarism of our own ancestry, and which they possess among the savages of Central and South America and Africa to-day. It is no solution of this difference in disposition to call him stupid, for it is not true ; he is far more alert, mentally, than the ghost-ridden Australian, or fetich-slave of Africa ; and, again, the sun-worshipping and intensely superstitious Incas were far superior, intellectually, to him.

Most of the Innuits give hardly a thought to the subject, yet they are exceedingly vivacious and social among themselves ; much more so than the Indians. They relate a great many supernatural stories, but it is only in amusement, and it seldom ever provokes serious attention.

CHAPTER V.

COOK'S INLET AND ITS PEOPLE.

Cook's "Great River."—The Tide-rips, and their Power in Cook's Inlet.—
The Impressive Mountains of the Inlet.—The Glaciers of Turnagain Canal.
—Old Russian Settlements.—Kenai Shore of the Inlet, the Garden-spot
of Alaska.—Its Climate best Suited to Civilized Settlement.—The Old
"Colonial Citizens" of the Russian Company.—Small Shaggy Siberian
Cattle.—Burning Volcano of Ilyamna.—The Kenaitze Indians.—Their
Primitive, Simple Lives.—They are the Only Native Land-animal Hunt-
ers of Alaska.—Bears and Bear Roads.—Wild Animals seek Shelter in
Volcanic Districts.—Natives Afraid to Follow Them.—Kenaitze Archi-
tecture.—Sunshine in Cook's Inlet.—Splendid Salmon.—Waste of Fish
as Food by Natives.—The Pious Fishermen of Neelshik.—Russian Gold-
mining Enterprise on the Kaknoo, 1848–55.—Failure of our Miners to
Discover Paying Mines in this Section.

THAT volcanic energy and amazing natural variation of the region
known as Cook's Inlet, and the Peninsula of Alaska, endow it
with a certain fascination which it is hard to adequately define
in words, and difficult to portray. The rugged, uninviting boldness of the Kenai Mountains turn us abruptly, after our departure
at Noochek, to the southward, where, in an unbroken frowning
cordon of one hundred and fifty miles in length, they bar us out
from the waters of that striking estuary—the greatest on the northwest coast, which is so well exhibited by the map to everybody as
Cook's Inlet. But it is known only in name—not by the faintest
appreciation, even, of its real character and of its strange belongings.

Two and three hundred miles still farther north than Sitka it
does not in itself present that increased wintry aspect at any season
of the year which would be most naturally looked for—but it does
offer, in physical contour and phenomena, a most marked contrast
to the Alexander archipelago and its people. It is an exceedingly
dangerous and difficult arm of the sea to navigate, and prompts an

involuntary thought of admiration for the nautical genius, skill, and courage of Captain Cook, who sailed up to the very head of this entirely unknown gulf, in 1778, seeking that mythical northwest passage round the continent—his dauntless exploration to the utter limit of Turnagain Canal—his extraordinary retreat in his clumsy ships, and safe threading of his way out and through the hundreds of then absolutely nameless and chartless islets and reefs to the shoals of Bering Sea—all this, viewed to-day, seems simply marvellous, that he should have escaped all these dangers which the best sailor now hesitates to undertake, even with excellent courses laid down and determined for him.

The ship's entrance to this great land-locked gulf, which the Russians named, for many years, the Bay of Kenai, lies between the extreme end of that peninsula called Cape Elizabeth, and Cape Douglas, which is a bold promontory jutting out from the Alaskan mainland. Nearly half-way between the two points is a group of bleak, naked islets, the Barren Islands : around them the tide-rips of this channel, which they obstruct, boil in savage fury, and are the dread of every navigator, civilized or Innuit, who is brought near to them ; these violent and irregular tidal currents here, even in perfectly calm weather, will toss the waters so that the wildest fury of a tempest elsewhere cannot raise so great a disturbance over the sea, or one which will so quickly wash a vessel under.

When your ship, bound in, passes this Alaskan "Hell Gate," she enters into a broad and ample expanse of water caused by the widening effect of two large bays which are just opposed to each other on the opposite shores. The coast of the Kenai Peninsula is low, the mountains contiguous are not high, though toward the interior the ridges become much loftier ; but everywhere between them and this coast-line is that characteristic marshy tundra of the Arctic—a low, flat, broad strip, varying in width from forty to fifty miles, through which sluggishly flow a multitude of streams and brooks, wooded with birch, poplar, and spruce everywhere on the banks, but bare of timber over the great bulk of its expanse. As the inlet contracts still further, especially at the point between the two headlands of East and West Foreland, the tide again increases in velocity and violence of action until it attains a speed of eight and nine knots an hour, with an average vertical rise and fall of twenty-four to twenty-six feet. The northeastern extremity of this large arm of the sea, which Cook entered with the confident

hope of finding a watery circuit of a. continent, and, being disappointed, applied to it the name of "Turnagain," presents a tidal
phenomena equal to that so well recognized in the Bay of Fundy.
Here the tide comes in with a thundering roar, raising a " bore "
wave that advances like an express train in rapidity, carrying everything before it in its resistless onward, upward sweep. High
banks of clay and gravel, which at low-tide seem as though they
were far removed from submersion, are flooded instantly, to remain
so until the ebb takes place. The natives never fail to remember
the angry warning of this incoming tide ; they always hurriedly
rush out of their huts, scan quickly everything surrounding, lest
some utensil, some canoe, or basket-weir be thoughtlessly left
within the remorseless rush of that swift-coming flood.

Those glacial sheets which fill countless ravines and cañons in the
mountain ridges at the head of Cook's Inlet, especially of Turnagain
Canal, and avalanches of snow, from their lofty cradles thereon, all
sweep down together upon the wooded flanks below, and are thus
destroying great belts of forest and piling up innumerable heaps of
rocky débris to such an extent as to often change the superficial
aspect of an entire section of country from season to season ; meanwhile the tide rushing up and down over this drift of avalanches
and glaciers, carries the débris hither and thither, so as to constantly alter the channels, and the very outlines of the coast itself.

One of the oldest and best of Russian posts was early established on the Kenai Peninsula, a few miles to the southward of that
narrowing of Cook's Inlet, caused by the two Forelands. On the
low banks of the Kinik River, and facing the gulf, the ruins of the
"Redoubt St. Nicholas" are still to be plainly seen, though at the
time of the transfer of the Territory, this old post was yet fortified
with a high stockade and octagonal bastions. But both stockade
and bastions have disappeared since then ; a number of new frame
buildings have been erected close by, and quite a colony of Russian
half-breeds are living here now, trading, and growing, to better advantage than anywhere else in Alaska, fair crops of potatoes and
turnips. They keep a few hardy cattle, and it is said that as much
as ten or twelve acres of ground are under cultivation by them.

The aspect of the country surrounding this settlement is much
more suggestive of farming and cattle-raising than is that presented
anywhere else in the Alaskan Territory. The land is rolling and
hilly, the higher eminences being covered with thick spruce forests ;

but as you advance into the interior, great swamps of tangled heather, fir, jungle, and sphagnum are prevalent. The soil everywhere, not covered with grass and forest, is mossy, with a little grass and many bushes. The trees are large, fifty to sixty feet high, and eighteen inches to twenty-four in diameter, mostly spruce —no cedar or hemlock. That district adjoining the East Foreland Head is, perhaps, the best with reference to dry, fertile soil, for, in its vicinity, there are broad plains where wild timothy and red-top grasses grow to the height of your waist and shoulders. An extended experience of the Russians taught them to locate their agricultural operations here ; that the coast-line belt of the Kenai Peninsula, between the Forelands and Kooshiemak Bay, a belt of low and semi-prairie uplands some eighty miles in length, and varying in depth from ten to twenty, was the most eligible base of agricultural effort afforded anywhere in Alaska, the quality of the crops always being best near the coast, the soil being drier, and the danger of little nipping summer-frosts wholly abated.

The several small settlements which we find upon this pastoral strip to-day have a curious history, as to the origin of their inhabitants. About the period of 1836–38, the expenses of the Russian American Company in maintaining their trading stations in Alaska were increasing to an alarming degree, while the receipts remained stationary, or fell off. An enquiry into its cause revealed it. The fact was, that hundreds of superannuated employés were drawing their salaries and subsistence, rendering no adequate return for the same. These persons had grown old, and had lost their health in serving the company ; were, nearly all of them, infirm survivors of Shellikov and Baranov's parties, whose daring and energy had established the company. It would be inhuman to discharge these aged and crippled Russians, and throw them upon their own resources in such a region. After much deliberation the company was authorized by the Crown to make the following terms of settlement and relief, and thus locate them as permanent pensioners and settlers in the country. Therefore all of the old employés who had married or lived with native or half-breed women, and who were unable to successfully engage in the trading avocations of the company, by reason of age and other infirmities, were, upon their written or witnessed request, after being stricken from the pay-rolls, provided for in this manner.

The company was obliged to select and donate a piece of ground, build a comfortable dwelling, furnish agricultural tools, seeds, cattle and fowls, and supply the pensioner receiving all this with provisions enough to support him and his wife for one year. These "old colonial citizens" (as they were called), thus established, were then exempted from all taxation, military duty, or molestation whatsoever, and a list of their names was annually forwarded in the reports of the company. The children of those settlers were at liberty to enter or not, as they pleased, the service of the company at stated salaries. The company, furthermore, was commanded to purchase all the surplus produce of these pensioners, furs, and dried fish, etc. This order of the Crown, thus fixing the status of those old servants, also included the half-breeds who were equally infirm by reason of such service. Such whites, or Russians, were officially designated "colonial citizens," the half-breeds were styled "colonial settlers."

The descendants of these pensioned servants of the Russian Company are the men and women you observe to-day in those little hamlets scattered along the east coast of Cook's Inlet, or the Kenai Peninsula. They are bright, clean, and, though very, very poor, still appear wholly independent. They are engaged in small trading with the Kenaitze savages and in their limited agricultural efforts, whereby they have potatoes, turnips, and other hardy vegetables. The cattle, of which they have a few in each settlement, are of the small, shaggy Siberian breed, not much larger than Shetland ponies, and capable of living in the rigors of a winter which would destroy or permanently injure our breeds of neat cattle. These people make butter by laboriously shaking the milk in bottles.

They are obliged to shelter their cattle during winters from the driving fury of heavy snow-storms, and when the herd ranges in the grass-season, the boys and old men always have to guard it from the deadly attention of the big brown bears which infest the entire region. They have a regular "round-up" in each hamlet every night.

Everywhere on the west coast of Cook's Inlet the mountains rise steeply and rugged from the sea, a wild and uninviting contrast with the park-like terraces of the Kenai coast just opposite. Here are the same lofty ridges and smoking peaks which startled and oppressed the brave heart of Captain Cook, as they muttered and trembled in volcanic throes when he sailed by. The two cones

THE VOLCANO OF ILYAMNA: 12,060 FEET

which rise dominant are the summits of Mount Ilyamna and the " Rédoute," from which columns of brownish smoke ascend by day and ruddy fire-glowings by night. So precipitous is this mainland shore of Cook's Inlet that at only two small points of the most limited area is there any low land to be found, and these spots have been promptly utilized by the Kenaitze Indians as sites for their villages of Toyonok and Kustatan. The dense, sombre coniferous forest which we have become so familiar with, clothes the flanks of those grim mountain walls with the thickest of all coverings to a height of one thousand feet above the beaches below. Here and there we glance into the recesses of a cañon or a gorge where the naked, mossy surface of immense rocky declivities arrests and fixes the eye, while the glittering caps of ice and snow far away above fit down snugly upon long, rough, treeless intervals, covered with heather, lichens, and varied arctic sphagnum.

The upper waters of Cook's Inlet are said to be quite remarkable for their barrenness of fish—salmon only being plenty in the running season, ascending all the numerous rivers and rivulets ; the reason most likely is due to the turbid upheaval of the bottoms everywhere by that violent tidal bore which prevails, recurring twice every twenty-four hours. The Indians here employ a curious trestle or staging of poles, which they use in spearing salmon, and netting them from its support.

An extensive spread of the largest fresh-water lake in Alaska just over the divide from Cook's Inlet, early led the Russians to explore it, and to find a portage via its waters to the sea of Bering. But, though this barrier can be passed by an active man in a single day, yet it has divided, and continues to absolutely separate, two distinct races of savages—the Innuits from the Indians ; for the Kenaitze are Indians, as we understand them, based upon our types of the great plains and foot-hills of the Rocky Mountains ; and, living here as they do on the shores of Cook's Inlet, they live, perhaps, in the most romantic and picturesque region of Alaska. Burning volcanoes, smoking and grumbling, a large inland sea rolling for miles and miles therein, and lay at their feet ; wide watery moors, tundra, timber and lakes, and rivers rising in the snow-white peaks everywhere visible, all combine to make the most striking lights and shades of natural scenery that human thought can realize in fancy.

These natives of Cook's Inlet are strongly defined from those of

Kadiak as a separate people, both in language (which no white man
has ever been able to repeat), in appearance, and in disposition.
They are true Athabascans, or exactly like the meat-eating Indians
of our great North American interior.   An average man here is an
Indian of medium height, say five feet seven or nine inches, well
built and symmetrical, lithe and sinewy.   The cold glint of his
small, jet-black eyes is not relieved by any expression of good humor

A Kensitze Chief: Cook's Inlet.

in his taciturn features and physical bearing.   His nose will present, as a rule, the full aquiline or Julius Cæsar outline.   Their
skin is darker than that of the Innuit, though now and then a
comely young person will show perceptible blood-mantlings to the
cheeks.   The mouth is large—lips rather full; beardless faces are
the rule.   Their women are much better-looking than either the
Siwash squaws of the Sitkan region, or the females of the Aleutian
and Innuit races.   Their hair is worn in clubbed bunches and

braids, hanging upon their backs, thickly larded over with grease, and often powdered with feathers and geese-down.

In the immediate vicinity of the shores of Cook's Inlet the primitive habits of these savages have been very much changed by their daily intercourse with the Creoles ; but at the head of the gulf, especially in the Sooshetno and Keknoo valleys, they are still dressed in their deer-skin shirts and trousers, men and women · alike. They work those garments with a great variety of beads, porcupine quills stained in bright colors, and grass plaitings.

These Kenaitze are the only real hunters in all Alaska. They place little or no dependence on fish like the other tribes, unless we except the walrus-eating Eskimo, who hunt, however, in water-craft entirely. And were they not natural Nimrods, the abundance of game which abounds in their district would stimulate such ambition alone in itself. The brown bear * of Alaska is found almost everywhere ; but it seems to prefer an open, swampy country to that dense timber most favored by its ursine relative, the black bear. It attains its greatest size, and exhibits the most ferocity, on the Kenai Peninsula. It should be called the grizzly, because it is frequently shot here fully as large, if not larger,† than those examples recorded in Oregon and California.

This wide-ranging brute is found away up beyond the Arctic Circle, though never coming down to the coast of the icy ocean except at Kotzebue Sound. It is a most expert fisherman, and a terror to the reindeer and cariboo of those hyperborean solitudes. It frequents, during the salmon season, all the Alaskan rivers and their tributaries which empty into Bering Sea and the North Pacific, as far as the fish can ascend. When the run for the year is over, then the animal retires into the thick recesses of semi-timbered uplands and tundra, where berries and small game, deer especially, are most abundant.

Everywhere throughout this large extent of Alaska the foot-paths, or roads, of that omnipresent ursine traveller arrest your attention. The banks of all streams are lined by the well-trodden trails of these heavy brutes, and offer far better facilities for progress than those afforded by the paths of men. Not only are the swampy

* *Ursus richardsonii.*

† One shot at Kenai Mission in 1880 measured nine feet two inches in length.

plains intersected by such well-worn routes of travel, but the
mountains themselves and ridges, to the very summits thereof, are
thus laid out ; and the judgment of a bear in traversing a rough,
mountainous divide is always of the best—his track over is sure to
be the most practicable route. On the steep, volcanic uplands of the
mountainous coast of the west shore in Cook's Inlet, groups of
twenty, and even thirty, of these huge bears can be seen together
feeding upon the berries and roots which are found there in
season. Their skins are not valuable, however, being "patched"

Bear "Roads" over the Moors of Oonimak Islands.

and harsh-haired. Then they are very fierce, so that they are not
commonly hunted anywhere except by the Kenaitze, who, like all
other aboriginal hunters, respect them profoundly, and invariably
address a few eulogistic words of praise to a bear before killing or
attempting to kill it.*

A peculiar dread which all the natives of this region have, of
visiting those areas where volcanic energy manifests itself, is taken
advantage of by those dumb beasts upon which the savage wages
relentless warfare ; the immediate vicinity of craters, of steaming

---

* Perhaps fully half the brown-bear skins taken by the Alaskan natives are
retained by them, used as bedding, and hung up as portières over the entrance-
holes or doors to their houses ; the smaller skins are tanned and then cut into
straps and lines to use in sledge-fastenings, snowshoe network bottoms, because this leather does not stretch when moist like deer and moose skin.

hot springs and solfataras, will always be a rendezvous for game, especially bears, which seem to fully understand that in staying there they will never be disturbed. But the Kenaitze are ardent hunters, nevertheless, and spend most of their time and energy in the chase of land animals—making long journeys into the interior, and gloomy recesses of mountainous cañons and defiles, to follow and find the fur-bearing quarry peculiar to their country.

They have regular tracks of main travel, where, like stage stations on our frontier post-roads, at intervals they have erected shelter-huts, in which they often live with their families for months of the year at a time ; they make birch-bark canoes for their river and lake transit, but in navigating Cook's Inlet, they buy skin bidarkas of the Kadiak model and use them altogether. They are fairly independent of salt water, and seldom pass many hours upon it, except in travelling and trading one with another, and the Creoles ; they are, however, very expert at fresh-water fishing through holes in the ice for trout in the thousand and one lakes, large and small, which are so common in their country.*

As these natives live in their permanent settlements, we find them distinguished by a peculiar architecture. Their houses are fashioned out of logs, and set above ground resting upon its surface ; the logs are hollowed out on one side so as to fit one upon the other in true spoon-fashion, and make a really air and water-tight wall ; an enclosure of these walls will hardly ever be larger than 20 feet square, and most of them never go over 12 or 15 feet ; they have regularly laid cross-rafters, with a low, or half-pitch, over which spruce-bark shells are so spread as to shed rain and drifting snow ; these shingle slabs are kept in place by a number of heavy poles, lashed transversely across ; a fireplace is always in the centre with a very small smoke-hole opened in the roof just above it ;

---

* The greatest number of different mammals found wild in any one region of Alaska is to be recorded here : bears, brown and black ; deer, reindeer and the woodland cariboo ; big-horn mountain sheep, a long-haired variety. These animals are all shot. The trapped varieties are : beaver, land-otter, porcupines, whistling marmot or woodchuck, large gray wolves, lynx, wolverine, marten, mink, ermine, weasels, and muskrats. Wild-fowl : grouse both white and ruffled, geese, ducks, sandhill cranes, and the great northern swan. Berries : whortleberries, salmonberries, gooseberries, and cranberries ; all gathered in season and mixed with the everlasting rancid oil used by every native in every section of Alaska.

the door is a square aperture cut through the logs at the least exposed front, about large enough to easily admit the ingress and
egress of a crouching Indian.   It is stopped in stormy weather by
a bear-skin, hung so as to fall directly over it from the inside.
When the door is thus closed the naturally dark interior becomes
almost wholly so ; but the howling of a tempest, laden with rain, sleet
or snow, as the case may be, renders this gloomy indoor perfectly
radiant to the senses of its sheltered inmates, and they loll in robes
and blankets and doze away the time on the rude wooden platform
which surrounds the walls and keeps their bodies from the cold
damp earth.   Upon this staging they spread grass mats and skins,
and, in fact, it is a catchall for everything.

The Bedroom Annex of a Kenaitze Rancherie.

An odd feature seen in some of the most pretentious houses of
those inlet savages, is the presence of a little kennel-like bedroom
annex, which many of the most wealthy or important have built up
against the main walls. These boxlike additions are tightly framed
and joined to the houses, the only entrance being from the inside
of the main structure by a small hole cut directly through the
logs of the wall ; they are sleeping chambers, and are furnished
with a rough plank floor, and sometimes a window made of a piece
of translucent bladder-gut.   They are also reserved and special
apartments during the occasion of those visits of ceremony which
Indians often pay, one to each other.   But the main idea is to
have these tight little dormitories so snug and warm that they will
insure the comfortable rest of the owner therein without much
burdensome bed-clothing—in many cases the Kenaitze can sleep

here in the coldest weather without any covering at all, and do. Such a bed is a great and priceless luxury to them.

No furniture annoys the Kenai housekeeper, unless the small square blocks of wood used occasionally as stools or seats can be so styled; the grease and fire-boxes which we have seen in Sitkan households are also duplicated here, but though made of wood they are not so neatly put together. The traders recently have introduced a very novel feature to the interior of nearly every Kenaitze house; it is the common, cheap, box-imitation, in miniature, of a Saratoga trunk with lock and key. Those oddly contrasted articles will be found everywhere among these people, who keep in them all their valuables, such as charms, and toys for the children, flashy handkerchiefs, small tools fashioned out of bits of iron and steel, bags of thread and stripped sinews, needles, ammunition, and their percussion-caps, which are to them as pearls without price—nothing so precious. Outside of this trunk-craze, and their odd sleeping-rooms, these Indians do not live together or act differently from the usual habit and manner of savages proper, so familiar to us by reason of repeated descriptions published of our own meat-eaters who live near by. They crave nothing from the white trader save powder, lead, good rifles, percussion-caps, tobacco, calico, and the sham trunks alluded to.

The sun shines out over Cook's Inlet much more than it does in the Sitkan region and the Aleutian Islands. The proportion of fair, bright weather is larger than that experienced anywhere else in all Alaska or its coast. The winter months here are not excessively cold; snow falls in December—sometimes as late as 3d of January before the first flakes of the season arrive. By the first to middle of May it has usually melted away on the lowlands, and the grass springs up anew, green and luxuriant. Summer, and even winter storms, are drawn along the lofty ranges of the Kenai Peninsula when all is serene and pleasant at the same time on the moors and lowlands of the inlet shores. Often, too, the people of that coast can look up to a continued falling of heavy rain and snow on the mountain summits of the steep ridges across the inlet, while they bask in unclouded sunshine, and have no interruption of its comfort.

We ourselves have as yet made but slight use of the natural resources and advantages of Cook's Inlet. A party of San Francisco merchants have established at the mouth of the Kassilov

River a salmon cannery, which has been worked to the full limit of demand ; and a smaller, similar factory is located at the head of this inlet, in the Kaknoo estuary.

The finest salmon known to man, savage or civilized, both in flavor and size combined, is that giant fish which runs in especial good form and number into Cook's Inlet, and which the Russians called the " chowichah ; " * they are most abundant during the summer neap-tides, but they are not as numerous as are the several other varieties of smaller and far less palatable salmonidæ, which also run up here with them. The average length of these superb chowichah fish is four feet, and a weight of fifty pounds is a low medium. They appear regularly on the 20th and 22d of every May, running in pairs, refusing the hook, though hugging the shore lines. Our people catch them in floating gill-nets, and in weirs of brush and saplings of wicker-work woven with spruce-roots and bark, which are erected on the mud-flats at the river mouth, during low tide.

The king salmon, however, is erratic in running to any one spawning spot, and in this respect differs from all the rest of its family, which is remarkably constant in annually returning to the same spawning ground. But the abundance of salmon which we see in their reproductive periods of each year, ascending every river and possible rivulet that communicates with the sea in Alaska south of Bering's Straits, is a never-failing source of wonder and delight to the white visitor and a measure of infinite creature-comfort to his physical being while sojourning here. Also, the pleasant thought constantly arises that when we shall have a populous empire on the Pacific slope, as we have now in the Mississippi Valley and east of the Alleghanies, what a handsome use we will make of this waste of fish-food wealth † which we now observe in the vast

---

* *Oncorynchus chouicha*—examples of this species have a recorded weight of one hundred pounds each, and six feet in length ; it is also abundant in the Yukon and Kuskokvim Rivers.

† Dr. T. H. Bean, who, as a trained ichthyologist, passed the season of 1880 investigating the fish of Alaska by cruising throughout its waters, says : " The greatest fish-wealth of Alaska, so far as the shore fisheries are concerned, lies in the abundance of salmon of the genus *Oncorhynchus*, which is represented by five species—*chouicha, keta, kisutch, nerka,* and *gorbuscha.* The first three of these are the largest, the whole series being named in the order of their size. *O. chouicha* is the giant of the group, and is the most important commercially ; it attains to its greatest size in the large rivers,

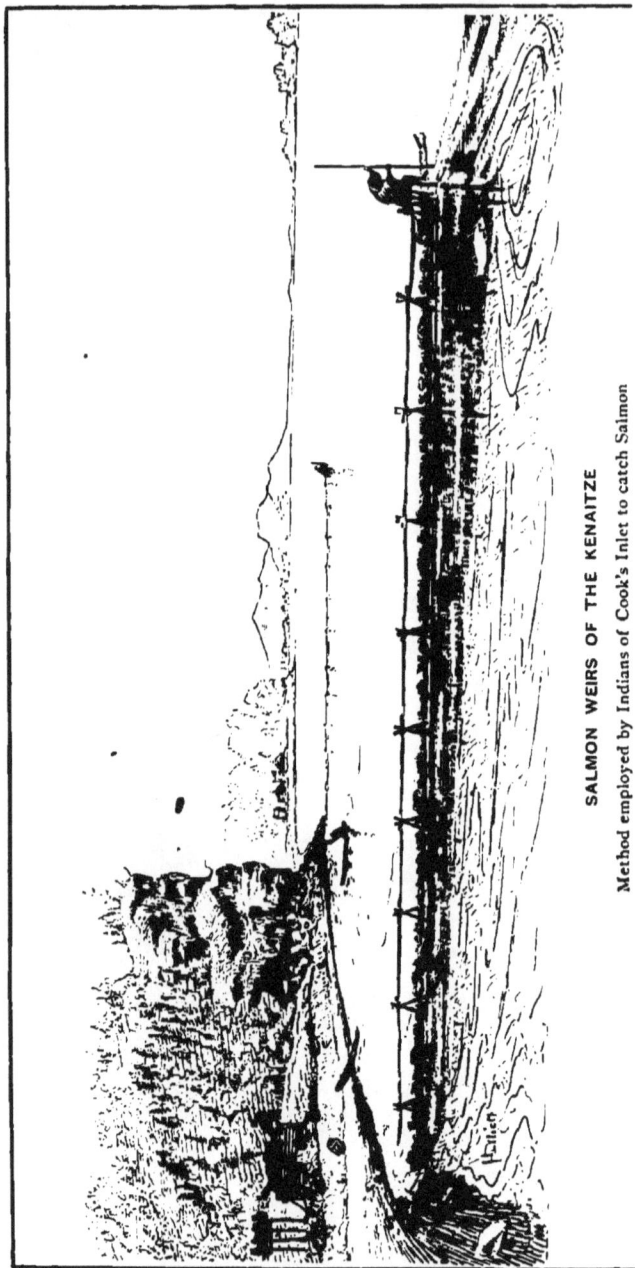

SALMON WEIRS OF THE KENAITZE

Method employed by Indians of Cook's Inlet to catch Salmon

realm of its indulgence throughout Alaska. Also in another, but wholly correct sense, the natives themselves shamefully waste the flesh of those fine salmon. To illustrate the extraordinary nature of this suggestion, let the following statements of fact be recalled : The native population of Cook's Inlet is not large—it is embraced in about one hundred and sixty-eight families, averaging four souls to each household ; everyone of these families prepares at least seven hundred and fifty to eight hundred pounds of dried salmon for its own specific consumption during the winter months. That amount of cured fish, therefore, is about one hundred and twenty-six thousand pounds, and as every pound weight of dried meat is equal to an original weight of at least eight or nine pounds of fresh, or un-dried, then this cured total gives us an immense aggregate of

---

which it ascends long distances in its spawning season. In Alaska it is known to extend as far north as Bering Strait, and it is especially abundant in Cook's Inlet and in the Yukon. Individuals weighing nearly one hundred pounds are occasionally reported from these waters, and even in the Columbia. The finest product of this salmon is the salted bellies, which are prepared prin-cipally on the Kenai, Kassilov, and Yukon Rivers ; the fame of this luxury once extended to the centre of government in Russia. The well-known ' quinnat salmon ' is the same species ; its importance, as evidenced by the efforts of the United States Fish Commission and other commissions toward its propagation and distribution, is too well understood to require additional mention. The great bulk of the salted salmon exported from Alaska are the small ' red fish,' *O. nerka ;* and this species is sought after simply on account of the beautiful color of the flesh and not for its intrinsic value, which is far below that of most of the other species. All the salmon extend northward to Bering Strait, but only one, *gorbuscha,* is reported as occurring north of the Arctic Circle ; *gorbuscha* is said by trustworthy parties to reach the Colville River. In the early part of its run the flesh of this little ' humpback ' seems to me to be particularly good. Other members of the family of *Salmonidæ,* and very important ones, are the species of *Salmo* (*purpuratus* and *gairdneri*) and *Salvelinus malma,* two of which reach a large size in Alaska. The first two are not known to exist much to the northward of Unalashka, while *malma* is believed to extend to the Colville. *S. gairdneri* resembles the Atlantic salmon in size and shape, but its habits are different ; it is found filled with mature eggs in June. I have not seen any very large examples of *S. pur-puratus* from the Territory, but the species is extremely abundant and valuable for food. The red-spotted char, *S. malma,* is everywhere plentiful and is highly esteemed as a food-fish ; it grows much larger in Northern Alaska than in California, and has some commercial value as an export in its sea-run con-dition under the name of ' salmon trout.' Natives of Alaska make water-proof clothing from the skins of this fish."

1,000,000 pounds of fresh salmon; this, figured down, shows that a single Indian uses, during the winter solstice—five months—the enormous amount of 1,430 pounds of this rich-meated article of diet, or about ten pounds every day, in addition to the bear-meat, deer, and sheep-meat, seal and beluga oil, berries and roots which he is constantly consuming, at the same time, in the greatest freedom, and which are always in abundant supply. The full thought of my presentation will be better understood when it is remembered that a pound of fresh salmon has more nourishing and sustaining quality than the same amount dried. The salt-dried codfish with which we are so familiar is very different in its texture, and weighs many times more than it would if it were cured by the air and smoke-exposure to which the natives of Alaska are driven in preserving their fish.

An exceedingly happy illustration of the singular force of habit which the salmon have in returning every recurring season to the exact localities of their birth was afforded near the Creole settlement of Neelshik on the Kenai Inlet coast. A small stream runs down to the gulf from the mountains and moors of the interior. Its mouth had been closed by a barrier of surf-raised sand and gravel during storms in the winter of 1879–80, and through which the sluggish stream filtered in its course without overflowing. When the salmon, which had descended the year previously from the upper waters of the stream in the course of their reproductive circuit, again returned to renew such labors in the following season, this unexpected wall barred their ingress. They did not turn away, but actually leaped out upon this sandy spit, and many of them succeeded, by spasmodic springs and wriggling, while on the dry gravel, in getting across and into the river-water beyond! the Creoles, in the meantime, having nothing to do except to walk down from their houses and gather up the self-stranded salmon as they fancied their size and condition. Inasmuch as these "old colonial settlers" are very pious, as well as very indolent, they were profuse in giving thanks to their patron saints for this unexpected bounty.

The color of gold everywhere found by washing the sands of Cook's Inlet on the Kenai shore early aroused the cupidity of the Russians. They made systematic examinations here under the lead of experienced men, between 1848 and 1855, and the Russian American Company spent a great deal of money in the same time by sustaining a large force of forty miners, directed by Lieutenant Doroshin, iu active operations at the head of the inlet on the Kaknoo River, and in the Kenai Mountains and Prince William or Choogatch Alps. Gold was found, but in such small quantities, compared with the labor of getting it, that the ardor of the Russians soon cooled, and nothing as yet has resulted from the prospecting of our own miners in this district, who have been all over these Slavonian trails since the transfer.

7

CHAPTER VI.

THE GREAT ISLAND OF KADIAK.

Kadiak the Geographical and Commercial Centre of Alaska.—Site of the First Grand Depot of the Old Russian Company.—Shellikov and his Remarkable History, 1784.—His Subjection of the Kaniags.—Bloody Struggle.—He Founds the First Church and School in Alaska at Three Saints Bay, 1786, One Hundred Years ago.—Kadiak, a Large and Rugged Island.—The Timber Line-drawn upon it.—Luxuriant Growth of Annual and Biennial Flowering Plants.—Reason why Kadiak was Abandoned for Sitka.—The Depot of the Mysterious San Francisco Ice Company on Wood Island.—Only Road and Horses in Alaska there.—Creole Ship and Boat Yard.—Tough Siberian Cattle. Pretty Greek Chapel at Yealovnie.—Afognak, the Largest Village of "Old Colonial Citizens."—Picturesque and Substantial Village.—Largest Crops of Potatoes raised here.—No Ploughing done; Earth Prepared with Spades.—Domestic Fowls.—Failure of Our People to Raise Sheep at Kolma.—What a "Creole" is.—The Kaniags or Natives of Kadiak; their Salient Characteristics.—Great Diminution of their Numbers.—Neglect of Laws of Health by Natives.—Apathy and Indifference to Death.—Consumption and Scrofula the Scourge of Natives in Alaska ; Measles equally deadly.—Kaniags are Sea-otter Hunters.—The Penal Station of Ookamok, the Botany Bay of Alaska.—The Wild Coast of the Peninsula.—Water-terraces on the Mountains.—Belcovsky, the Rich and Profligate Settlement.—Kvass Orgies.—Oonga, Cod-fishing Rendezvous.—The Burial of Shoomagin here, 1741.—The Coal Mines here Worthless.

THE boldest and the most striking cape in this wilderness of bluffy headlands and jutting promontories is that point which marks the dividing line between the Kadiak region and Cook's Inlet—Cape Douglas. It is a lofty alpine ridge or spur, abruptly thrust out at a right angle to the coast, and into and over the sea for a distance of three miles, where it drops suddenly with a sheer precipitous fall of over one thousand feet into the waves that thunder on its everlasting foundations. Baffling winds here, and turbulent tide-rips distress that navigator who, coming down from the inlet, seeks the harbor of St. Paul's village. He hardly regards this scared and rugged headland with that admiration which the geologist and the artist always will. The "woollies," which blow fiercely off from it, worry him and challenge all his nautical skill.

Kadiak Island is the centre, geographical and commercial, of a most interesting and wide-extended district, perhaps the most so, of the Alaskan Territory, and Kadiak village, or Saint Paul Harbor is, in turn, the central and all-important settlement of this district.* It was the site of the first grand depot of the old Russian American Company, and also the location of the first missionary establishment and day-school ever founded on the northwest coast of the continent. From the quiet moorings of this beautiful Kadiak bay hundreds of shallops and vessels bearing courageous monks and priests have set out in every direction over all Alaska, carrying scores of them to preach the gospel among its savage inhabitants, who then were savage indeed to all intents and purposes.

The first visit ever made by white men to the great Island of Kadiak was the landing here in the autumn of 1763, at Alikitak Bay, of Stepan Glottov, a Russian sea-otter trader, who went into winter-quarters at the southeastern extremity of the island, on a spot now called Kahgooak settlement. The natives were ugly, hostile, refused all intercourse, and kept the Russians in a chronic state of fear. Scurvy broke out in their camp and nearly destroyed the invaders, leaving less than one-third of them alive in the spring. They managed then, with the greatest effort, to launch their vessel and get away, the savages meanwhile constantly attempting to finish that destruction which bodily disease had so well-nigh effected.

The beginning of the eighth decade of the eighteenth century is a true date of the real epoch of Russian domination in Alaska. All history of white exploration in this country prior to that is simply the cruel legend of an eager, heartless band of outraging Muscovites, doing everything just for the gain of the present moment, sowing so badly that they dared not remain and reap. One of those big-brained, cool, and indomitable Russians, who gave then as they give now, the stamp of high character to the race, was for several years prior to 1780 prominently engaged in the American fur trade. Grigor Ivan Shellikov was this man. He was a citizen of the Siberian town of Roolsk. He resolved to survey in person those scenes

---

* With the exception of Prince of Wales Island in the Sitkan archipelago, Kadiak is the largest Alaskan island. There is not much difference between these two islands in landed area ; the former, however, is the bigger.

of rapid demoralization and ruin to the profitable prosecution of his Alaskan business, and, if possible, to attempt a change for the better. An evident decrease in furs, together with the hostile attitude of the natives, provoked altogether by their inhuman treatment at the hands of the "promishlyniks," called for reform in the most emphatic manner. After a carefully deliberated plan of action had been determined upon between himself and his partners, the brothers Gollikov, he at once proceeded to the Okotsk Sea and fitted out three small vessels for his expedition.* He did not reach Kadiak until 1784, two years after starting out, when two of his vessels came to anchor in the harbor now known, as it was then christened by him, Three Saints Bay. Shellikov was a ready and willing correspondent. His numerous letters to his Siberian partners and his own published "Journeys" give us a clear idea of the hardihood of his enterprise, and they have a rare ethnographic value. From them we learn of the great liking which Shellikov's party took to the Island of Kadiak, and how they resolved, soon after making a short reconnoissance, to establish themselves permanently if they could gain the confidence of its savage inhabitants.

Shellikov sent out a scouting party and captured a Kaniag, brought him into camp, and loaded the bewildered native with presents and kindness, then sent him back to his people; but the native, though won wholly over himself, † could not prevail upon his hostile countrymen, who soon gave the Russians ample evidence of their enmity. A party of the latter in two of the ships' boats were exploring and hunting, when they were disturbed by the appearance of a "perfect cloud" of natives that were encamped on rough and precipitous uplands of Oogak Island, a short distance from the main island itself. Shellikov resolved to proceed himself to the spot and endeavor to win them over to amity and trade. He ex-

---

* These "galiots" where characteristically named by Shellikov's spiritual advisers, viz.: *The Three Saints; The Archangel Michael and Simeon, the Friend of God;* and *Anna the Prophetess.* Bad weather and poor navigation caused the vessels to separate, so that Shellikov was compelled to winter on Bering Island; but during the following year the little fleet was reorganized, and it reached Oonalashka, where repairs again were necessary.

† Shellikov says that this man returned the following day and refused to leave the Russian camp; that he not only accompanied and served him in all his voyages thereafter, but often warned the party of hostile ambuscades and hidden dangers by land and sea.

hausted every art of pacification that his ready wit could suggest
without making the slightest favorable impression upon these men,
who treasured up in the liveliest recollection those outrages and indignities which they had hitherto suffered from the arms and vices
of Shellikov's Muscovitic predecessors. The only answer that they
made to the trader now was that he at once embark and leave the
island, and a few arrows and bird-spears were discharged and
thrown at him by way of clinching the argument. The Russians
retired to their camp, and wisely erected over day a rude stockade—
none too quick, for these Kaniags approached the harbor in the
middle of the night, unobserved, and threw themselves with frenzied fury upon the slightly fortified Russians. The battle lasted
until daylight. The necessity and instinct of self-preservation
caused the whites to fight with desperate coolness and intrepidity.
The slaughter was great among the natives, and, considering the
vastly inferior numbers of the Russians, their loss, too, was heavy.
In spite of the bravery of the whites in this terrible midnight struggle, they would have been overpowered and exterminated ere the
dawning, had it not been for the consternation which the reports
of their small iron two-pounders created in the assailing ranks of
those dusky hosts.

Recognizing the fact that now the only hope of peace and commercial intercourse with these natives lay in their complete subjugation, Shellikov, immediately after the sullen retreat of the hostiles,
armed one of his vessels and followed them up to their rocky for-
tresses in Oogak, where they had taken up a position that was well-
nigh impregnable, and to which savage reinforcements were rapidly
flocking from the main islands. Unable to reach the entrenched
camp of those defiant natives with the small ship's-cannon, Shellikov picked a party of sixty men out of his company, went ashore
with them, and, with his little iron two-pounders, he stormed the
enemy with such impetuosity that the rapid discharges of these guns
and small fire-arms of the charging Russians utterly demoralized
an immensely superior force of the savages, who became panic-
stricken, and actually jumped by scores off the high bluffs of
Oogak into the sea, hundreds of feet below ; the rest of them,
more than a thousand souls, surrendered to the Russians, who took
and located them on a rocky islet, several miles from the harbor of
Three Saints, and temporarily provided them with provisions ; and
then, with hunting-gear, they were set to work and liberally paid for

their peltries. Twenty or thirty of their leaders were kept as host-
ages on the vessels, and the result was entire submission everywhere afterward to the Russians in this region. Occasional attacks and massacres would now and then be made upon far-distant
hunting parties of the Russians, it is true, but the moral effect of
the Oogak victory and slaughter was such among the Kadiakers
that no further combined organized resistance or opposition was
ever given again.

Shellikov soon realized that he was in no further danger from
savage attacks. He began a most extensive and thorough exploration of the great island, and organizations of trading-posts at every
eligible point. He sent a large party around to the north side and
located it at Karlook, where we now find quite a salmon-canning
establishment. Here, during the winter of 1785–86 fifty-two Russians and as many natives ranged all over the water of Shellikov
Straits in eager search of the sea-otter ; in the meantime the whites
under Shellikov's immediate command were actively examining
the recesses and fiörds which are so numerous and deep on the
south side of Kadiak. So well and so thoroughly was this work
carried out, that by the beginning of 1786 Shellikov had made himself well acquainted with the whole region—had established his
trading-posts at every point between Shooak, in the north, and
Trinity Islands, at the extreme south; and had even made himself
tolerably familiar with the coast of Cook's Inlet, having chastised
the ugly Kenaitze in a most summary manner.

Again, this remarkable man is distinguished by the successful
and sensible effort which he made in substituting for the orgies of
Kadiak demonology the practices of the Greek Church, which, he
wisely foresaw, if effected, would bind the natives closer to the
Russians than any other power. He was aided in this by the personal labors and example of his wife, who accompanied him at the
outset. She instructed the girls and women in needle-work, and
acquired an influence over them that was very great. Feeling
certain that he had established his trade on a secure foundation,
Shellikov and his wife sailed for home on May 22, 1786, in the
same vessel which brought them out, leaving an impress of enduring character upon Alaska of the greatest good and worth.

During the first years of the existence of the Shellikov Company,
thus established here in Kadiak, it enjoyed the partial protection
of the Crown and many exclusive privileges, by which advantages

nearly all the smaller trading companies had been fairly crowded out of the country. But it was not always the power conferred upon a great firm by its favor at Court and larger capital that gained supremacy in Alaska during those early days—it frequently occurred that the employés of one association resorted to physical force of arms in dispossessing those of another, and then, this order initiated, the strongest organization was sure to eventually dominate the coveted region. This commercial anarchy led to the autocratic monopoly of the Russian American Company in a very few years—it was the best thing that could then have transpired for Alaska and its people.

The approach to Kadiak from the ocean is striking, because it and the numerous islets and islands that join it closely are mountainous and hilly, with many lofty peaks that have plateaux and ravines full of eternal snow. It is not often seen clearly, however, along its full extent of wild topography, on account of clouds, fog and boisterous weather, which terrifies the navigator, driving him from its vision. It is, however, an island that affords the greatest number of safe and snug harbors, and has no rival as the most enjoyable place for the traveller to visit. It so justifies us in our mind to-day, just as it warranted the Russians in expressing their preference for it a full hundred years ago.

Nature has drawn across Kadiak in a firm line, the ultimate limit of timber growth to the westward. It seems to be as arbitrary and capricious as if traced there by the humor of a human ruler. Only one-third of the island itself, its northern extremity, is covered with spruce-forest;. the invisible barrier to the west seems to be a perfectly straight line over from the heads of Orlova Bay on the south side to that of Ooganok on the north coast. Here the change from a vigorous growth of spruce-forest to bare hills and grassy tundra is most abrupt and astonishingly sharp in definition ; you pass from the jungle of the woods, at a single step, into heather of the moor. This line, with a slight curve to the westward only, strikes the same definition over on the mainland of the peninsula opposite, and runs right up north to Bering's Straits' latitude, avoiding the coast everywhere except at Cape Denbigh, Norton's Sound.

There is scarcely any lowland, indeed none at all, on the large island itself; it is everywhere mountainous and abruptly rolling, with spaces here and there in which the grasses flourish to a great extent. A legion of small streams rush down to the bays from their mountain sources, but none of them are navigable—they are mere rapids and cascades in their entire length. A growth of the characteristic circumpolar annuals and biennials on the slopes of these hills of Kadiak is of exotic luxuriance, and of the most varied beauty of floral display in June, July, and August. Willows and alders fringe the borders of the streams in their range throughout the woodless area, while stunted birch and green grasses reach to the very summits of the hilly ridges of the interior.

Although Shellikov had established the headquarters of his Russian Company on Three Saints Bay with good reason at the time, yet when the entire Alaskan region went into the control of a single organization, it became necessary to have a grand central depot of supplies. Therefore Baranov promptly removed to the site of the present village of Kadiak. Upon that wooded island in the offing he procured the lumber and timbers necessary for the erection of those huge warehouses and numerous dwellings of many sizes required to house the merchandise, furs, and his employés. The harbor, too, is ample, and so situated that sailing-craft can come and go in all winds. Sadly, indeed, did the Russians, a few years later, abandon Kadiak for Sitka, as the numerous letters and protests still on file show ; but the menacing encroachments of foreign traders in the far-distant Alexander archipelago were too grave in their portents of loss and usurpation of vested rights to allow of any other action.

To-day many of the ancient Russian structures are still pointed out in the village here which commands the harbor of Saint Paul, and in which some three hundred Creoles are living in well-built log and frame houses. Everything is clean* and orderly, but very, very quiet, inasmuch as no commerce, no monthly steamer, no tramping miners invade the solitude of its location. It supports a large Greek church and the priest attendant. Its people, as a rule, are wholly engaged in the business of trading fur and hunting sea-otters. Small codfish schooners often rendezvous here, and the

---

* Cleanliness and comfort, however, were but little regarded by the Russian fur-traders, who gave their surroundings of residence no sanitary attention whatever. Even Baranov himself was supremely indifferent, and when the Imperial Commissioner, Resanov, called on him at Sitka in 1805, the chief manager of the Russian American Company was living in a mere hut, " in which the bed was often afloat," and a leak in the roof too small a matter to notice !

natives also cut considerable cord-wood, for the use of such fur-traders who ply to the treeless districts westward, and fuel for fishing canneries at Karlook and Kassilov. Several little mountain rivulets flow through the limits of this settlement, which is everywhere well drained, and therefore dry in the streets by rea-son of its position on the rising slopes of the lofty hills which make a bold background, when the picture is viewed from the ship's deck as you sail up to the anchorage. The presence here of some thirty white men, pure Russian Creoles, and several of our own people who have really settled in the country, many of them mar-ried, and who call the place home, makes Kadiak unique in this re-spect. Elsewhere, if we find a white man living at the trading-posts, or plying his vocation as a cod-salmon fisherman, or miner, he always draws himself up and emphatically denies any idea of permanent residence in Alaska.

Looking down the bay, we observe a thickly timbered and a somewhat more level island than usual—it is the famous Wood Island, where the largest spruce-trees in all this section grow ; upon it is a small village of one hundred and fifty-six souls, living in thirteen log houses, thickly clustered together ; they are all sea-otter hunters during the summer. This village is also the depot of that mysterious San Francisco corporation which has regularly cut up and stored tons of ice here every winter since 1856, and never has shipped a pound of it away ! and when the bright, hearty agent of this corporation asks you to come out with him to the stable and advises you to mount one of the three or four horses sheltered therein, so that you can gallop round the island with him, your astonishment is perfect.

Sure enough, there is a road, incredible as it first seemed ; for, in order that the horses might be exercised, a good track has been made upon the entire tide-level circuit of the island, about twelve miles in length, over which the ice company's stock is trotted every summer at frequent intervals ; in the winter these unwonted animals are busy hauling ice. You may well improve this oppor-tunity, for it will not occur again as you travel in Alaska—you will not be able to ride elsewhere on a road worthy of the name.

A number of small trading-sloops and schooners have been built here in a boatyard, fashioned by the skill of some Creole ship-carpenters, who were trained in the yards at Sitka when Russian authority was dominant, and who have taken up their permanent abode in this "Leesnoi" settlement. A few small, tough Siberian cattle, such as we saw at Neelshik, Cook's Inlet, are roaming about here, cared for by the natives who prize milk ; also several of these same bovines are to be seen at Kadiak, where they are limited also to a few head, on account of the trouble of winter attendance and loss from bears in the summer pasturage.

An odd, weather-beaten faded little building is pointed out by the natives with pride and animation, as the house in which a "soul-like man"—a Russian monk made his abode for thirty consecutive years, teaching the children of the village and those of the neighboring towns, who flocked here in great numbers to be instructed. He taught the Russian alphabet, so that the church service might be intelligible ; also rudimentary art-principles, gardening and divers useful habits for such youth. This unique shrine is in the heart of the next village closely adjoining, and which is located on Spruce Island, or "Yealovnie," as the seventy odd Russian Creoles who live there call it. It is a little hamlet of only fifteen small log houses, very neat and clean ; and the prettiest of flower-pots within the scant windows give you a far-away thought as you observe them. Here is also one of the tiniest of Greek chapels, in which the natives are regularly joined by the small number of those of Oozinkie village (a little way off) and just across the straits ; these people, who have no church, are also pure Creoles, and unite in perfect accord with those of Spruce town.

Near by, on the southern shore of Afognak Island, is the largest settlement of the "old colonial citizens" in the Territory ; three hundred and thirty of these people are living here in a very picturesque and substantial village ; a large chapel, which is also used as a school-house, is the distinguishing architectural feature, while a number of newly-built row-boats for fishermen, on the stocks, in a miniature shipyard, point to an industry worthy of attention. The town is spread over a large landed extent, which in many places between the dwellings is devoted to vegetable gardens. More land is under cultivation here than all the rest so treated in Alaska to-day ; the crops of potatoes, cabbages, turnips, and garden-salads, like radishes, etc., seldom fail except in very backward years. No ploughing* is done ; the earth prepared for potatoes is

---

* On Wood Island, however, a small field of rye, oats, or barley, is planted every year for the use of the horses kept there ; here a plough is employed.

thrown with spades, picks, and hoes up as small ridges or tumuli, into the surface of which the seed is planted. A few of those shaggy little bulls and cows, which we have noticed before at Wood Island and Kadiak, are also roaming about, and a great many domestic fowls, such as chickens and ducks, are raised by the women and children, who take the poultry into the attics or lofts above their living rooms during the inclemencies of winter.

The desire of the Russians to have beef, milk, and butter, led to a very general importation of Siberian cattle from Petropaulovsk so that every post in Alaska, at one time, had at least a pair of these useful animals to start with. The greatest care was given to them at first, everywhere ; they were especially fostered at Sitka, where the demand for their flesh and milk was most urgent, but at Kadiak and the Kenai mission on Cook's Inlet, the only partial success in causing an increase to the stock was achieved. Impressed with an idea that certain sections of the Kadiak region would serve admirably for sheep-husbandry, a San Francisco merchant-firm shipped a flock of rams and ewes—one hundred and fifty of them—sheep of the hardiest breed, to Kolma, a spot not far from St. Paul's Harbor, Kadiak. They were in charge of a trained Scotch shepherd ; but while the flock did remarkably well in the summer, yet most of them perished during the following winter, not from exposure nor want of food, but the long-continued and frequent intervals when the sheep are obliged to be shut up tightly from the fury of wintry gales laden with sleet and rain and snow, causes their wool to "sweat" and fall from the skin in large patches, producing an emaciation and debility which the animal seldom fully recovers from. Also, the general dampness everywhere under foot during the summer season in many good grazing sections of Alaska, is such as to cause an abnormal increase of the hoofs, so that the horny toes turn and grow upward, destroying the peace and comfort of a sheep and literally confine its movements and destroy its thrifty life.*

These cereals never ripen, but are cut green, and fed as fodder. Corn is a total failure everywhere, even as fodder. No cereals have been ripened in Alaska ; the attempt, however, has been made a thousand times.

* The first cattle brought into Alaska were taken to Kadiak in 1795, and from this central station the stock was distributed—so that by 1833 it had increased to a herd of over two hundred and twenty. At the present writing it is very doubtful whether there are sixty head in the whole region. Every

Since these little villages of Kadiak, Leesnoi, Yenlova, and Afognak embrace within their limits a large majority of the sixteen or seventeen hundred Creoles who are residents and natives of Alaska, it may be interesting if a sketch be given of the physical and mental characteristics which distinguish them broadly from the aboriginal types. The original Creole was the offspring of a Russian father and an Aleutian or Kaniag mother. He inherited the strong thickset frame and bushy, curly beard and brown hair of his father; in many cases his eyes were as blue (and his hair sometimes red), his skin as white, and his bearing just as good as was his Russian progenitors'. The aggressive energy, however, of the sire seldom was transmitted, the Creole being indolent and very pacific in disposition. If this original Creole, in his time and turn, married a full-blooded Aleutian or Kaniag girl, then the offspring would show a marked dominance of the mother's race—indeed, the child would be as much like other Aleutian babies as they are related in looks among themselves; but if this original Creole marries an original Creole girl, sired like himself, then we have a type which cannot be distinguished at all from the full-blooded Slavonian, only much less demonstrative, alert, and pugnacious. Most of these old colonial citizens of this district of Kadiak are therefore full-blooded Russian quadroons and octoroons, and in every physical aspect are as much like Russians as if of pure origin. Those early Creoles, male and female, who mated, as they matured, with the native males and females, in so doing caused all their offspring, long ago, to revert to the savage types, and we cannot distinguish them to-day.

Some of the Creole girls and women whom we observe in these settlements are exceedingly handsome, modest, and the only fault we can find with them is their absolute speechlessness—they cannot be induced to chat with us, though they seem to enjoy our presence. Most of them live in scrupulously clean houses, the floors scrubbed and sanded like a well holystoned ship's deck, walls papered and decorated with pictures of saints and other pious subjects; old Russian furniture, chairs, settees, bureaus, and

season it is the habit of traders and others to send upon steamers as they go, a few head of beef-steers, which are turned out at Sitka, Kadiak, and Oonalashka to fatten during the summer, and then are slaughtered when winter ensues. Pigs thrive here, but live too much on the sea-refuse for the good of their flesh. So they are not favored.

CREOLES AND ALEUTES

Pencil Portraits of typical Alaskan faces, selected from the Author's Portfolio

1. Luka Mandriggan, an Aleute,   49 years of age
9. Philip Vollkov,   a Creole,   49 years of age
8. Matroona Vollkov,   "   "   50   "   "
10. Anoorka Meeseekin,   "   "   girl, 15   "   "
7. Fevronia Eevanov,   "   "   "   11   "   "
4. Natalia, an Aleute girl, 12 years of age
11. Deemietri Yeatkin, a Creole,   49 years of age
6. Aggie Kooshing,   "   boy, 18   "   "
3. Ivan,   an Aleute boy, 7 years of age
5. Domian,   "   "   12   "   "
2. Paraskeeva,   "   girl, 13   "   "

clocks of our own make ; the bright, omnipresent "samovar" in which the boiling water for tea is never allowed to get cool ; little curtains over the small windows, and big curtains puckered around the beds—everything is usually clean, tidy, and quiet within the Creole's home.

The wants of the Creole are very few outside of what the country in which he lives affords him. He manages to so deal in sea-otter, and fox and bear skins as to get from the trader's store what tea, sugar, flour, and cloth are required for his family. Beyond this exertion and that displayed in his gardening he rests wholly at peace with himself and all the rest of the world.

The Kaniags or Kadiakers, who are the natives of this island and contiguous islands, are in much greater numbers, and are to be found everywhere here in small hamlets that nestle in the deep fiörds and bays of Kadiak. They resemble the Aleutes so closely in outward form and characteristics that the full description given in a following chapter of those people will cover the whole ground of this inquiry, only let it be remembered that the Kaniag is a trifle taller than his Aleutian cousin, has a fairer skin, a somewhat broader face, and is considerably more muscular. Like the Aleutes, he has small feet and hands, small black eyes set in deep sockets, little or no beard, and an abundance of coarse, straight, black hair, which he cuts off roughly just above his shoulders ; he has a trifle more beard and a better mustache, but this is a very fine distinction. He is lighter-hearted, freer, and more jovial, but has less patience during seasons of privation or epidemic disease.

When the Kaniags gather together they are exceedingly talkative, abounding in jokes, in the recitation of funny legends, and stories of every imaginable nature associated with their simple lives. As they paddle their bidarkas and bidarrahs in making long journeys, they enliven the labor by continuous songs, snatches from church tunes, or lively airs taught them by the Russians and later by our soldiers and traders. They are in every respect much more susceptible of emotional impulses than are the Aleutes. This greater sociability is well exhibited by the invariable erection, in every settlement, of a "kashima," or public dance and work-house, or, in fact, a town-hall as we have it :—the Aleutes have nothing of the sort. They pass a good deal of their time on the land, traversing mountain trails in quest of bears, wolves, foxes, the land-otter, and the marmot, or "yeavrashkie," which is made into that famous

skin coat called the " parkie " all over this Alaskan country outside
of the Sitkan archipelago.

As these natives exist.to-day there are only eighteen hundred,
a few more or less, of them, which is an immense shrinkage from
the Russian enumeration of six thousand five hundred made by
actual count of Baranov. They seem to be declining even now,
year by year, even as the Koloshians of the Sitkan region do, so
that the native population of the Kadiak district, if decreased * in
the next two decades as it has in the last, will hardly have a living
representative. No one can well avoid a train of fast-crowding
thought when he stops in contemplation of sickness and death as
it appears and is treated in savage settlements—the only medical
counsel that they ever have is their own individual instinct. Ignorant as they are of the simplest anatomical details of their structure,
it is not surprising that they should surrender to disorders and
disease with that remarkable passive apathy which is so distinctive
of the sick everywhere in such communities.

Indians, and these Aleutes and Kaniags, as they grow up, have
no parental supervision whatever as to details of diet, of warm or
cool clothing, or of any of those many attentions which our children
receive from their parents. For the first ten or twelve years of
their lives they literally run wild, and are semi-naked or wholly so,
both male and female ; † this is their condition, then, at all seasons
of the year. Exposed as they are, in their manner of living, to
draughts, to insufficient covering, and damp, cold nooks for slumber, in which the air reeks with odors too vile for the power of
language to express, naturally they lay a foundation, at the very
outset of their existence, for pulmonic troubles in all the varied
degrees of that dread disease. Consumption is, therefore, the
simple and broad term for that single ailment which alone destroys
the greatest number of these people, every season, in Alaska ; all
the natives, the Eskimo, the Aleut, the Kaning, and the Indians
suffer from it alike, and they all exhibit that same stolid indifference
to its stealthy but fatal advancement—no extra care, no attempt to

---

* The church records show that the people of the Kadiak district have
decreased as follows : 1796—6,510 ; 1818—3,430; 1819—3,252 ; 1822—2,819 ;
1863—2,217 ; 1880—1,813. Small-pox, measles, and other imported diseases
have caused this.

† The little girls, as a rule, receive the earliest garments, generally nothing
but a cotton shift and a torn blauket.

shelter, to protect or to ward off in the slightest manner this trouble, until the very moment of supreme dissolution calls in a shaman and the sorrowing relatives.

After lung diseases, the next destroying factor of greatest power is embodied in the virulence of scrofulous affections, which take the form of malignant ulcers that eat into the vitals and slough away the walls of the large arteries. This most loathsome blood-poisoning renders a few settlements entirely leprous, especially so to our startled eyes when we visit them. And in this regard it is hard to find a village in the whole Alaskan boundary where at least one or more of the families therein has not got upon some one of its members the singularly prominent scars that attest this disease. Often a comely young girl or man will, in turning suddenly, reveal under the jaws or on the neck and throat, a disgusting, livid eruption which a scrofulous ancestry has cursed the youth with. Since most of this complaint is on the surface, as it were, we naturally would look for some care on the part of the afflicted native, even if for no other end than self-contentment and the ready alleviation of this cutaneous misery ; but we will look in vain, the patient never gives it. On the contrary, it is utterly neglected, and by reason of the filthy habit of these people, it is immensely aggravated and made infinitely more violent. In regard to consumption this apathy on the part of the victim is not, in contrast, so very remarkable, since it is more concealed and not near so disagreeable both to the native and his associates.

Though consumption and scrofula are the two great indigenous sources of disease and death among the natives, yet there is still a list—quite a long one—of other ills, such as paralysis, inflammatory rheumatism and peritonitis, fits, and an abrupt ending of life in the middle-aged, called most graphically "general debility." As might be inferred from the method and exigencies of aboriginal life in Alaska, these natives do not survive to any great age ; rarely, indeed, will an authenticated case of the full limit of sixty years be recorded or observed—an overwhelming majority of them are old at thirty-five and forty. When a man or a woman in a settlement rounds the fiftieth year of his or her life, a noted example of the tribe is afforded ; but should this age be attained, and the man then be free from rheumatic troubles or the death-grasp of scrofulous or pulmonic disease, he is sure to be afflicted with injured and defective vision, if not totally blind ; the glint of snow and the intensely smoky interiors of every style of native dwelling so affect the eyes of these people that those organs of sight, in the middle-aged, are seldom without signs of decay—showing some one of the various stages of granular ophthalmia, as a rule.

Snow-blindness can be remedied and its pain abated by the use of peculiar goggles, which the savages know well how to make and use, but the greater evil of smoke-poison to the optic nerve is not obviated at all by any action on their part, though it would be easy so to do. They actually seem determined to live on so as to live as wretchedly in the future as they have in the past.

Another singular characteristic of these Alaskan savages is the fact that none of the many tribes have any medicine whatever ; nor have they any knowledge, so far as we can find out, of any medicinal herb or mineral, and this again is the most extraordinary item of it all. Every less or great indisposition is treated by a universal resort to the sweat-bath ; this is the sole specific, and this is the only relief, except when the shaman is called in to worry the last hours of the unhappy patient to death, or, perhaps, in rare cases, to prolong his wretched existence for a longer period, by stimulating an undue or extra nervous tension, which then causes, at times, the usually languid and resigned sufferer to rally, as it were, before the flame flickers out. Truly these people are predestinarians ; they are wonderful in their patience when suffering long and acutely, as they lie stretched out or squatted in their gloomy, noisome hovels.

All the traders, and every vessel that sails in Alaskan waters, have medicine-chests, and to their credit be it said that, as far as they can, they do everything in their power to aid the natives when sick ; but the aborigines have not the right idea of taking physic, since they appreciate nothing but forcible treatment—large doses of something that acts immediately, or nothing at all. For instance, if the trader gives an Indian a dose of Epsom-salts, the amount given must be at least four or five times as much as would do for himself, or there will be no effect on the patient whatever. Consequently, the simplest remedies known are the only ones which the white man dare give to these people, and they have, as a matter of course, very little power to relieve them. During the last six or seven years a violent form of typhoid-pneumonia has been wasting whole settlements on the Kadiak and Aleutian coasts ; the Creoles and the natives alike yield at once to the disease, making scarcely ·

an effort to save themselves. The traders everywhere became seriously alarmed, as the force of sea-otter hunters was rapidly decreasing, and exerted themselves to their utmost in staying the epidemic, which seemed to be carried from one village to the other in vessels and by canoes. But the only medicines which can be used in the safe and successful treatment of this complaint were regarded as unworthy of notice by the suffering natives, who, not feeling immediately relieved after taking them, would then totally ignore their further use.

Bad enough are the indigenous ills of the savages in Alaska. They were, however, nothing to the horrors which followed the importation of small-pox by the Russians in 1838-39. This terrible scourge swept like wildfire up from its initial point at Sitka, over the whole length of the Alaskan mainland and island coast, until it faded out in the far north where it had nothing to prey upon. It actually carried in its grim grasp one-half of the whole population then living in that large area to an abrupt and violent death—several districts were ·so afflicted that not a soul escaped—every human being was exterminated ; it was exceedingly fatal and virulent in the Sitkan archipelago. We, knowing the filth and exposure of the lives of these people, can readily understand how they fell down and were crushed under the march of this disease.* As might be supposed the Russians lost no time in thoroughly vaccinating the survivors ; and they have been faithfully followed, in this duty, by our own sailors and traders who now live in the country.

· Another imported evil, the measles, is almost as deadly up here among the natives as small-pox. While it is a simple trouble arousing no especial anxiety with us, yet in this climate, together with the careless methods of life, it assumes a black form and becomes malignant and fatal. The last extended attack took place principally in the villages of the Kadiak district in the winter of 1874-75, where it so alarmed and impressed the sojourning members of an Icelandic Commission as to shake their desire to emigrate to

---

* La Pérouse, who touched on this coast in 1786 at Litooya Bay, under the flanks of Mount Fairweather, declares that he saw marks of the small-pox on the savages who were there then ; most likely what he saw was the scar of scrofulous sores. In 1843-44 another small-pox outbreak on the Aleutian Islands took place, but the people had been vaccinated in the meantime, and nothing serious came of it.

8

that region—at least, when they returned to their country, they were never heard from in favor of Alaska.

A very natural question arises in this connection as to whether or no the savages of Alaska will ever increase in numbers or diminish to actual extermination as time advances. It appears very plain, however, that the inhabitants of the Aleutian chain, the Peninsula, and Cook's Inlet are nearly as numerous to-day as they have been ever since the small-pox decimation of 1838–39. But all authorities agree in declaring that these people have never regained their numerical force represented in the settlement prior to the advent of the scourge which depopulated them. As to the Eskimo of the Bering Sea coasts and the Koloshians of the Sitkan region, it seems well established, from what we can learn, that they have regained their former strength in part, and were they only provident they might live by hundreds where they now exist in tens. Indifferent, wholly indifferent when living, they are as apathetic when they face death.

After reading the quaint yet strong narrative of the ferocity and strength of the Kaniags which Shellikov * has given us, it is hard, indeed, to realize that bold pioneer's feeling as we now look in upon the steep slopes of Three Saints Bay, where, at the head of it, within the sweep of a sand-spit, he erected the first permanent white habitation ever planted on Kadiak with the aid of the one hundred and fifty or sixty Russians who formed his company. Here, to-day, we see a cluster of sod-walled barraboras and two small, frame trading-houses, in which live one hundred and ninety of the descendants of those hardy savages who terrified and nearly annihilated the party of Shellikov one hundred years ago in this very spot. Nothing else is left, for Baranov in 1796 removed the post itself to the present site of Kadiak village. As we scan the settlement of Three Saints we notice that the most prominent object is the rough-hewn walls and thatched roof of an old Greek chapel, in front of which is a rude trestle; from the upper frame of this a bell hangs. Now a stooping figure emerges from the church door; he seizes the clapper, or bell tongue, with both hands and swings it vigorously. Promptly the villagers emerge from their huts; trotting and shambling in single file, they all troop into the chapel.

---

* Grigoria Shellikova Stransvovania, or Shellikov's Journeys, from 1783 to 1787. Published, St. Petersburg, 1792–93. 12mo. 2 vols.

Meanwhile the dusky sub-deacon still tolls and chimes away long after every inhabitant has been gathered in. These men and women who, with bowed heads and fervent crossings, bend and kneel as they enter that place of worship, are the children of the "blood-thirsty and implacable" Kaniags of whom Shellikov gave so vivid a picture to the Empress of all the Russias just a century ago.

They are hunting sea-otters, however, just as they did then, and living in precisely the same manner, save the variations of outward demeanor and intercourse due to the teachings of the Greek Church. But if you go among them and strive to have them tell you of the heroic battle made by their ancestors on the Oogak "kekour," you will be rewarded by either a stupid stare of vacancy or a muttered "Bogue ezniet" (God knows) !

The deep recess of Eagle Harbor, which lies between this point of earliest Russian occupation and Kadiak village, affords the location of another large native village, and its region is called the best grazing ground in all Alaska. On the surf-beaten islets at the mouth of the inlet a great many sea-lions are always found, and thus yield to these hunters of Orlova a rich return in hides and sinews so essential for the construction of the "bidarka." A few families of Creoles also reside here, who attend to a small herd of cattle, keep fowls, and generally look after their commissions as middle-men in the sea-otter revenues.

From the earliest colonial time to the present the little village of Karlook, on the north side of the island, has been the busiest spot in the country. Here is a salmon-fishing settlement right on the coast at the mouth of a small river, where from the ancient date of Russian occupation there has been a salt house and packing establishment, in which the salt and dried fish used throughout the entire Alaskan region was annually secured and prepared. To-day we find two large canning establishments set up and sustained by San Francisco merchants. The run of salmon into this river of Karlook at the height of the season is so great that it interferes with the free movement of canoes in crossing the stream ; while the fishermen of long experience in such matters say that twenty thousand barrels of the red-meated flesh could be easily secured and packed away at Karlook every summer and autumn. This salmon,* so

---

* *Oncorynchus nerka*. The fishing is done entirely with seines, floating across the river twenty to twenty-five fathoms in length, three fathoms in

abundant here, is much smaller in average size than is the one common in Cook's Inlet—it does not average ten pounds in weight. But the rich red color of its flesh is an object of the canner, who soon finds out what public taste prefers.

The rough, rocky islands of Trinity, which constitute the extreme southern limit of the Kadiak influence, are the chosen resort of sea-lions and many of the rare sea-otter. Their capture lures a few hardy natives to live in close juxtaposition to the favored haunts of these much-prized animals, and they have a most extended hunting range, reaching far away down to the westward and southward as low as that remarkable barren island of Ookamok, where the celebrated "Botany Bay," of the old Russian régime, was established. That lonely, isolated, desolate spot was the point where the old-time criminals who were guilty of murder, arson, and other capital offences, were always shipped, and left largely to their own devices for a livelihood. They were literally entombed alive on this islet, where nothing but moss and lichens and scant sphagnum could exist upon the rough, rocky surfaces, where the soil was barely appreciable—elsewhere there was none. But, strange to say, upon this island great numbers of that lively little ground squirrel, Parry's marmot, were found, and still continue to be found, which were characterized then as now by a peculiar bluish ground-tint to their fur. This color is most popular and the one so highly prized in those universal coats or cloaks used by the natives, and called "parkies" by them.

Therefore the convicts were obliged, in order to get food of their liking and many small luxuries, to diligently hunt these little animals, which they did, not only for this reason alone, but in self-defence to kill time as well.

In 1870 the descendants of the original convicts, and survivors of recent transportation by Russian order, learned in some way or other that they might lead a free life; so they then actually removed *en masse* in two large skin bidarrahs, loaded to the gunwales, and made in safety that long sea-voyage which intervenes between Ookamok and Kadiak Island. They had about one chance in a hundred of getting over the route alive, for the least of those chronic gales and storms that prevail here would have swept

depth, with a three-and-a-half-inch mesh. The whole native population is also employed in this fishery during the summer.

them to the bottom of the ocean had it arisen in the time of their passage.*

A great expanse of tide-troubled and wind-tossed water is bound between the northern coast of Kadiak and the volcanic ridges of the mainland opposite. The Straits of Shellikov are fair to see on the chart, but the mariner who has once sailed into them, lured there by the false promise of a sheltered passage, never fails to avoid the track afterward—he gladly makes the open detour of the broad Pacific. That same precipitous mountain range which we have gazed upon as it rose in sullen grandeur from the waters of Cook's Inlet, still fronts us, just as boldly, as it sweeps down the entire three hundred miles of the peninsula, forming the southern coast of that land. The sombre green and blue timber-cloak, so characteristic of its northern range, is here replaced by the russet-grays and brownish-yellows of that sphagnum and moss which now supplant the coniferous forests of Cook's Inlet, giving to the picture a much richer tone. Several of these peaks in this chain of mountains thus extended down the south coast of the peninsula are five and seven thousand feet in altitude, their summits much eroded and broken. They hold in their lofty solitudes a great many little glaciers, that, however, never come down to the sea as they habitually do in the Choogatch and Elias Alps. The feet of these peninsular mountains are washed by the direct roll of the ocean waves, which dash into innumerable fiords and coves, studded with small,

* The true reason for this hegira of the convicts is a most amusing one. It is as follows: Shortly after the transfer, in 1869, General Thomas made an extended inspection of the Alaskan posts on a steamer detailed for that work. He was accompanied by a certain representative of a Protestant Board of Missions. The vessel accidentally ran across Ookamok Island when making her way to the westward from Kadiak and touched there, where, ignorant of the fact that the people were convicts and their descendants, moved by their pitiful tales of privation, a large amount of ship's stores were landed upon the beach to satisfy the "suffering" natives: they ate, drank, and were merry, and lived sumptuously for several months afterward. But an end to these good things came at last ; the reaction in the settlement was terrible. So, urged by its pangs, the penal colony determined to pack up and move to the nearest point possible, where, when living, they could again meet, and often too, their kind benefactors ! Hence that startling journey to find those generous Americans. Lately, however, the traders at Kadiak have taken many of these people back to Ookamok, where they begged to be allowed to go and end their lives. This is the most desolate island, perhaps, in all the range of that vast Aleutian archipelago.

rocky islets and reefs awash. A beautiful geological demonstration of the effect which surf-beating waves of the ocean have made upon these mountains ages ago, is shown by the plainly evident lines traced on their flanks fully one thousand feet above the present level of the tide ; and again, another terrace is sculptured in parallel relief just above it, some five hundred feet higher—a silent, but conclusive showing of the truth that the entire Aleutian chain has been lifted out, at two successive periods, and up from the sea.

This range of the peninsula is in itself quite peculiar from the others which we have hitherto noticed thus far. It differs from their physiognomy in one respect—the mountains and ridges themselves are interrupted in one continuity down the line of their extension by abrupt depressions. These passes, as they appear to be, are not so in fact, but are either low or elevated marshy plains, which extend clear across the peninsula ; they create an impression in the mind of the observer that at a not very remote period, geologically speaking, the peaks of this peninsula range were then islands, and the marshy portages, now elevated, were the bottoms of the straits then between them. The natives are continually going to and fro between the waters of Bering Sea and the Pacific Ocean over these areas of swampy level, engaged in hunting reindeer, bear, or in friendly intercourse with the settlements. The most signal mountain groups on the peninsula are those of Morshovie, of Belcovsky, and the Pavlosk volcanic cluster—all joined by low, wet isthmian swales. The Shoomagin volcano of Veniaminov is also a noteworthy peak. The peninsula is almost bisected between Moller and Zakharov Bays, where the natives cross from water to water in a half-day's portage, and again at Pavlov Harbor. All these isolated or nearly detached mountain sections have a striking resemblance in every respect to the first large island, Oonimak, that is separated from the mainland by the narrow and unnavigable Krenitzin Straits.

The Bering Sea coast-line of the great Alaskan Peninsula presents a most radical contrast to that of the Pacific—the unbroken, rocky abruptness and roughness there is here suddenly transformed right at the very turn in the Straits of Krenitzin, to low, sandy reaches and slightly elevated moorland tundra, which cover a wide interval between the mountains and the waters of Bristol Bay and Bering Sea. The huge masses of lava, of breccia and conglomerate tufa, that everywhere rear their black-ebony shoulders above the Pacific surf, disappear entirely and suddenly here. At Oogash-ik, where we find a small settlement of Aleutes from Oonalashka, hunting walrus and sea-lions, reindeer and bears, the first rocks of granite and quartz-porphyries appear, every evidence of that character to the westward being purely and essentially igneous.

Belcovsky is the metropolis of the Alaskan Peninsula. It is the chief settlement of the sea-otter hunters, and the seat of the greatest rivalry and traffic in that fur-trade, based wholly upon the costly skins of the " bobear," * and which constitutes the only traffic worthy of mention in which the inhabitants of the entire Aleutian and Kadiak districts can engage. Here we observe from our an-

The Walrus-hunting Village of Oogashik.

chorage a little town perched upon the summit of a bluff and clinging to the flanks of a precipitous mountain that looms up behind it, usually so wreathed in fog that its summit is seldom seen. Some two hundred and sixty or seventy Aleutian sea-otter hunters and their families are living here in an oddly contrasted hamlet of frame houses and earthen barraboras; the freshly painted red roof and yellow walls of a large, new church, in the tower of which a pleasing chime of bells (but rudely struck, however), arrests the ear and the eye as the most attractive single object within the limits of the place.† The rival traders have run up their flags very

* Literally "beaver." The Russians always called this animal the "sea-beaver," but shortened from "morskie-bobear" to the simple name.

† This church was finished in 1882—begun in 1880, it cost $7,000, every cent being freely contributed by the natives.

smartly on the poles that are erected before their doors as we swing
to anchor in the offing, and a great bustle is evident among the
inhabitants when our boat pulls away for the landing, which is a
sheltered surf-eddy right under the blackest and most forbidding
of bluffs. Two rival trading-firms have each erected a landing
warehouse for the reception of their stores upon the rocky beach
where we step ashore. The ascent to the village above is steep,
but over a sloping slide of mossy earth and rocks. A clear, brawling brook runs down through the town, and we cross it by a little foot-bridge on our way. We observe cord-wood piled upon the
beach, which the traders have brought from Kadiak, and several
heaps of coal that had been brought up as ballast from Vancouver's
Island. This fuel is regularly sold to the natives here, who have
none, unless it be a stray stick of drift-wood or the "chicksa" *
vines, which the women gather on the hill-sides.

Sea-otter hunting is the sole industry and topic of conversation,
for within a radius of fifty miles from the site of Belcovsky fully
one-half of the entire Alaskan catch of these valuable peltries is secured. Were they not hopelessly improvident, shiftless and extravagant, they would be a really wealthy community ; but the notoriety
of the debauches here has become a by-word and a reproach over
the whole region between Cook's Inlet and Attoo. Every dollar of
their surplus earnings is squandered in orgies, stimulated by the
vile "quass" or beer which they make. They dress, however, in
suits of every-day clothing, such as we wear ourselves, when lounging about the village, and their women wear cloth garments and
hats cut after a fashion not very remote in San Francisco.

The neatness of the villages which we have just visited at Kadiak
and Cook's Inlet has no counterpart in Belcovsky, where, in spite of
its much greater trade and wealth, the filth and neglect everywhere
manifested among the barraboras and their interiors, are in harsh
and disagreeable contrast, while the taciturn, swelled heads of the
inmates speak volumes for the strength of that carousal during the
night prior to your arrival. A small frame house is pointed out
as the school, where it seems that those natives actually sustain a
teacher and send a large percentage of their children. It declares
that these people are not vicious at heart, though they cannot resist intemperance. They read and write, however, principally in the

---

* Trailing tendrils of the *Empetrum nigrum.*

Aleutian dialect, using an alphabet prepared for their race by the Greek Catholic missionaries in 1810-25. But, while the large capture of sea-otters and consequent flow of the traders' money and supplies into this settlement brings these people greater wealth than that showered elsewhere, yet the real physical misery of those natives of Belcovsky proves the truth and points the moral of a very old saying which declares that riches alone do not bring contentment to the human mind, be it ever so high or ever so low.

A strong south wind is springing up, and you are told by the skipper that you must get aboard as quickly as possible, for it is sure destruction to his vessel if she lies long at anchor in the offing, since the sunken rocks and open roadstead are dangerous. The little schooner is rapidly put under way, "beating out" in the freshening gale and headed for Oonga, which is the next settlement in importance, about fifty miles east. Sailing-vessels never come into Belcovsky, except those of rival traders, because it is the most risky port that the mariner has to make in all these waters of Alaska.

Before leaving the sea-otter emporium it is well to call attention to the fact that at a small indentation of this same peninsula, twenty-nine miles to the northward, is a settlement made up entirely of the poor relatives of these Belcovsky people, some forty or fifty souls, who, however, take a great pride in their superior health and morality. They have a little chapel, and enjoy much better opportunities for hunting bear and reindeer. These animals, the reindeer leading, always followed by the bears, come down at regular intervals in large herds from a great moorland to the northeast, travelling on a well-beaten "road" or track, which leads clear to the westernmost end of the peninsula, where those bovine roadmakers plunge into and cross the narrow Krenitzin Straits to renew their land, march and scatter all over the rugged and extended tundra and mountain sides of Oonimak Island.

With a line of dissipation and general misery which the rich commerce of Belcovsky causes in that settlement, we ought not to fail to include the Protassòv or Morserovic village which is located on the far end of the peninsula—the extreme west end, where a much smaller community exists, though equally opulent and just as dissolute. Here is a settlement of nearly a hundred natives, who have an annual average income of about $1,000 to each family. Yet, in spite of this small fortune in such a region, when visited by an agent of the Government in 1880, they shocked him by their aspect of abject physical misery and that excessive debauchery which had stamped them more wretchedly than it had even their cousins of Belcovsky. These people, in addition to their fine natural advantages of position for hunting sea-otters, enjoy a location in close juxtaposition to walrus-banks and sea-lion spits and islands elsewhere on the Bering shore, where they find these pinnipeds in great numbers at certain seasons of the year. The flesh, skins, blubber, and sinews are both articles of essential use and of luxury to them. Also, the same reindeer and brown-bear road, which we have just noticed, passes close by the village, so that those desiderata of food-supply and trade are very accessible.

Near by the village, less than half a mile, as if planned especially by a merciful providence, there are a number of hot sulphur-springs which would afford the diseased and sickly natives infinite relief, if they could only be induced to make the necessary exertion to go to them and bathe therein. Yet this officer of the Government declares that not one of them could be induced by him to try the efficacy of the healing waters—"It was too far to walk!"

When our little vessel comes to anchor in Delarov Harbor, Oonga Island, of the Shoomagin group, we see a flag flying from the summit of a grassy knoll which caps an irregular but bluffy headland. The village lies directly over, and under the shelter of that ridge, and it opens quickly on our view as we pull around the point and land with our dingy in a deeply indented cove upon a smooth sand and pebbly beach. The town is just above, in its full extent, but it is a thickly clustered mass of fourteen frame houses, twenty or twenty-one bar-rabkies, and the ever-present church. It does not make near as much of a spread as does Belcovsky, although it is quite as large. This is the chief codfishing rendezvous for the white fishermen who annually come up to the Shoomagin banks from San Francisco in six or seven small schooners. The location and surroundings of the little hamlet are exceedingly picturesque, but, unfortunately, though in a somewhat less disagreeable extent, the people here are also given over to those Belcovsky orgies, inasmuch as they, too, are great and successful otter-hunters, and have an income of over six hundred dollars for each family, which wealth seems to demoralize far more than it comforts their existence.

The strong southerly and southeast winds that prevail here during the summer season are the most severe, and, strange to say, they are the ones which are the coldest and the chilliest—a north wind is always warmer! These south winds bring to Oonga its foggiest weather, its heaviest rains, and raise such a ground swell in the village harbor that the craft therein are often compelled to go to sea for safety, and it always drives the fishermen from the banks outside. Those cod-banks are best, off the southerly range of the islands, and hence, when a southeaster blows, the schooners are on a most dangerous lee-shore. They seldom ever take the risks of riding out such a gale. Old skippers who have fished for forty years on the Grand Banks and "Georges," for the Gloucester and Boston markets, declare that the fury of the sea and wind is greater off the Shoomagins in a southeaster than anything of the kind experienced on the Atlantic. These wild gales become stronger, loaded with sleet and snow, as winter approaches, so that by the middle or end of November, until next April, all sailing-craft are practically driven from the fishing grounds.

The same method of catching cod is employed here as practised by our Gloucester men, in only one respect, however: the long, buoyed lines are not set out and regularly under-run, but instead, small boats and dories, with two men in each, are put off from the schooners, and fish with hand-lines, using what is known as "11-inch" and "12-inch" hooks. Halibut, and "squid," or cuttle-fish, make the best bait. A good, smart man, if he is fortunate, will haul up four hundred codfish in a day's steady labor, but this is an extraordinary streak of luck. An average of three hundred every fair day is one that gives the highest satisfaction. These fish are taken on board of the schooner, salted, and not touched again until the cargo is broken for re-drying and curing at several points chosen for that purpose in California. At first our people were disposed to hire the natives up here to do this hand-line fishing, and they did so; but a patient trial has demonstrated the fact that it pays to employ our own men instead, even at greatly advanced wages. The Aleutes are docile, and do exceedingly well in spurts, but they do not like to work in steady, well-sustained periods of any great length at a time.

Were it not for the intense physical discomfort of the rapidly recurring fog, sleet, and rain-laden gales, Oonga would undoubtedly be a site well chosen for a neat New England fishing village. Many of those white men now employed up there in the cod-fishery declare that they would bring their wives and children into the country, to permanently settle, if they thought that they could be

happy under the conditions of climate which prevail. But they
argue that where they themselves cannot peacefully exist the year
round, it would be idle to suppose that a civilized settlement could
be well established. We will find, however, quite a number of
genial, sociable fellows, men of our race, who are well educated,
and who have had excellent opportunities, and who to-day are
roaming here, there, and everywhere in Alaska, hunting, fishing,
and trading, or prospecting. They appear to be entirely happy,
not a bit cynical, and never express the slightest desire to return
with us to the world which they have left behind them voluntarily.
Alaska to them is a perfect Mecca of peace, and they have no desire to see it changed. They unite usually in saying that their
wants are few, easily supplied, and they scarcely remember what
care was—it does not trouble them now.

The cod-fishermen do not make their working headquarters in
this village, but across, over the bay on Popov Islet, at a spot which
is called Pirate Cove. They are not annoyed by idle villagers there,
and are also somewhat nearer to the fishing-resorts which are just
outside. They are most likely not far from that spot where Bering
landed, August 30, 1741, to bury one of his seamen named Shoomagin, and to refill his water-casks. The exact locality, or even the
precise islet of the many that form this Shoomagin group, on
which the then sick and sadly demoralized explorer and his crew
interred the remains of their dead comrade, will never be satisfactorily established; the cross of wood set up was immediately
pulled down, after his departure, by the natives, who were then
decidedly hostile, and who eyed him and his vessel with unaffected
dislike and apprehension.* When the *St. Peter*, six days later,
hauled off from those islands and turned her prow for Kamchatka,
perhaps that gloomy, timid Dane commanding her may have had
an astral premonition of the wreck of this vessel, which soon followed—and his own death too, in a self-made sand grave beneath
the black shadows of the bluffs at "Kommandor"—this may have
caused him to earn that reproach which has been so lavishly laid
upon his conduct of a most remarkable and disastrous voyage.

The Shoomagins are all bold and bluffy, with high uplands and

* From the record made in the ship's log it would seem most likely that
he landed on either Popov Island, or else Nogai; the description will fit either
locality.

lofty ridges; on Oonga the most elevated summits are to be seen. Bare of timber, but covered with sphagnum and mosses and clumps of dwarfed crab-apples and willows, they stand as rock-ribbed break-waters against the full sweep of the mighty uninterrupted roll of a vast ocean. The surf that dashes foaming and booming upon their firm foundations is of unrivalled force, and fear-inspiring.

Oonga Island has also been the base of a very.extended and thorough attempt to develop a large vein of coal which is found cropping out on the face of a bluff in a small inlet of its north shore. The oldest coal-mine in the region of Alaska is located in Cook's Inlet near its mouth, at a spot still indicated on maps as Coal Harbor. Here the Russians, eager to be able to obtain fuel for the use of their steam-vessels, began, in 1852, a most active and systematic series of mining operations; they brought machinery and ran it by steam-power; experienced German miners were engaged to superintend and direct a large force of Muscovitic laborers sent up from Sitka. In 1857 the work had been so energetically pushed that shafts had been sunk, and a drift run into the vein for a distance of one thousand seven hundred feet; during this period, and three following years, two thousand seven hundred tons of coal were mined, the value of which was forty-six thousand rubles, but the result was a net loss. The thickness of the vein was found to vary from nine to twelve feet, and its extent was practically unlimited. But the Russians found out then, as our people at Oonga did afterward, that this Alaskan lignite was utterly unfit for use in the furnaces of the steamers—that it was so highly charged with sulphur as to burn like a flash and eat out, fuse and warp the grate-bars—even melting down the smoke-stacks! Steam-vessels now bring their own coal with them from San Francisco, Puget Sound or Nanaimo, or have it sent up from there by sailing-tenders to depots previously designated.*

As we leave the sheltering bluffs of Oonga, our course seems to

* Captain F. W. Beechey in his voyage of the *Blossom*, 1825-27, discovered and located at Cape Beaufort, in the Arctic Ocean and on the Alaskan coast, a vein of coal; this has been subsequently revisited and mined to a small extent by the officers of the Revenue marine cutters of our Government, who pronounce it very satisfactory for steaming purposes. Its situation, however, is so remote that it has no economic significance, and no harbor is there for a vessel of any kind.

be laid directly south ; so much so, that for once we express our surprise to the skipper, who, feeling sure that he understands our dread of losing time in reaching Oonalashka, spreads out his chart and calls us to the table. A moment's inspection shows the wisdom of the roundabout course, for a forest of rough, rocky islets studs the ocean directly to the west and many to the south. To sail through the intricate passages of the Chernaboors and the reefs of Saanak would be to invite certain destruction. Therefore, as we make a long detour to clear the path of our progress from all danger, we will give the reader some interesting facts relative to the chase of the sea-otter, which is the sole object of those natives who hunt in this district.

CHAPTER VII.

THE QUEST OF THE OTTER.

LITTLE does my lady think, as she contemplates the rich shimmer
of the ebony sea-otter trimming to her new sealskin sacque, that
the quest of the former has engaged thousands of men during the
last century in exhaustive deeds of hazardous peril and extreme daring, and does to-day—that the possession of the the sea-otter's coat
calls for more venturesome labor and inclement exposure on the
part of the hunter than is put forth in the chase of any other fur-
bearing or economic animal known to savage or civilized man. No
wonder that it is costly ; what abundant reason that it should be
rare !

The rugged, storm beaten resorts of the sea-otter, its wariness
and cunning, and the almost incredible fortitude and patience, skill
and bravery, of its semi-civilized captor, have so impressed the
writer that he feels constrained to rearrange his notes and touch up
his field-sketches made upon the subject-matter of this chapter several years ago, while cruising in Alaskan waters, so that he may give
to the readers of this work the first full or fair idea of the topic ever
put into type and engraving.

Feodor Altasov, with a band of Russian Cossacks* and Tartar "promishlyniks," were the pioneers of civilized exploration in Eastern Siberia, and finally arrived at the head of the great Kamchatkan Peninsula, toward the end of the seventeenth century. Here they found, first of all their race, the rare, and to them the exceedingly valuable, fur of the sea-otter. The animal bearing this pelage then was abundant on that coast, and not prized above the seals and sea-lions by the natives who displayed their peltries to the ardent Russians, and who in barter asked little or nothing extra from the white men in return. The feverish eagerness of the Slavonians, quickly displayed, to secure these choice skins, so excited the natives as to result very soon in the practical extirpation of the "kahlan," as they termed it, from the entire region of the Kamschadales. The greedy fur-hunters then rifled graves and stripped the living of every scrap of the precious object of their search, and, for the time being, searched in vain for other haunts of the otter.

Along by the close of 1743, the survivors of Bering's second voyage of exploration and Tscherikov brought back to Petropaulovsk an enormous number of skins which they had secured on the Aleutian and Commander Islands, until then unknown to the Kamchatkans or the Russians. In spite of the rude appliances and scanty resources at the command of these eager men, they fitted out rude wooden shallops and boldly pushed themselves over dark and tempestuous seas to the unknown and rumored resorts of the sea-otter. In this manner and by this impulse the discovery of the Aleutian Islands and the mainland of Alaska was fully determined, between 1745 and 1763. In this enterprise some twenty-five or thirty different individuals and companies, with quite a fleet of small vessels and hundreds of men, were engaged; and so thorough and energetic

---

* The Cossacks who came with Altasov were rough-looking fellows of small size, lean and wiry, with large, thin-lipped mouths and very dark skins. Most of them were the offspring of Creole Russian Tartars and women from the native tribes of Siberia. They were filthy in their habits. Naturally cruel, they placed no restraint upon their actions when facing the docile Aleutes, and indulged in beastly excesses at frequent intervals. The custom of the Cossack hunters after establishing on an island, was to divide the command into small parties, each of which was stationed in or close by a native settlement. The chief or head Aleut was induced by presents to assist in compelling and urging his people to hunt. When they returned, their catch was taken and a few trifling presents made, such as beads and tobacco-leaf.

were they in their search and stimulated capture of the coveted animal, that, along by the period of 1772-74, the catch of this unhappy beast had dwindled down from thousands and tens of thousands at first, to hundreds and tens of hundreds at last. When the Russian traders opened up the Aleutian Islands they found the natives commonly wearing sea-otter cloaks, which they willingly parted with at first for trifles, not placing any especial value upon the otter, as they did upon the bodies of the hair-seal and sea-lion, the flesh and skins of which were vastly more palatable to them and serviceable. But the fierce competition and raised bidding of the greedy traders soon fired the savages into hot and incessant hunting. During the first decade or two of pursuit the numbers of these animals taken all along the Aleutian chain and down the entire northwest coast as far as Oregon, were so great that they appear fabulous in comparison with the exhibit made now.*

The result of this warfare upon sea-otters, with ten hunters then where there is one to-day, was not long delayed. Everywhere throughout the whole coast-line frequented by them, a rapid and startling diminution set in ; so much so, that it soon became difficult to get from places where a thousand were easily taken, as many as twenty-five or thirty. When the region known as Alaska came into our possession, the Russians were taking between four and five hundred sea-otters annually from the Aleutian Islands and South of the Peninsula and Kadiak, with perhaps one hundred and fifty more from Cook's Inlet, Yakootat, and the Sitkan district, the Hudson Bay traders and others getting some two hundred more from the coasts of Queen Charlotte's and Vancouver's Islands, and Gray's Harbor, Washington Territory.

Now during the last year, instead of less than seven hundred skins taken as above specified, our traders have secured more than four thousand. This immense difference is not due to the fact of a proportionate increase of sea-otters—that is not evident—but it is due to the keen competition of our people, who have reanimated the organization anew of old-fashioned hunting-parties, after the

* In 1804 Baranov (the Colonial Governor) went from Sitka to the Okotsk with fifteen thousand sea-otter skins, that were worth as much then as they are now, viz., fully $1,000,000. Last year the returns from Alaska and the northwest coast scarcely foot up four thousand skins; but they yielded at least $200,000 directly to the native hunters, being ten times better pay than they ever brought under Russian rule to these people.

9

style of Baranov's bateaux. As matters are now conducted, the hunting-parties do not let the sea-otter have a day's rest during the whole year : parties relieve each other in orderly, steady succession, and a continual warfare is maintained. Stimulated by our people, this persistence is rendered still more deadly to the kahlan by the use of rifles of our best make, which, in the hands of the young and ambitious natives, in spite of the warnings of their old sires, must result in the virtual extermination of that marine beast.*

This is the more important because all the world's supply comes from the North Pacific and Bering Sea, and upon its continuance between four thousand and five thousand semi-civilized natives of Alaska depend absolutely and wholly for the means by which they are enabled to live beyond simple barbarism ; its chase and the proceeds of its capture furnish the only employment offered by their country, and the revenue by which they can feed and clothe themselves as they do, and, by so doing, appear to all intents and purposes much superior to their Indian neighbors of Southeastern Alaska, or their Eskimo cousins of Bering Sea.

The sea-otter, like the fur-seal, is another striking illustration of an animal long known and highly prized in the commercial world, yet respecting the life and habits of which nothing definite has been ascertained or published. The reason for this is obvious, for, save the natives who hunt them, no one properly qualified to write has ever had an opportunity of observing the *Enhydra* so as to study it in a state of nature, inasmuch as of all the shy, sensitive beasts upon the capture of which man sets any value whatever, this creature is the most keenly on the alert and difficult to obtain ; · and, also, like the fur-seal, it possesses, to us, the enhancing value

---

* It is a fact, coincident with the diminution of the sea-otter life under the pressure of Russian greed, that the population of the Aleutian Islands fell off at the same time and in the same ratio. The Slavonians regarded the lives of these people as they did those of dogs, and treated them accordingly. They impressed and took, under Baranov's orders, in 1790–1806, and his subordinates, hunting-parties of five hundred to one thousand picked Aleutes, eleven or twelve hundred miles to the eastward from their homes at Oonalashka, Oomnak, Akoon, and Akootan. This terrible sea-journey was made by these natives in skin "baidars" and bidarkies, traversing one of the wildest and roughest of coasts. They were used not only for the drudgery of otter-hunting in Cook's Inlet and the Sitkan archipelago, but forced to fight the Koloshians and other savages all the way up and down those inhospitable coasts. That soon destroyed them—very few ever got back to the Aleutian Islands alive·

and charm of being principally confined in its geographical distribution to our own shores of the Northwest. A truthful account of
the strange, vigilant life of the sea-otter, and the hardships and the
perils of its human hunters, would surpass, if we could give it all,
the novelty and the interest of a most weird and attractive work of
fiction.

The sea-otter is widely removed from close relationship to our
common land-otter. Unlike this latter example, it seldom visits the
shore, and then only when the weather is abnormally stormy at sea.
Instead of being a fish-eater, like *Lutra canadensis*, it feeds almost
wholly upon clams, crabs, mussels, and echinoderms, or "sea-
urchins," as might be inferred from its peculiar flat molars of den-

The Kahlan or Sea-otter.

tition. It is, when adult, an animal that will measure from three
and a half to four and a half feet in length from nose to root of its
short, stumpy tail. The general contour of the body is strongly
suggestive of the beaver, but the globose shape and savage expression of the creature's head are peculiar to it alone. The small,
black, snaky eyes gleam with the most wild and vindictive light
when the owner is startled ; the skin lies over its body in loose
folds, so that when taken hold of in lifting the carcass out of the
water, it is slack and draws up like the elastic hide on the nape
of a young dog. This pelt, when removed in skinning, is cut only
at the posteriors, and the body is drawn forth, turning the skin
inside out, and in that shape it is partially stretched, air-dried, and
is so lengthened by this process that it gives the erroneous impression of having been taken from an animal the frame of which was

at least six feet in length, with the proportionate girth and shape of a mink or weasel.

There is no sexual dissimilarity in color or size, and both male and female manifest the same intense shyness and aversion to man, coupled with the greatest solicitude for their young, which they bring into existence at all seasons of the year, for the natives capture young pups in every calendar month. As the hunters never have found the mothers and their offspring on the rocks or beaches, they affirm that the birth of a sea-otter takes place on the numerous floating kelp-beds which cover large areas of the ocean south of the Aleutian chain and off the entire expanse of the northwest coast. Here, literally "rocked in the cradle of the deep," the young kahlau is brought forth and speedily inured to the fury of fierce winds and combing seas. Upon these algoid rafts the Aleutes often surprise them sporting one with the other, for they are said to be very playful, and one old hunter told the writer that he had watched a sea-otter for half an hour as it lay upon its back in the rollers and tossed a bit of sea-weed up into the air from paw to paw, apparently taking great delight in catching it before it fell into the water.

The sea-otter mother clasps her young to her breast between her fore-paws, and stretches herself at full length on her back in the ocean when she desires to sleep, and she suckles it also in this position. The pup cannot live without its mother, though frequent attempts have been made by hunters to raise them, for the little animals are very often captured alive and wholly uninjured; but, like some other animals, they seem to be so deeply imbued with fear or dislike of man that they invariably die of self-imposed starvation. The enhydra is not polygamous, and it is seldom, indeed, that the natives, when out in search of it, ever see more than one animal at a time. The flesh is very unpalatable, highly charged with a rank taste and odor. A single pup is born, as the rule, about fifteen inches in length and provided with a natal coat of coarse, brownish, grizzled hair and fur, the head and nape being rather brindled, and the nose and cheeks whitish-gray, with the roots of the hair everywhere much darker next to the skin. From this poor condition of fur at birth the otter gradually improves as it grows older, shading darker, finer, thicker, and longer by the time they are two years of age. Then they rapidly pass into prime skins of the most lustrous softness and ebony shimmering, though the creature is not full-grown until it has passed its fourth season. The rufous-white nose and mustache of the pup are not changed in the pelage of the adult, but remain constant through life. The whiskers are short, white, and fine. So much for the biology of the sea-otter. Now we turn to the still more interesting one of its captors.

The typical hunter is an Aleutian Islander or a native of Kadiak. He is not a large man—rather below our standard—say five feet five or six inches in stature. There are notable exceptions to this rule, for some of them are over six feet, while others are veritable dwarfs—resemble gnomes more than anybody else. He wears the peculiar expression of a Japanese more than any other. His hair is long, coarse, and black; face is broad; high, prominent cheek bones, with an insignificant flattened nose; the eyes are small, black, and set wide in his head under faintly marked eyebrows, just a faint suggestion of Mongolian obliquity; the lips are full, the mouth large, and the lower jaw square and prognathous; the ears are small, likewise his feet and hands; his skin in youth is often quite fair, with a faint flush in the cheeks, but soon weathers into a yellowish-brown that again seams into deep flabby wrinkles with middle and old age. He has a full, even set of good teeth, while his body, as might be inferred from his habit of living so much of his life in the cramped "bidarka"* or skin boat, is well developed in the chest and arms, but decidedly sprung at the knees, and he is slightly unsteady in his pigeon-toed gait.

The mate of this hunter was when young a very good-looking young woman, who never could honestly be called handsome, yet she was then and is now very far from being hideous or repulsive.

---

* The "bidarka" is a light framework of wooden timbers and withes very tightly lashed together with sinews in the form indicated by my illustrations. It is covered with untanned sea-lion skins, which are sewed on over it while they are wet and soft. When the skins dry out they contract, and bind the frame, and are as taut as the parchment of a well-strung bass-drum. Then the native smears the whole over with thick seal-oil, which keeps the water out of the pores of the skin for quite a long period and prevents the slackening of the taut binding of the little vessel for twenty-four to thirty hours at a single time. Then the bidarka must be hauled out and allowed to dry off in the wind, when it again becomes hard and tight. Most of them are made with two man-holes, some have three, and a great many have but one. The otter-hunters always go in pairs, or, in other words, use two-holed bidarkies.

She partakes, somewhat mellowed down, of the same characteristics which we have just sketched in the face and form of her husband. As they live to-day, they are married and sustain this relation, sheltered in their own hut or "barrabkie." They have long, long ago ceased to dress themselves in skins, and now appear in store clothes and cotton gowns, retaining, however, their characteristic waterproof garment known as the "*kamlayka*," and the odd boots known as "*tarbosars*," in which they are always enveloped in wet weather, or whenever they venture out to sea in their bidarka. They dress themselves up on Sundays, when at home, in boots and shoes and stockings of San Francisco make. He wears a conventional "beaver" or plug hat often, and she affects a gay worsted hood, although, on account of the steady persistence of high winds, he prefers a smart marine-band cap, such as our soldiers on fatigue-duty wear. He is, however, inclined to be quite sober, not giving much attention to display or color, as is the habit of semi-civilized people everywhere else ; but he does lavish the greatest care and labor over the decoration of his bidarka, and calls upon his wife to ornament the seams of his water-repellant kamlayka and tarbosars with the gayest embroidery, and tufts of bright hair and feathers, and lines of cunning goose-quill work.

Mrs. Kahgoon, however, is a true woman. She naturally desires all the bright ribbons and cheap jewellery which the artful trader exhibits to her longing eyes in his store, that stands so near and so handy to her barrabkie, and her means only limit the purchase which she makes of these prized desiderata. She dresses her hair in braids, as a rule, and twists them up behind. She seldom wears a bonnet or hat, but has a handkerchief, generally of cotton, sometimes silken, always tied over her head, and when she goes, as she often does, out to call on a neighboring spinster or madam, or to the store, she throws a small woollen shawl over her head and shoulders, holding it drawn together under her chin by one hand. As we have intimated, she dresses principally in cotton fabrics, with skirts, overskirts, white stockings, etc. ; but when she was a girl, and much more than that, she usually went, with her legs and feet bare, into the teeth of biting winds and over frosty water and wood-paths.

The domestic life of this hunter and his wife is all bound up within the shelter of their "barrabkie." This hut or house of the Aleutian hunter is half under ground, or, in other words, it is an

excavation on the village site of a piece of earth ten or fifteen feet
square, three or four feet deep, which is laid back up and over upon
a wooden frame or whalebone joisting, which is securely built up
within and above this excavation, so that a rafter-ceiling is made
about six feet in the clear from the earthen floor. A wall of peaty
sod is piled up around outside, two and three feet thick. The native architect enters this dwelling through a little hall patched on
to that leeward side from the winds prevalent in the vicinity. The
door is low, even for Kahgoon, and he stoops as he opens and

A Barrabkie.

(*The characteristic dwelling of Aleutes and Kadiakers.*)

closes it. If he has been a successful hunter, he will have the floor
laid with boards secured from the trader ; but if he has been unlucky, then nothing more to stand upon than the earth is afforded.
This barrabkie is divided into two rooms, not wholly shut out one
from the other, by a half-partition of mats, timber, or some hanging
curtains, which conceal the bedroom or " spalniah " from the direct
gaze of the living and cook-room. They are very fond of comfortable beds, having adopted the feather-ticks of the Russians. Soap
is an expensive luxury, so Kahgoon's wife is economical of its use
for washing in her laundry ; and, though she may desire to spread
over her sleeping couch the counterpane and fluted shams of our
own choice, she has nothing better than a colored quilt which the

traders bring up here especially to meet this demand. A small deal-table, two or three empty cracker-boxes from the store, and a rude bench or two constitute all the furniture, while a little cast-iron stove, recently introduced, stands in one corner, and the heating and cooking is created and performed thereon. The table-ware of a hunter's wife and the household utensils do not require much room or a large cupboard for their reception. A few large white crockery cups, plates, and saucers, with gaudy red and blue designs, and several pewter spoons, will be found in sufficient quantity to entertain with during seasons of festivity. She manifests a marked dislike to tin dishes, probably due to the fact that it is necessary to take care of this ware, or it rusts out. Then, above all the strange odors which arise here in this close, hot little room, we easily detect the smell of kerosene, and, sure enough, it is the oil which is burned in the lamp.

Such a barrabkie built and furnished in this style and occupied by Kahgoon, his wife, two or three children, and a relative or so, is a warm and a thoroughly comfortable shelter to him and his, as long as he keeps it in good repair. It is true that the air seems to us, as we enter, oppressively close, and, in case of sickness, is positively foul ; yet on the whole the Alaskan is very comfortable. He never stores up much food against the morrow—the sea and its piscine booty is too near at hand. Whatever he may keep over he does not have in a cellar, but hangs it up outside of his door on an elevated trestle which he calls a " laabas," beyond the reach of the village dogs, while there is no thought of theft from the hands of his neighbors. He lives chiefly upon fresh fish—cod, halibut, salmon and other varieties, which he secures the year round as they rotate in the sea and streams. He varies that diet according to the success of his hunting, by buying at the store tea, sugar, hard-bread, crackers, flour and divers canned fruits or vegetables. Nature sends him in season the flesh and eggs of sea-fowl, geese, ducks, and a few land birds like willow-grouse.

In this fashion the sea-otter hunter appears to us as we view him now ; his children come, grow up, and branch out, to repeat his life and doing, as they show themselves capable of living by their own exertions as hunters and fishermen. He is a peaceful, affectionate, and thoroughly undemonstrative parent, a kind husband, and he imposes no burden upon his wife that he does not fully share, unless he becomes a drunkard, when, in that event, a sad change is made in the man. He gets drunk, and his wife too, by taking sugar, flour, and dried-apples, rice or hops, if he can get them, in certain proportions, puts them into a barrel or cask, with water, bungs it up and waits for fermentation to do its work. Before it has worked entirely clear he draws off a thick, sour liquid which intoxicates him most effectually—he beats his spouse and runs her and children from the house, smashes things, and for weeks afterward the barrabkie is desolate and open as the result of such orgies. If he continues, his health is shattered, he rapidly fails as a hunter, and he suffers the pangs of poverty with his family. It is said upon good authority that the brewing of this liquor was taught to these people by the earliest Russian arrivals in their country, who made it as an anti-scorbutic ; but it certainly has not proved to be a blessing in disguise, for it has brought upon them nearly all the misery that they are capable of understanding.

In concluding this brief introduction to the life of the otter-hunter, we may fitly call attention to the fact that Kahgoon and his family are devout members of the Greek Catholic Church, as are all of his people, without a single exception, between Attoo and Kadiak Islands—nearly five thousand souls to-day, living in scattered hamlets all along between.

The subtle acumen displayed by the sea-otter in the selection of its habitat can only be fully appreciated by him who has visited the chosen land, reefs, and water of its resort. It is a region so gloomy, so pitilessly beaten by wind and waves, by sleet, rain and persistent fog, that the good Bishop Veniaminov, when he first came among the natives of the Aleutian Islands, ordered the curriculum of hell to be omitted from the church breviary, saying, as he did so, that these people had enough of it here on this earth! The fury of hurricane gales, the vagaries of swift and intricate currents in and out of the passages, the eccentricities of the barometer, the blackness of the fog enveloping all in its dark, damp shroud, so alarm and discomfit the white man that he willingly gives up the entire chase of the sea-otter to that brown-skinned Aleut, who alone seems to be so constituted as to dare and wrestle with these obstacles through descent from his hardy ancestors, who, in turn, have been centuries before him engaged just as he is to-day.

So we find the sea-otter-hunting of the present, as it was in the past, entirely confined to the natives, with white traders here and there vieing in active competition one with the other in bidding for the quarry of those dusky captors. The traders erect small frame dwellings as stores in the midst of the otter-hunting settlements, places like Oonalashka, Belcovsky, Oonga, and Kadiak villages, which are the chief resorts of population and this trade in Alaska. They own and employ small schooners, between thirty and one hundred tons burthen, in conveying the hunting parties to and from these hamlets above mentioned as they go to and return from the sea-otter hunting-grounds of Saanak and the Chernaboor rocks, where five-sixths of all the sea-otters annually taken in Alaska are secured. Why these animals should evince so much partiality for this region between the Straits of Oonimak and the west end of Kadiak Island is somewhat mysterious, but, nevertheless, it is the great sea-otter hunting-ground of the country. Saanak Island, itself, is small, with a coast-circuit of less than eighteen miles. Spots of sand-beach are found here and there, but the major portion of the shore is composed of enormous water-worn boulders, piled up high by the booming surf. The interior is low and rolling, with a central ridge rising into three hills, the middle one some eight hundred feet high. There is no timber here, but an abundant exhibit of grasses, mosses, and sphagnum, with a score of little fresh-water ponds in which multitudes of ducks and geese are found every spring and fall. The natives do not live upon the island, because the making of fires and scattering of food-refuse, and other numerous objectionable matters connected with their settlement, alarm the otters and drive them off to parts unknown. Thus the island is only camped upon by the hunting-squads, and fires are never made unless the wind is from the southward, since no sea-otters are ever found to the northward of the ground. The sufferings—miseries of cold, and even hunger, to which the Aleutes subject themselves here every winter, going for weeks and weeks at a time without fires, even for cooking, with the thermometer below zero in a wild, northerly and westerly gale of wind, is better imagined than portrayed.

To the southward and westward of Saanak, stretching directly from it out to sea, eight or ten miles, is a succession of small, submerged islets, rocky, and bare most of them, at low water, with numerous reefs and stony shoals, beds of kelp, etc. This scant area is the chief resort of the kahlan, together with the Chernaboor Islets, some thirty miles to the eastward, which are identical in character. The otter rarely lands upon the main island, but he is, when found ashore, surprised just out of the surf-wash on the reef. The quick hearing and acute smell possessed by this wary brute are not equalled by any other creatures in the sea or on land. They will take alarm, and leave instantly from rest in a large section, over the effect of a small fire as far away as four or five miles distant to the windward of them. The footstep of man must be washed from a beach by many tides before its trace ceases to scare the animal and drive it from landing there, should it approach for that purpose.

The fashion of capturing the sea-otter is ordered entirely by the weather. If it be quiet and moderately calm, to calm, such an interval is employed in "spearing surrounds." Then, when heavy weather ensues, to gales, "surf-shooting" is the method ; and if a furious gale has been blowing hard for several days without cessation, as it lightens up, the hardiest hunters "club" the kahlan. Let us first follow a spearing party ; let us start with the hunters, and go with them to the death.

Our point of departure is Oonalashka village ; the time is an early June morning. The creaking of the tackle on the little schooner out in the bay as her sails are being set and her anchor hoisted, cause a swarm of Aleutes in their bidarkies to start out from the beach for her deck. They clamber on board and draw their cockle-shell craft up after them, and these are soon stowed and lashed tightly to the vessel's deck-rail and stanchions. The trader has arranged this trip and start this morning for Saanak, by beginning to talk it over two weeks ago with these thirty or forty hunters of the village. He is to carry them down to the favored otter-resort, leave them there, and return to bring them back in just three months from the day of their departure this morning. For this great accommodation the Aleutes interested agreed to give the trader-skipper a refusal of their entire catch of otter-skins—indeed many of them have mortgaged their labor heavily in advance by pre-purchasing at his store, inasmuch as the credit system is worked among them for all it is worth. They are adepts in driving a bargain, shrewd and patient. The traders know this now, to the grievous cost of many of them.

If everything is auspicious, wind and tide the next morning, after sailing, bring the vessel well upon the ground. The headlands are made out and noted ; the natives slip into their bidarkies as they are successively dropped over the schooner's side while she jogs along under easy way, until the whole fleet of twenty or thirty craft is launched. The trader stands by the rail and shakes the hand of each grimy hunter as he steps down into his kyack, calling him, in pigeon-Russian, his " loobaiznie droög," or dear friend, and bids him a hearty good-by. Then, as the last bidarka drops, the ship comes about and speeds back to the port which yesterday morning she cleared from, or she may keep on, before she does so, to some harbor at Saanak, where she will leave at a preconcerted rendezvous a supply of flour, sugar, tea, and tobacco for her party.

If the weather be not too foggy, and the sea not very high, the bidarkies are deployed into a single long line, keeping well abreast, at intervals of a few hundred feet between. In this manner they paddle slowly and silently over the water, each man peering sharply and eagerly into the vista of tumbling water just ahead, ready to catch the faintest evidence of the presence of an otter, should that beast ever so slyly present even the tip of its blunt head above for breath and observation. Suddenly an otter is discovered, apparently asleep, and instantly the discoverer makes a quiet signal, which is flashed along the line. Not a word is spoken, not a paddle splashes, but the vigilant, sensitive creature has taken the alarm, and has turned on to its chest, and with powerful strokes of its strong, webbed hind feet, has smote the water like the blades of a propeller's screw, and down to depths below and away it speeds, while the hunter brings his swift bidarka to an abrupt standstill directly upon the bubbling wake of the otter's disappearance. He hoists his paddle high in the air, and holds it there, while the others whirl themselves over the water into a large circle around him, varying in size from one-quarter to half a mile in diameter, according to the number of boats engaged in the chase.

The kahlan has gone down—he must come up again soon somewhere within reach of the vision of that Aleutian circle on the waters over its head ; fifteen or twenty minutes of submergence, at the most, compel the animal to rise, and instantly as its nose appears above the surface, the native nearest it detects the movement, raises a wild shout, and darts in turn toward it ; the yell has sent the otter down again far too quickly for a fair respiration, and that is what the hunter meant to do, as he takes up his position over the spot of the animal's last diving, elevates his paddle, and the circle is made anew, with this fresh centre of formation. In this method the otter is continually made to dive and dive again without scarcely

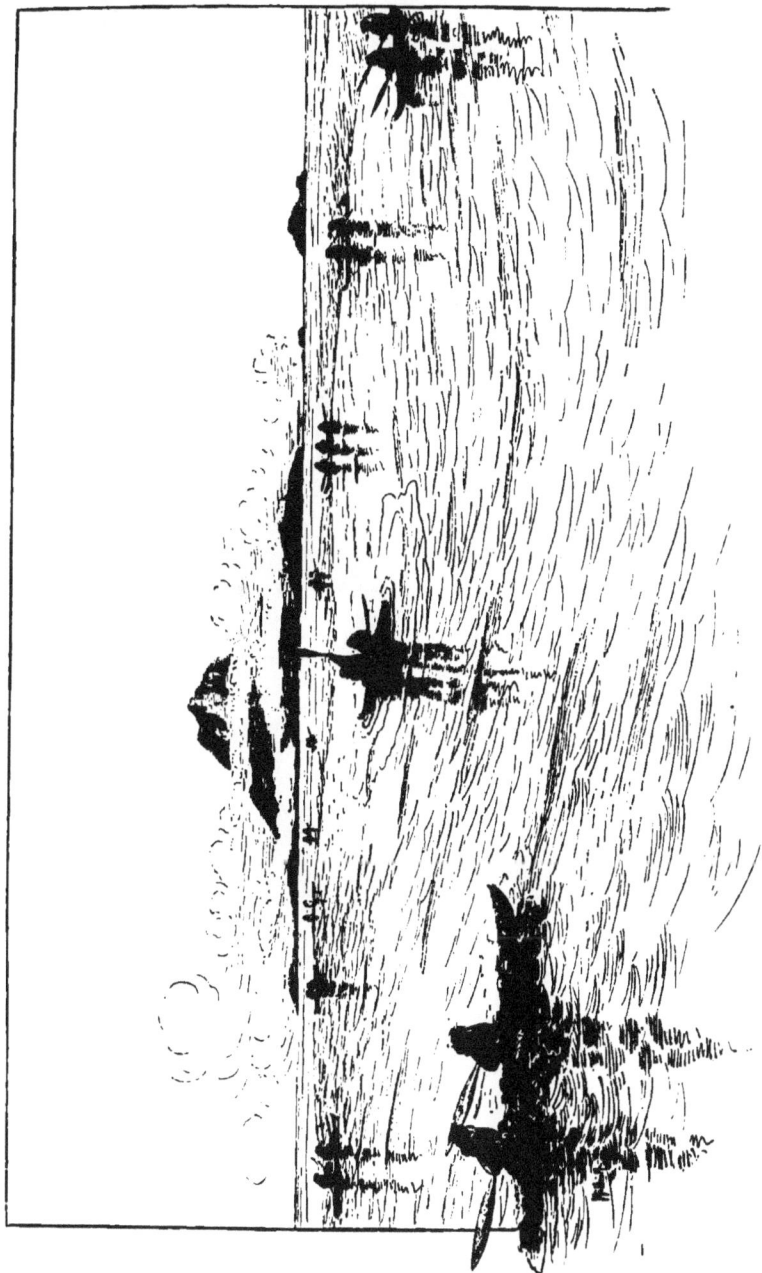

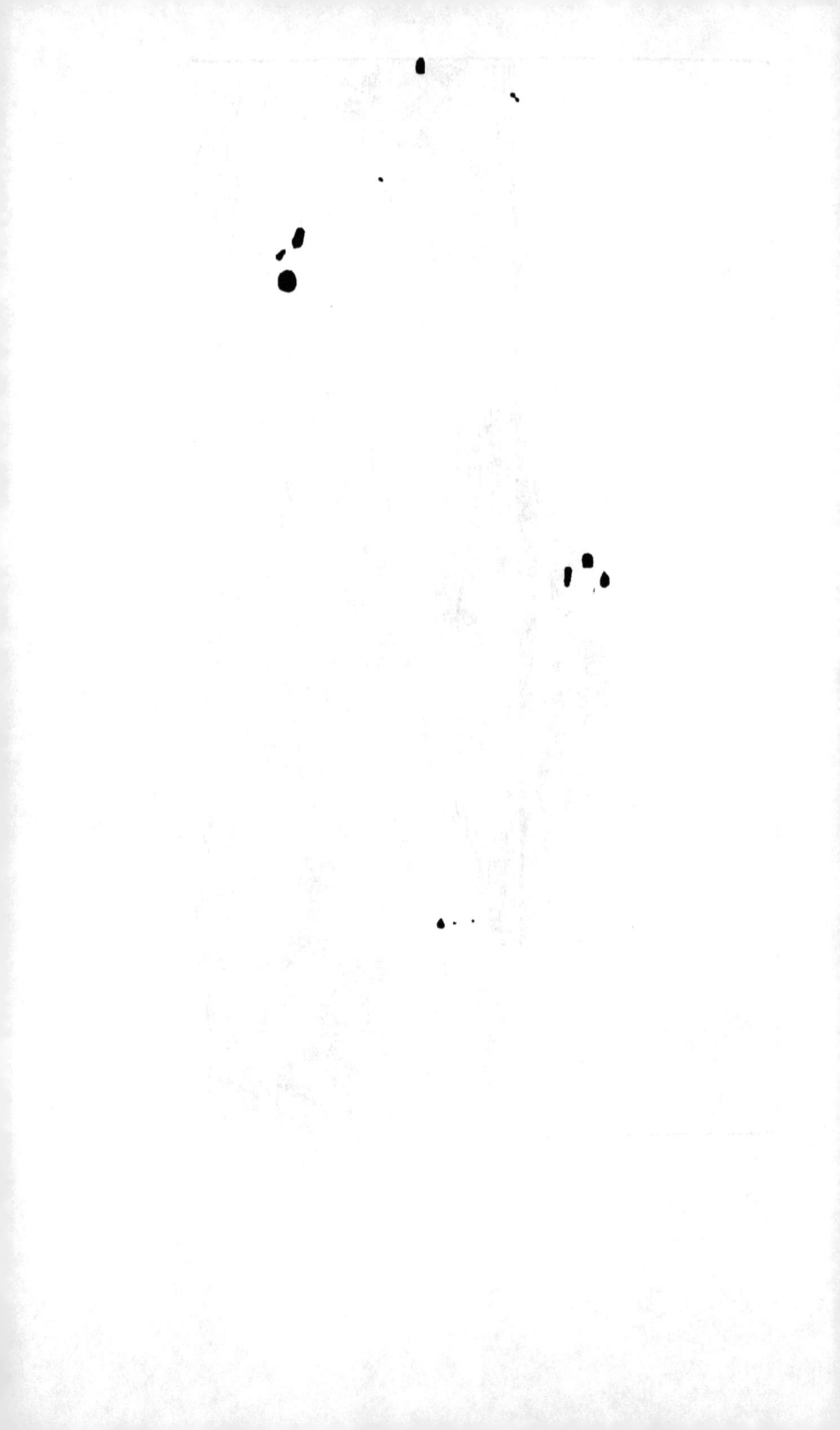

an instant to fully breathe, for a period, perhaps, of two or even three hours, until, from interrupted respiration, it finally becomes so filled with air or gases as to be unable to sink, and then falls at once an easy victim. During this contest the Aleutes have been throwing their spears whenever they were anywhere within range of the kahlan, and the hunter who has stricken the quarry is the proud and wealthy possessor, beyond all question or dispute.

In this manner the fleet moves on, sometimes very fortunate in finding the coveted prey; again, whole weeks pass away without a single surround. The landings at night are made without any choice or selection, but just as the close of the day urges them to find the nearest shore. The bidarkies are hauled out above surf-wash, and carefully inspected; if it is raining or very cold, small Λ-tents are pitched, using the paddles and spears for poles and pegs, into which the natives crowd for sleep and warmth, since they carry no blankets or bed-clothes whatever, and unless the wind is right they dare not make a fire, even to prepare the cherished cup of tea, which they enjoy more than anything else in the world, not excepting tobacco. After ninety or a hundred days of such employment, during which time they have been subjected to frequent peril of life in storm, and fog-lost, they repair to the rendezvous agreed upon between the trader and themselves, ready and happy to return for a resting-spell, to their wives, children, and sweethearts in the village whence we saw them depart. They may have been so lucky as to have secured forty or fifty otters, each skin worth to them at least fifty to sixty dollars, and if so, they will have a prolonged season of festivity at Oonalashka, when they get back. Perhaps the weather has been so inclement that this party will not have taken a half-dozen pelts; then gloomy, indeed, will be the reception at home.

While the "spearing surround" of the Aleutian hunter is orthodoxy, the practice, now universal, of surf-shooting the otter, is heterodoxy, and is so styled among these people, but it has only been in vogue for a short time, and it is primarily due to our traders, who, in their active struggle to incite the natives to a greater showing of skins, have loaned and have given, to the young hunters in especial, the best patterns of rifles. With these firearms the shores of many of the Aleutian channels, Saanak, and the Chernaboors, are patrolled during heavy weather, and whenever a sea-otter's head is seen in the surf, no matter if a thousand yards out, the expert, patient marksman shoots seldom in vain, and if he does miss the mark, he has a speedy chance to try again, for the great distance, and thunderous roar of the breakers prevent the kahlan from hearing or taking alarm in any way until it is hit by the rifle-bullet : nine times out of ten, when the otter is thus struck, it is in the head, which is all that the creature usually exposes. Of course such a shot is instantly fatal, so that the hunter has reason to sit himself down with a long landing-gaff and wait serenely for the surf to gradually heave the prized carcass within his ready reach.

Last, but most exciting and recklessly venturesome of all human endeavor in the chase of a wild animal is the plan of "clubbing." You must pause with me for a brief interval on Saanak to understand, even imperfectly, the full hazard of this enterprise. We cannot walk, for the wind blows too hard—note the heavy seas foaming, chasing and swiftly rolling by, one after the other—hear the keen whistle of the gale as it literally tears the crests of the breakers into tatters, and skurries on in sheets of fleecy vapor, whirring and whizzing away into the darkness of that frightful storm which has been raging in this tremendous fashion, coming from the westward, during the last three or four days without a moment's cessation. Look at those two Aleutes under the shelter of that high bluff by the beach. Do you see them launch a bidarka, seat themselves within and lash their kamlaykas firmly over the rims to the man-holes? And now observe them boldly strike out beyond the protection of that cliff and plunge into the very vortex of the fearful sea, and scudding, like an arrow from the bow, before the wind, they disappear almost like a flash and a dream in our eyes !

Yes, it looks to you like suicide ; but there is this method to their madness. These men have, by some intuition, arrived at an understanding that the storm will not last but a few hours longer at the most, and they know that some ten or twenty or even thirty miles away, directly to the leeward from where they pushed off, lies a series of islets, and rocks awash, out upon which the long-continued fury of this gale has driven a number of sea-otters that have been so sorely annoyed by the battle of the elements as to crawl there above the wash of the surf, and, burying their globose heads in heaps of sea-weed to avoid the pelting of the wind, are sleeping and resting in great physical peace until the weather shall change : then they will at once revive and plunge

ALEUTES CLUBBING SEA OTTERS

During a furious gale on the Chernaboor Rocks

back into the ocean without the least delay. So our two hunters, perhaps the only two souls among the fifty or sixty now camped on Saanak, who are brave enough, have resolved to scud down on 'the tail of this howling gale, run in between the breakers to the leeward of this rocky islet ahead of them, and sneak from that direction over the land and across to the windward coast, so as to silently and surely creep up and on to the kelp-bedded victims, when, in the fury of the storm, the fast falling footsteps of the hunter are not heard by the active yet somnolent animal ere a deadly whack of his short club falls upon its unconscious head. The noise of such a tempest is far greater than that made by the stealthy movements of these venturesome natives, who, plying their heavy, wooden bludgeons, despatch the animals one after another without alarming the whole number. In this way, two Aleutian brothers are known to have slain seventy-eight otters in less than one hour!

If these hardy men, when they pushed off from Saanak in that gale, had deviated a paddle's length from their true course for the islet which they finally struck, after scudding twenty or thirty miles before the fury of wind and water, they would have been swept on and out into a vast marine waste and to certain death from exhaustion. They knew it perfectly when they ventured, yet at no time could they have seen ahead clearly, or behind them, farther than a thousand yards! Still, if they waited for the storm to abate, then the otters would all be back in the water ere they could even reach the scene. By doing what we have just seen them do they fairly challenge our admiration for their exhibit of nerve and adroit calculation, under the most trying of all natural obstruction, for the successful issue of their venture.

In conclusion, the writer calls attention to a strange habit of the Aleutian otter-hunters of Attoo, who live on the extreme westernmost island of the grand Alaskan archipelago. Here the kahlan is captured in small nets,* which are spread out over the floating kelp-beds or "otter-rafts," the natives withdrawing and watching from the bluffs. The otters come out to sleep or rest or sport on these places, get entangled in the meshes, and seem to make little or no effort to escape, being paralyzed, as it were, by fear. Thus they fall an easy prey into the hands of the captors, who say that

---

* Sixteen to 18 feet long, 6 to 10 feet wide, with coarse meshes; made nowadays of twine, but formerly of seal and sea-lion sinews.

they have caught as many as six at one time in one of these nets, and that they frequently get three. The natives also watch for surf-holes or caves awash below the bluffs: and, when one is found to which a sea-otter is in the habit of going, they set this net by spreading it over the entrance, and usually capture the creature, sooner or later.

No injury whatever is done to these frail nets by the sea-otters, strong animals as they are ; only stray sea-lions and hair-seals destroy them. There is no driving an otter out upon land if it is surprised on the beach by man between itself and the water ; it will make for the sea with the utmost fearlessness, with gleaming eyes, bared teeth and bristling hair, not paying the slightest regard to the hunter. The Attoo and Atkah Aleutes have never been known to hunt sea-otters without nets, while the people of Oona-lashka, and those eastward of them, have never been known to use such gear. Salt-water and kelp appear to act as disinfectants for the meshes, so that the smell of them does not repel or alarm the shy, suspicious animal.

CHAPTER VIII.

THE GREAT ALEUTIAN CHAIN.

The Aleutian Islands.—A Great Volcanic Chain.—Symmetrical Beauty of Shishaldin Cone.—The Banked Fires in Oonimak.—Once most Densely Populated of all the Aleutians ; now Without a Single Inhabitant.—Sharp Contrast in the Scenery of the Aleutian and Sitkan Archipelagoes.—Fog, Fog, Fog, Everywhere Veiling and Unveiling the Chain Incessantly.—Schools of Hump-back Whales.—The Aleutian Whalers.—Odd and Reckless Chase.—The Whale-backed Volcano of Akootan.—Striking Outlines of Kahlecta Point and the "Bishop."—Lovely Bay of Oonalashka.—No Wolf e'er Howled from its Shore.—Illoolook Village.—The "Curved Beach."—The Landscape a Fascinating Picture to the Ship-weary Traveller.—Flurries of Snow in August.—Winds that Riot over this Aleutian Chain.—The Massacre of Drooshinnin and One Hundred and Fifty of his Siberian Hunters here in 1762-63.—This the Only Desperate and Fatal Blow ever Struck by the Docile Aleutes.—The Rugged Crown and Noisy Crater of Makooshin.—The Village at its Feet.—The Aleutian People the Best Natives of Alaska.—All Christians.—Quiet and Respectful.—Fashions and Manners among Them.—The "Barrabkie."—Quaint Exterior and Interior.—These Natives Love Music and Dancing.—Women on the Wood and Water Trails.—Simple Cuisine.—Their Remarkable Willingness to be Christians.—A Greek Church or Chapel in every Settlement.—General Intelligence.—Keeping Accounts with the Trader's Store.—They are thus Proved to be Honest at Heart.—The Festivals, or "Prazniks."—The Phenomena of Borka Village.—It is Clean.—Little Cemeteries.—Faded Pictures of the Saints.—Attoo, the Extreme Western Settlement of the North American Continent.—Three Thousand Miles West of San Francisco !—The Mummies of the "Cheetiery Sopochnie."—The Birth of a New Island.—The Rising of Boga Slov.

AFTER "lying-to" in a fierce southwester for three whole days and nights, in which time the fury of the gale never abated for an hour, our captain had so husbanded his resources that, when the weather moderated, he was able to clap on sail and get under swift headway ; then we quickly left the watery area of our detention and soon opened up a splendid vista of Oonimak Island, in the early dawning of a clear June day. This is the largest one of that long-

10

extended archipelago which stretches as an outreaching arm for Asia from America; it presents to our delighted gaze a sweep of richly-colored, rolling uplands, which either slope down gently to the coast at intervals, or else terminate in chocolate-brown and reddish cliffs abruptly stopped to face the sea breaking at their feet. Very high ridges, with summits entirely bare of vegetation, traverse the centre of the island from east to west, while the towering snowy cone of Shishaldin and the lower, yet lofty, head of Pogromnia—two volcanoes—rear themselves over all in turn.

There are a multitude of huge and cloud-compelling mountains in Alaska, but it is wholly safe to say that Shishaldin is the most beautiful peak of vast altitude known upon the North American continent; it rears its perfectly symmetrical apex over eight thousand feet in sheer height above those breakers which thunder and incessantly roll against its flanks, as these precipitous slopes fall into the great Pacific Ocean on the south, and Bering Sea to the north. A steamy jet of vapor curls up lazily from its extreme summit, but it has not been eruptive or noisy at any time within the memory of the Russians. No foothills, that crowd up against and dwarf the presence of most high mountains, embarrass your view of Shishaldin; from every point of the compass it presents the same perfect cone-shape; rising directly from the water and lowlands of Oonimak, it holds and continues long to charm your senses with its rare magnificence; the distance of our vessel, ten or twelve miles away, serves to soften down its lines of numerous seared and blackened paths of prehistoric lava overflow, so that they now softly blend their purplish tones into those of the rich-hued mantle of golden-green mosses and sphagnum which cover the rolling lower lands.

As we draw into Oonimak Pass—it is the gateway for all sailing vessels bound to Bering Sea from American ports, we, in closing up with the land, almost lose sight of Shishaldin, and come into the shadow of the rougher and less attractive volcano of Pogromnia. It shows ample evidence of its origin by the streams of blackened frost-riven basalt and breccia which are ribbed upon its rugged sides; great masses of eruptive rock and pumice lie here and there scattered all over the broad-stubbed head of the mountain; tons and tons of this material have rolled from thence in lavish profusion and disorder, clear down for miles to the very waters of the sea and straits, strewing that entire route with huge débris. Seams of snow and ice lacquer, in white, thread the bold black crown of this, the "booming" or "noisy" volcano of the Russians. It has not been in action since 1820, when it then threw showers of ashes and pumice; but those fires in its furnace are only banked, as it has been smoking in inky brown and black clouds at irregular and frequent intervals ever since; loud mutterings, deep rumblings and wide-felt tremors of land and sea are aroused by it constantly. This Island of Oonimak has been always regarded by the Russians as the roof of a subterranean smelting furnace with many chimneys through which telluric forces ascended from the molten masses beneath. It has been, and is still, the theatre of the greatest plutonic activity in Alaska. Russian eye-witnesses have described violent earthquakes here where whole ridges of the interior and coast have been rent asunder, cleft open, from which torrents of lava poured and columns of flame and clouds of ashes, steam and smoke, have risen so as to be viewed and noticed for a circuit of hundreds of miles around. These manifestations were always accompanied by violent earthquakes, and tidal-waves which often submerged adjacent villages on the sea-level, and also whole native settlements were swept away in mountain floods caused by the sudden melting of those big banks of ice and snow on such volcanic summits and their foothills, upon which the hot breccia from a vomiting crater fell.*

This great island in olden times was the one most densely populated by the Aleutes. The excesses and terrible outrages of Russian promishlyniks, followed by the wholesale work of death wrought by small-pox, have utterly eliminated every human settlement from the length and breadth of Oonimak, upon which no one has resided since 1847. Ruins on the north shore show the abandoned sites of numerous large hamlets; one was over four thousand

* Bishop Veniaminov, who witnessed one of these eruptions in 1825, describes the occurrence: "On the 10th March, 1825, after a prolonged subterranean noise resembling a heavy cannonade, that was plainly heard on the islands of Oonalashka, Akoon, and the southern end of the Aliaska Peninsula, a low ridge at the northwestern end of Oonimak opened in five places with violent emissions of flames and great masses of black ashes, covering the country for miles around; the ice and snow on the mountain tops melted and descended in a terrific torrent five to ten miles in width, on the eastern side of the island. The Shishaldin crater, which up to that time had also emitted flames, continued to smoke only. '

two hundred feet in its frontage on the beach. The fear and
superstition which those tragedies of early Russian intercourse
produced in the simple minds of the natives, who belonged by
birth to this great island, became at length so potent as to cause
the entire and permanent abandonment of their desolated villages,
which were once so populous and well satisfied.

The craters, and outflow therefrom, on Oonimak have been, from
time immemorial, resorted to by the natives as their storehouses for
sulphur, and that shining obsidian with which they tipped their
bone-spear and arrow-heads ; of it, also, they made their primitive
knives, and traded the surplus stock to those Aleutes living elsewhere. They used the sulphur with dried moss in making fires,
which they started with the fire-stick and by rocky concussions.*

Before entering the straits of Oonimak, we had a fine view of the
entire sweep of the Krenitzin group, that presents a succession of
the wildest and most irregular peaks and bluffs, everywhere seeming to jut up and fall into the sea, without a gentle slope for a
human landing, as they face the Pacific billows dashing so incessantly upon their basaltic bases ; the extreme eastern islet of the
group is Oogamok, and it forms the opposite land from Cape Hect-
hook on Oonimak, directly across the straits. A swarm of sea-
parrots fly out from its rocky bluffs on the south shore, stirred into
unwonted activity and curiosity by the near approach of our vessel,
while a dozing herd of sea-lions suspiciously stretch their long necks
into the air, smell us, then simultaneously and precipitately plump
themselves into the foaming breakers just below their basking-place
above the surf-wash.

It is very difficult to adequately define or express those varying
impressions which are inspired by a panorama of these Aleutian
Islands, such as unfolds itself to your eye when rapidly sailing along.
under their lee on a clear day. The scene is one of rare beauty.
The water is blue and dancing until it strikes in heavy waves upon
the rocky curbing of the islands, dashing up clouds of spray in
white, fleecy masses against the dark-brown and reddish cliff-walls
rising over all. The slopes and the summits of everything on land,

* A flat, flinty rock—upon it a layer of dried moss or eider-down was
spread, then a sprinkling of powdered sulphur was cast over the moss or
feathers, then a large quartzite stone was grasped in the native's hand, who
struck it down with all his force upon this preparation. The concussion produced fire, and, when feathers were used, a terrible smell.

save the very highest peaks, are clothed in an indescribably rich green and golden carpet of circumpolar sphagnum ; exquisitely-colored lichens* adorn the stony sea-bluffs and precipices inland. Every minute of the ship's progress in a free, fair wind shifts the fascinating scene—a new peak, another bold headland, a narrow pass, unfolds now between two islets that just before apparently were solid and as united as one island could be ; a steamy jet of hot-spring vapor rises from a deeper, richer mass of green and gold than that surrounding it, and a dark-brownish column of smoke that issues from a lofty, cloud-encircled summit in the distance is the burning crater of Akootan.

Everything is so open here, is so plain to see, that when you try to find some points of resemblance to that picture which has challenged your admiration in the Sitkan archipelago, you find nothing—absolutely nothing—in common effect. It is, nevertheless,

---

* The range and diverse beauties of the numerous mosses and lichens on these islands must serve as an agreeable and interesting study to anyone who has the slightest love for nature. They undoubtedly formed the first covering to the naked rocks, after these basaltic foundations had been reared upon and above the bed of the sea—bare and naked cliffs and boulders, which with calm intrepidity presented their callous fronts to the powerful chisels of the Frost King. Rain, wind, and thawing moods destroyed their iron-bound strongness ; particles larger and finer, washed down and away, made a surface of soil which slowly became more and more capable of sustaining vegetable life. "In this virgin earth," says an old author, "the wind brings a small seed, which at first generates a diminutive moss, which, spreading by degrees, with its tender and minute texture, resists, however, the most intense cold, and extends over the whole a verdant velvet carpet. In fact, these mosses are the medicines and the nurses of the other inhabitants of the vegetable kingdom [in the North]. The bottom parts of the mosses, which perish and moulder away yearly, mingling with the dissolved but as yet crude parts of the earth, communicate to it organized particles, which contribute to the growth and nourishment of other plants. They likewise yield salts and unguinous phlogistic particles for the nourishment of future vegetable colonies, the seeds of other plants, which the sea and winds, or else the birds in their plumage, bring from distant shores and scatter among the mosses." Then the botanist needs no prompting when he observes the maternal care of these mosses, which screen the tender new arrivals from the cold and imbue them with the moisture which they have stored up, and "nourish them with their own oily exhalations, so that they grow, increase, and at length bear seeds, and afterward dying, add to the unguinous nutritive particles of the earth ; and at the same time diffuse over this new earth and mosses more seeds, the earnest of a numerous posterity."

just as attractive, just as grand ; but how different! All is laid
perfectly bare to inspection here—no dense forests and tangled
thickets to conceal the surface of the diversified uplands and mountain slopes, or to hide the innermost recesses of the deep ravines
and narrow valleys. While there is a vast variation in the islands,
yet there is, to the mind of him who views them for the first time,
the most helpless inability on his part to distinguish or even recognize them apart when he happens to revisit them. They are seldom
ever clearly defined, being more or less obscured in fog and heavy
rifts of cloud. The top of a headland peeps aloft, sharply outlined, while all below is lost in the mists and banks of fog that roll
up there from the sea. Then, in remarkable contrast, only a few
miles beyond, the rocks at sea-level and foothills of the next island
will be entirely plain to your sight ; while everything above is concealed, in turn, by a curtain of the same moist and vanishing
misty fog. Fog, fog, fog everywhere, rising and descending with
the force of wind-currents that bear it—now veiling, now revealing
the startling and impressive beauties of this vast sea-girt chain of
the Aleutian archipelago. These majestic blue swells of the great
Pacific join with those cold green waves of that lesser, shallower
ocean of the North in holding with firm embrace the most impressive range of fire-eaten mountains known to the geographer. This
cordon of smoking, grumbling, quaking hills and peaks, when
once surveyed, leaves an enduring image, grand and superb, on the
retina of that eye which has been so fortunate as to behold it.

As the little schooner bears up to the westward for our port of
Oonalashka, after we have well passed the Straits of Oonimak, we
sail into the shorter, choppy waters of Bering Sea—into its characteristic light gray-green hue of soundings. The precipitous walls
of Akoon Island, rising like so much Titanic sandstone masonry
everywhere abruptly from the surf, carry a broad green plateau,
that rolls and extends high above the surrounding tide-level.
Here, under their lee, on the north shore, we encounter one of
those large schools of humpback whales * which are so common
and so frequently met with in the Aleutian straits and passages.
These animals rise and sink alongside of the vessel, in utter disregard of its presence ; and even volleys and bullets of our breech-
loading rifles rapidly fired into their broad, glistening, gray-black

* *Megaptera versabilis.*

backs and sides do not seem to arouse or alarm them in the least. Down they lazily go, to soon rise again with a sonorous whistle as they " blow." A cloud of whale-birds hover over and settle on the watery area occupied by the feeding whales, ever and anon rising, to alight again as the cetacean fleet leaves its feathered convoy tossing behind on the wavelets of the sea.

Our skipper, who has been a whaler in his youth, tells us, with a quaint air of contempt for what we· so much admire, that these fish-like monsters are of no consequence in the eyes of a wise whaling captain, for though they are large enough, it is true, yet they are the wrong breed of whales—they are lean, fighting humpbacks, which, if struck with a harpoon, will run like an express engine for fifty miles or more, carrying a boat and crew of our species, either down in its rapid rush, or else diving in the shoals, over which it feeds, it rolls the death-dealing iron out or breaks it off on the bottom.

A stiff head-wind causes the course of the vessel to frequently lie close in to the shore where the massive bluffs of Akoon and Akootan rise in grim defiance, and from the shelves and interstices of which flocks of sea-parrots and little auks fly out in circling flights of curiosity and inspection around the schooner. As we watch the lazy motions of the whales, we recall the fact that on the summits of these bluffs and headlands now before us, the natives of Oonimak, as well as those to the country born, were in the habit of standing through long vigils of daily and nightly watch, as they went whale-fishing long ago after their own primitive fashion.

Nothing fit to eat is, or was, so highly prized by the Aleutes or Kaniags, as the. blubber and gristle of a whale. To secure this luxury these savages were in the habit of subjecting themselves to infinite hardship and repeated bitter disappointment. The chase of the "ahgashitnak"* and the little "akhoaks"† was the important business of their lives in times of peace. The native hunter used, as his sole weapon of destruction, a spear-handle of wood about six feet in length ; to the head of this he lashed a neatly-polished socket of walrus ivory, in which he inserted a tip of serrated slate that resembled a gigantic arrow-point, twelve or fourteen inches long and four or five broad at the barbs, and upon the point of which he carved his own mark.

---

* Yearling whale.   † Calf whales.

In the months of June and July the whales begin to make their first inshore visits to the Aleutian bays, where they follow up schools of herring and shoals of *Amphipoda*, or sea-fleas, upon which they love to feed. These bays of Akootan and Akoon were and are always resorted to more freely by those cetaceans than are any others in Alaska, and here the hunt is continued as late as August. When a calm, clear day occurs the natives ascend the bluffs and locate a school of whales; then the best men launch their skin-canoes, or bidarkas, and start for the fields. "Two-holed" bidarkas only are used. The hunter himself sits forward with nothing but his whale-spear in his grasp; his companion, in the after hatch, swiftly urges the light boat over the water in obedience to his order. Carefully looking the whales over, the hunter finally recognizes that yearling, or the calf, which he wishes to strike; for it is not his desire to attack an old bull or angry cow-whale. He calculates to a nice range where the "akhonk" will rise again from its last point of disappearance, and directs the course of the bidarka accordingly. If he is fortunate he will be within ten or twenty feet of the rising calf or yearling, and as it rounds its glistening back slowly and lazily out from its cover of the wavelets the Aleut throws his spear with all his physical power, so as to bury the head of it just under the stubby dorsal fin of that marine monster; the wooden shaft is at once detached, but the contortions of the stricken whale only assist to drive and urge the barbed slate-point deeper and deeper into its vitals. Meanwhile the canoe is paddled away as alertly as possible, before the plunging flukes of the tortured animal can destroy it or drown its human occupants.

As soon as the whale is thus wounded it makes for the open sea, where "it goes to sleep" for three days, as the natives believe; then death intervenes, and the gases of decomposition cause its carcass to float, and, if the waves and currents are favorable, it will be so drifted as to lodge on a beach at some locality not so very remote from the place where it was struck by the hunter. The business of watching for these expected carcasses then became the great object of everyone's life in that hunters' village; dusky sentinels and pickets were ranged over long intervals of coast-line, stationed on the brows of the most prominent headlands, where they commanded an extensive range of watery vision. But the caprices of wind and tides are such in these highways and byways of the Aleutian Islands, that on an average not more than one whale

ALEUTES WHALING

Natives of Akoon and Akootan killing Humpback Whales

in twenty, struck in this manner by native hunters, was ever secured; nevertheless, that one alone (when cast ashore) amply repaid the labor and the exposure incurred chiefly by watching day after day, in storm and fog, from the bluffs of Akoon and Akootan. The lucky hunter who successfully claimed, by his spear-head mark, the credit of slaying such a stranded calf or yearling, was then an object of the highest respect among his fellow-men, and it was remembered well of him even long after death.* Also, the greatest expression of respect for the size and ability of a native village and its people was the statement that it was so populous as to be able to eat all the meat and blubber of a large whale's carcass in a single day!

As we "put about" under the frowning walls at Cape North, of Akootan, our captain says that the next tack will carry us into Oonalashka Harbor. Meanwhile, as we stand out into the waters of Bering Sea, we have a superb vista of the rugged, seared, and smoking summit of Akootan itself, which rears its hot head high above the rough, rocky island that bears its name. The beaches are few and far between, and there is but little land upon this island to invite a pedestrian, since masses of dark basalt, vesicular and olivine, are scattered in wild profusion everywhere. Over the northeast side steamy clouds arise from the path of a hot spring, which gushes out of the mountain, so hot that meat and fish are cooked in its scalding flood by the natives. On the very crest, as it were, of this whale-backed volcano, are two small, deep lakes that once were the vent-holes of subterranean fires. In olden times seven settlements, with a population of more than six hundred Aleutes, lived on the coast of this island, which, with Akoon, was then the whale-hunter's paradise. To-day we find it utterly desolate, inhabited by a poverty-stricken hamlet of sixty-five natives, who are located on the southwest shore. The able-bodied men of

---

* Then it was the custom to cut up the dead body of a celebrated native whale-hunter into small pieces, each of which was kept by the survivors to rub over their spear-heads, being carefully dried and preserved for that purpose. Again, in ancient times, the pursuit of the whale was the prelude to many secret and superstitious observances by the hunters. These primitive whalers preserved the bodies of distinguished hunters in caves, and before going out on a whale-chase would carry those remains into the water of streams so as to drink of that which flowed over them. The tainted draught conveyed the spirit and luck of the departed!

this place spend the greater part of their time, however, far away
from home on the sea-otter grounds of Saanak, being carried, like
their brethren of Akoon and Avatanak, to and from that spot by
a trader's vessel.

Closely joined to them is the village of Akoon, in which fifty-five
or sixty of their countrymen live on the northwest shore, who hunt
and deport themselves as do those of Akootan. The Akoonites,
however, enjoy the satisfaction of being nearer than their neighbors
to that small, rugged islet of Oogamak, which stands in the path,
as it were, of the great Pass of Oonimak ; here on the low rocks a
comparatively large number of sea-lions repair, and the little hair-
seal also. For some reason or other, more of these last-named
seals are found here than elsewhere in the entire large extent of this
gigantic island chain. Akoon used to boast of many mighty whalers
among its prehistoric population of five or six hundred natives,
who, in fading away, have left the ruins only of eight settlements
to attest their previous proud existence.

While we have noticed the poverty of the Akootans, yet, as we
contemplate the wretched little village on Avatanak, close by and
facing the straits, we must call this the most abject human settlement, perhaps, that we shall or can find throughout the archipelago
—only nineteen souls living here in the most abandoned squalor and
apathy, principally upon the sea-castings of the beach and mussels.
Yet this island in olden days was the happy home for a busy little
fishing community which then had three settlements on the banks
of a beautiful stream that empties its clear waters into the sea on
its north side. The most revolting chapter in all the long story of
Russian outrage and oppression of Aleutian natives is devoted to
a recital of the savage brutality of Solovaiyah and Notoorbin, who
lived here during the winter of 1763.

Steam-vessels usually make the jagged headlands and peaks of
Tigalda Island as their first land-fall *en route* from San Francisco to
Oonalashka and Bering Sea. They then shape their course into
Akootan Straits very easily and safely. The currents and winds,
which always cause a variation of the ship's course, never carry the
vessel much to the right of Tigalda, or to the left of Avatanak, so
that an experienced Alaskan mariner has but little difficulty—even
though dense fog prevails, which only gives him fitful gleams of
the rude landscape—in recognizing some one of the characteristic
peaks or bluffs of these Krenitzin islands ; then, with a known

point of departure, he can literally feel his way into Oonalashka Harbor. He almost always has to do so, for seldom indeed does he enjoy as fair a sweep of these coasts of Avatanak and Tigalda as that viewed by the author, who scanned this rocky group in a calm, clear September afternoon of 1876.

To-day, Tigalda is an utterly abandoned island, given over during the summer to the undisturbed possession of foxes and those flocks of "tundra" geese which settle on the uplands to breed and preen in safety. When moulting here, they have the shelter of several lakes, upon which they swim in mocking security, even if crafty, lurking Reynard attempts to capture them. Near the largest lake on this island a settlement once throve. The inhabitants had control of a mine of red and golden-yellow chalk, which formed the base of a pigment highly prized by all Aleutes, far and near, for painting their ancient grass, and wooden hats, and other work of the same materials. On the north side of this island is a singular cluster of needle rocks which rise, as twenty-eight points, abruptly from the sea. On them, in positive security, the big burgomaster gull breeds, and the eagle-like pinions of this bird bear thousands of heavy bodies in stately flight over and around these nesting-places. The shrill, hawk-like screams of those "chikies" can be heard far out at sea, over the noise of the surf.

Oogalgan rock, which stands up boldly, and defies that fury of an ocean in the mouth of Oonalga Straits, is another striking headland which the mariner should be well acquainted with, for in times of arrival, when fog prevails, it is often the first land-fall made after leaving California or Oregon, when bound in for Oonalashka. It is a bleak, tempest-swept islet, presenting to the Pacific a black, reddish front of abrupt precipitous cliffs, without a sign of vegetation in the crevices ; but, from the inside passages of Akootan and Oonalga, it exhibits two or three saddle-backed slopes covered with green mosses and lichens. Flocks of those comical shovel-billed sea-parrots breed upon it, and skurry in their rapid, noiseless manner all around.

At last our little schooner "comes about," to make that "reach" which is to take us into the peace and quiet of a beautiful harbor, and, with every sail drawing hard, she fills away, and we glide swiftly ahead. That richly banded waterfall bluff on our right, and the striking outline of Kahlechta Point, over the "Bishop" rock under it, on our left, are eagerly scanned as we dash through the

heavy roll of Akootan Straits and its violent tide-rips, the surf breaking on the "Bishop" and the point beyond it most grandly. 'A short
hour, and the rough water is passed. We have entered Captain's
Harbor, and are "fanning" along over a glassy surface up to our
anchorage off from, but close by, the village of Oonalashka.*

What San Francisco is to California, so is Oonalashka to all
Alaska west of Kadiak. It is the point of all arrivals and all departures for and from this vast area. It is most fitly chosen, and
beautifully located. From earliest time, an Aleutian legend never
failed in its rendition to the dusky people then living in their
yourts and kazarmies to vividly impress upon the native mind a
full sense of those pleasures of life and hope at Illoolook ; not, however, as expressed so sadly by our own bard, whose inimitable
poem declares that the wolf howled long and dismally from this
lovely shore of Illoolook.

Cold on his midnight watch the breezes blow
From wastes that slumber in eternal snow,
And waft across the wave's tumultuous roar
The wolf's long howl from Ounalaska's shore.

If Campbell had only substituted "Akoon" for "Oonalashka"
in this much-admired verse descriptive of savage desolation, he
would not have marred a famous passage by the slightest error—
but, at Oonalashka, never, never was a wolf ever known to be. In
1830, however, two of these animals got over from Oonimak as far
west as Akoon—on drifting ice-floes, most likely. They were speedily noticed by the natives, who killed them at once, so Veniaminov
says, for they were cordially hated by the Aleutes, since these
beasts "kill foxes and spoil the traps."

The panorama of land and water here in summer is an exceedingly attractive one—in its effect fully as charming as is the lovely
spread of Sitka Sound ; but its character is widely opposed. If
we chance to view Oonalashka in clear sunshine during a day in the
summer months, we will recall this picture to our mind's eye often
with positive pleasure. Here, strung along for half a mile just
back of a curved and pebbly beach, is an irregular row of frame,
single-story cottages, a large Greek church, and a fine parsonage,

* The natives always called this settlement "Illoolook," or "curved
beach."

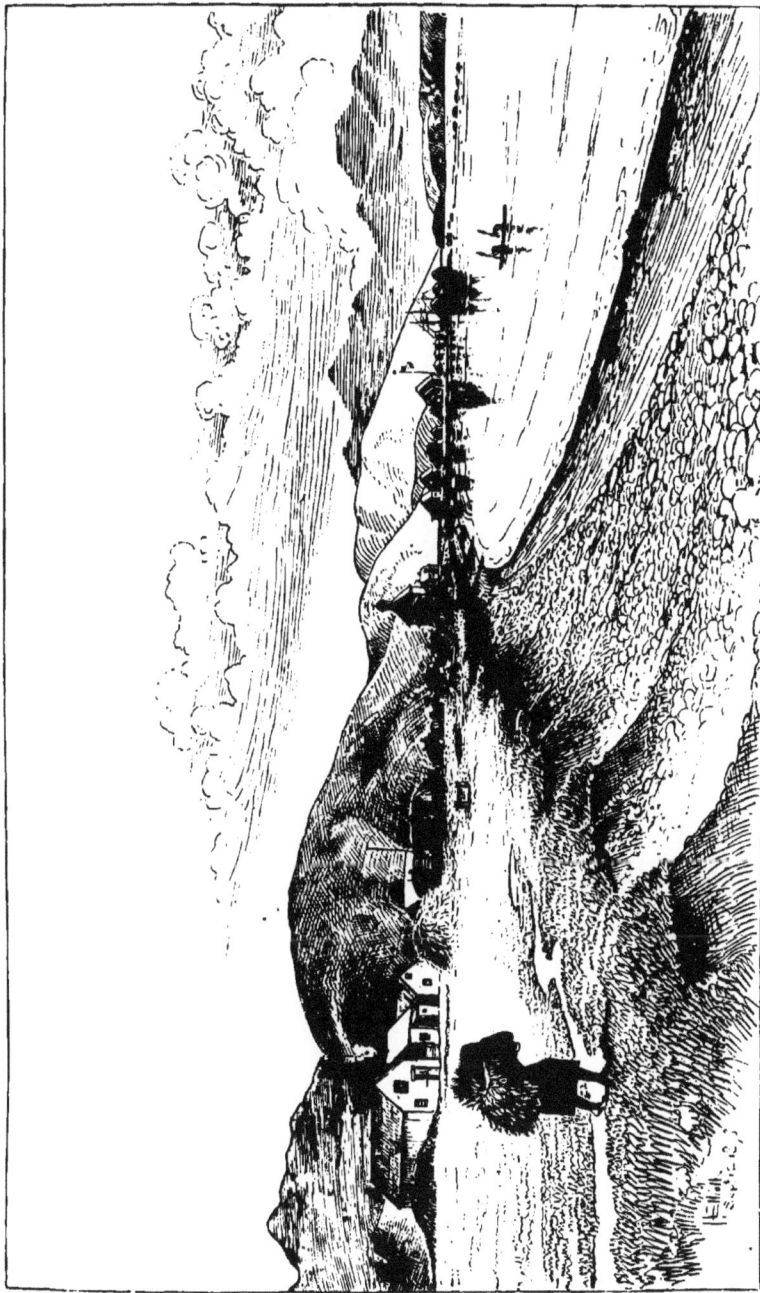

ILLOOLOOK, OR OONALASHKA

View of the Village looking West from the Cemetery

three or four big wooden warehouses with a wharf running well into the harbor, two or more trading-stores, one of them quite imposing in its size, and fifty or sixty barraboras—these constitute the abiding-places of the four hundred residents of Illoolook. They are placed upon a narrow spit of alluvium that divides the sea from the waters of a small creek which runs just back of the village right under these hills that abruptly rise there, to rise again, farther inland, to higher peaks in turn. A rich, dark, vivid green covers and clothes the mountain slopes, the valleys, and the hills, even to the loftiest summits, where only a light patch of glistening snow is now and then seen relieved thereon by the grayish-brown rocky shingle. These hills and mountains, rising on every hand above us from the land-locked shores of Captain's Harbor, bear no timber whatsoever, but the mantle of circumpolar sphagnum, interspersed with grasses and a large flora, makes ample amends for that deficiency and hide their nakedness completely—in their narrow defiles and over the bottom-land patches grass grows with tropical luxuriance, waist-high, with small clumps of stunted willow-bushes clinging to the banks of little water-courses and rivulets. This is the only growing timber found anywhere on the Aleutian chain. It never becomes stouter than the thickness of a man's wrist, and the tallest bushes in scattered thickets are never over six or seven feet high, rapidly dwindling in growth as they ascend the hillsides.

Especially gratifying is the landscape, thus adorned, to the senses of any ship-worn traveller, who literally feasts his eyes upon it. But if he should go ashore and step upon what appeared to him, from the vessel's deck, to be a firm greensward, he will find instead a quaking, tremulous bog, or he will slide over a moss-grown shingle, painted and concealed by cryptogamic life, where he fondly anticipated a free and ready path. The thick, dense carpet of crowberry* plants that is spread everywhere over the hill-sides, into which the pedestrian sinks ankle-deep at every step, makes a stroll very laborious when undertaken at any distance from the sea-beach.

If a wide survey is accomplished here of Oonalashka Island, the studies made will give a perfect understanding of every other island

---

* *Empetrum nigrum.* The natives call it "shecksa." It is their chief supply of fuel.

to the westward in this great archipelago, which is enveloped during the major portion of each year in fogs, and swept over by frequent gales. Such a combination of the elements, with mists and hidden sea-currents, make it a region dreaded by mariners ; yet there is enough sunshine now and then to make the life of our landsmen very comfortable, even though they cannot engage in any other profitable calling than that of sea-otter trading with the natives.

Summers are mild, foggy, and humid. The average temperature is about 50° Fahrenheit. Winters are also mild, foggy, and humid, with a slightly colder average of 30°. The thermometer nowhere in the Aleutian chain ever went much below zero at sea-level. There is no record even of a consecutive three or four weeks in winter lower than 3° or 5° above zero. The mercury seldom ever falls as low as 10°. There is no nice distinction of the four seasons here. We can notice only two. Winter begins in October and ends by May 1st to 5th, when summer suddenly asserts herself for the rest of the year not thus appropriated.

Flurries of snow sometimes fall in August and often in September. It never stays long on the ground or even on the hilltops then, and generally melts as fast as it comes, away into December ; but on the highest peaks it is seen all the year round. From January to May 1st or 5th, as a rule, snow covers everything in a spotless shroud from two to five feet deep. The high, blustering wintry gales make this snow intensely disagreeable to us, driving into and through air-tight crevices, and literally making the inmates of the village huts prisoners for weeks at a time. The dogs and sleds so common and characteristic elsewhere in the vast expanse of Alaska are never seen here. They would be a mere nuisance to these people, since the rugged inequalities of the Aleutian country simply prohibit their use.

This is, however, the chosen land for lingering fogs. The foggy cloudiness of the Aleutian Islands is most remarkable. There are not a dozen fogless days in the whole year at Oonalashka, though the sun may be seen half the time. Fifty sunshiny days in the year is a handsome average. Thunder is never heard, or seldom ever, while lightning is never seen, although the dark swelling clouds seem to constantly suggest it ; also the northern lights— these auroral displays are almost unknown, and when seen are very, very faint.

But the wind—ah, the winds that riot over this range of rocky islands! They are always stirring. A perfect calm has never been recorded at Oonalashka. They are strong and come from all points of the compass; they are freshest and most violent in October and November, December, and March. Gales follow each other in quick succession during these months every year, lasting usually about three days each.

All sides of Oonalashka Island are deeply indented by bays and fiörds; but the points on the southern coast are avoided and not well known. They are not safe to approach on account of reefs and rocks, awash and sunken, which extend out to sea a long distance, and upon them the heavy billows of the Pacific Ocean break incessantly, as well as against the cliff-beaches of this forbidding shore. But around the northern and eastern margins of the island more good harbors are located than can be found on all of the other islands of the Aleutian archipelago put together. They call the bay which we entered, as we sailed in from Akootan Pass, "Captain's Harbor." It is the same place where the natives first gazed upon a white man and his ship after the frightful massacres of 1762 and 1763. Here in 1769 Layvashava, with a crew of those Siberian promishlyniks, anchored during the whole of one autumn and engaged the astonished inhabitants in active trade; but it was a guarded and tedious barter, since the Aleutes had a lively recollection of the terrible past, so recent and so bloody.

The island of Oonalashka chanced to be the scene of that only real desperate and fatal blow ever struck by the simple natives of the Aleutian chain at their Cossack oppressors. By 1761 the Russians had advanced to the eastward as far as Oonimak, and up to this time the relations between the natives and the white invaders had been altogether of an outwardly friendly character, the former submitting, as a rule, patiently to the demands of the newcomers, but the Cossack Tartars, encouraged by their easy conquests, rapidly proceeded from bad to worse, committing outrages of every kind, so that in 1762 they had reduced the Aleutes to the verge of absolute slavery, and continued to act in this manner until the patience and the timidity of the simple race were exhausted. The arrival of a brutal, domineering, lustful party of over one hundred and fifty of these Cossack Russians at Chernovsky, on the north-west coast of this island, in the summer of 1762, under the nominal command of a Siberian trader named Drooshinnin, proved to be "the last straw laid upon the camel's back." At a given signal the

despoiled and ravished natives arose in every one of the then populous Oonalashkan settlements (twenty-four villages), flocked together, and unitedly fell upon their oppressors. They slaughtered
every man except four, who happened, luckily for them, to have
been absent from their vessels in Chernovsky Harbor, hunting
grouse in the mountains. They were secreted in the recesses of a
hot cave (that is still pointed out in the flanks of Makooshin Mountain), by the kindness of a charitable native, until they were able
to escape and join the expedition of Solovaiyah, which appeared at
the offing of Oomnak early in the following year. Fired by a
recital of the Drooshinnin slaughter, this fierce Cossack turned his
half-savage comrades, and worse yet, himself, loose upon the unhappy people of Oonalashka, and literally exterminated every male,
old and young, that he could find, visiting each settlement in swift
rotation of death and desolation. The men and boys fled to the
fastnesses of the interior, followed by many of the women, and
when the inclemencies of winter began to threaten their starvation,
they humbly sued for peace, and became the abject and submissive
vassals of the promishlyniks ever after. *

A smoking volcano that rears its ragged crown high above all
the surrounding hills and peaks is Makooshin ; it juts, alone and
unsupported, as a bold promontory, five thousand four hundred
and seventy-five feet above, and into the green waters of Bering
Sea. It is the chief point of scenic interest on Oonalashka Island,
and the objective one in particular, if the day be clear, as the
visitor sails up and into the harbor of Illoolook. While it is not
near so majestic in elevation, or perfect of outline, as the Shishaldin Mountain, yet it is wild and striking. It can be easily ascended
in July and August, when the winds do not blow their hardest, and
when there is the least snow. No one remembers, nor is there any
legend of any disturbance more serious than the shaking of the
earth and loud noises which Makooshin is charged with. In 1818
it made the whole island tremble violently during a period of several days, emitting, however, nothing but dense columns of smoke,
and fine ashes were sifted lightly everywhere with the winds. A
resounding cannonade that then burst from its bowels sorely alarmed
the people, however, who fled from their little hamlets clustered at
its base.   •

Immediately under the steep slopes and large proportions of this
quiescent volcano is a small settlement of sixty natives, housed in

those typical Aleutian barraboras, with a small chapel, of course. Here, in 1880, lived the oldest inhabitant of the Oonalashka parish, an Aleut who had an undisputed age of eighty-three years. These simple souls have that same faith in the good behavior of Makooshin which distinguished the citizens of Herculaneum and Pompeii with reference to the dangers of Vesuvius. But the most amusing indignation is expressed by them in speaking of the bad behavior of an Oomnak crater, just across the straits from them, which in 1878 broke out into earthquakes, smoke, fire, and mudshowers, that so frightened the fish all about in these waters as to literally cause a famine at Makooshin. The finny tribes seem to be driven off by a trembling of the rocky bottom to the sea.

It was at Makooshin that the first Russians landed under Stepan Glottov in 1757. These traders in their reports declared that the natives here then " were very numerous and warlike," and that they had a great deal of that peculiar trouble with them which we so thoroughly understand now in the light of their infamous record. Certain it is that a more innocent-looking, indolent group of Aleutes cannot be found in all this region to-day than are these descendants of the "blood-thirsty savages," which Glottov saw in council here. They trap cross-foxes on the flanks of the great mountain which overshadows their settlement, and do but little else. They are not at all impressed by the volcano, and cannot understand why we should walk over a long portage of eight miles from Oonalashka Harbor just to ascend it : because, they say truly, that the chances are ten to one against our seeing anything when we shall get up there, inasmuch as fog will surely shut down over everything. In spite, however, of their argument we ascended, and they were right. We could not see a rod beyond our footing in any direction, and had it not been for their guidance, as the fog continued, we would have had a very difficult matter in regaining the lowlands at all that day.*

When Makooshin is seen from Bering Sea, in the early autumn, the snow rests upon its peculiar form so as to make a most striking suggestion of its being extended as a huge corpse, with a sheet thrown over the upper part only of the body. The natives have

---

* But on two other occasions the author has had clear and unfogged glimpses of this singular mountain, which he made careful studies of; they are presented to the reader in this connection.

many folk-lore stories and legends which belong to the mountain ;
but these yarns are like the ballads of our sailor boys, they run on
forever, ending in the same manner as they began. A hot spring
sends a little rivulet of warm water across our path as we come
down, and we notice that most of the boggy places are tinged with
iron oxides.

In over-looking any of these islands from an interior view of
high altitude, you are impressed by the large number of fresh-water
lakes and ponds that nestle in the valleys, in the uplands, and even
in the depressions on the loftiest summits. One of the prettiest
pools of water which can be imagined is formed by the red, bowl-
shaped walls of an extinct crater that makes the top of Paistrakov
Mountain : this is a very prominent landmark just across the bay
from Oonalashka village, looking west.

A superb survey of Oonalashka Island can be made by the ascent of Mount Wood, which rears its sharp, syenitic peak two thousand eight hundred feet behind and right over the village and harbor of Illoolook. The path to the summit is not difficult, and the
panorama spread out under your eyes well repays the effort. It
gives you a better idea of what a singularly mountainous region the
island is, of the comparative absence of level or bottom-land areas
—everything seems to spring from the surrounding ocean mirror,
to hills—from hills, in turn, to mountains that end in sharp and
rugged peaks. Upon the rocky, frost-riven shingle of these summits nothing can grow except those tiny polar lichens which we find
existing, clinging to the earth and rocks of the uttermost limits of
the North as far as we have knowledge.

If the fog lifts its gray-blue curtain from the unruffled, clear
surface of Captain's Harbor, and rolls back and away from the red
and brown head of the cold crater of Paistrakov on the left, and
from the black, jagged outlines of the "Prince" on your right, you
will then have at your feet a picture of surpassing scenic beauty,
both of contour and color, before and under your delighted vision.
The rougher waters of Bering Sea have power no farther inland
than their foaming at the feet of Waterfall Head and the dark bases
of the Prince, for they rapidly fade into a smooth, still peace as the
queer, hook-like sand-spit of Oolachta Harbor is reached, and the
anchorage of Illoolook village is attained ; its houses and barraboras just peep out from the obscuring foothills of the mountain upon which we stand, and we can faintly discern a deli-

cate fringe of sea-foam along the border of a long-curved beach in front. Two schooners and a steamer lie motionless upon the glassy bay, like so many microscopic water-insects.

Turning right about and looking south, our eyes fall upon a radically different landscape—a bewildering, labyrinthian maze of Oonalashkan mountain peaks and ranges, rising in defiance to all law and order of position, with that lovely island-studded water of the head to Captain's Harbor in the foreground. Ridge after ridge, summit after summit, fades out one behind the other into the oblivion of distance, where the suggestion of a continuance to this same wild interior is vividly made, in spite of wreaths of fog and lines of snowy sheen, relieved so brightly by that greenish-blue of the mosses and sphagnum in which they are set. A few pretty snow-buntings flutter over the rocks' to the leeward of our position; their white, restless forms are the only evidence or indication of animal life in our rugged vista of an Oonalashkan interior. Yet, could we see better, we might notice a lurking red fox, and flush a bevy or two of summer-dressed ptarmigan, feeding as they do on the crowberries, the sphagnum, willow-buds and insect-life.

While gazing into the endless succession of valleys, and scanning the varied peaks, a puff of moist wind suddenly strikes our cheeks—we turn to its direction and behold it bearing in and up from Bering Sea—a thick and darkening bank of fog which rapidly envelopes and conceals everything that it meets. It ends our sight-seeing, and peremptorily orders a return to the village below from which we came.

When we look at the Aleutes we are impressed at once with their remarkable non-resemblance to the Sitkans. They constantly remind us of Japanese faces and forms in another costume. The average Aleut is not a large man ; he is below our medium standard—being about five feet six inches in stature, though, of course, there are a few exceptions to this rule, when examples will be found six feet tall, and many that are mere dwarfs. The women are in turn proportionately smaller. The hair is coarse, straight, and black ; the beard scanty ; cheek-bones are broad, high, and very prominent ; the nose very insignificant and almost flattened out at the bridge—the nostrils thick and fleshy ; the eyes very wide-set— very small, too, with a jet-black pupil and iris ; the eyebrows very faintly marked ; the lips are thick ; the mouth large ; the lower jaw is very square and prognathous ; the ears are small, set close to the head, and almost always pierced for brass or silver rings. The complexion is a light yellowish-brown; in youth it is often fair, almost white, with a faint blush in the cheeks; in middle age and to senility the skin always becomes very strongly wrinkled and seamed, with a leathery harshness. They all have full even sets of teeth, but never take the least care of them whatever. They have small, well-shaped hands and feet, but the finger-nails are exceedingly thin and brittle, bitten off, and never trimmed neatly. They walk in a clumsy, shambling manner, with none of that lithe, springy stepping so characteristic of the Rocky Mountain Indians. When we meet them as we saunter through the settlement, men, women, and children alike drop their eyes to the ground, and pass by in stupid humility, or indifference, as the case may be.

As we see these people at Oonalashka, so they are seen in every respect elsewhere, as they exist between Attoo and Bristol Bay and the Shoomagin Islands. They spend most of their time, men and women, in their skin-canoes, hunting the sea-lion and sea-otter—in codfishing and travelling to and from their favorite salmon-runs and berrying-grounds. Therefore, they have not enabled a symmetrical figure to develop—their legs are always sprung at the knees, some badly bowed, and all are unsteady in walking. While there is nothing about the countenances of the women or girls which will warrant the term of handsome, yet they are not so ugly as the squaws of the Sitkan archipelago. Many of them have very kindly expressions, and a gleam of true womanly instinct far above their surroundings.

No people are more amiable or docile than are these natives of the Aleutian Islands to-day. They are quiet and respectful in their intercourse with the traders, and are all duly baptized members of the Greek Catholic Church. A chapel is never absent from their villages. They hunt sea-otters and trap foxes for their means of trading for those simple luxuries and necessaries of their life which they cannot find in their own country. There are no other fur-bearing animals here, and no other industries whatever in which they can engage.

As they live here to-day, they are married and sustain very faithfully the relation of husband and wife. Each family, as a rule, has its own hut or barrabora. They have long, long ago ceased to dress in skins; but they still retain and wear the primitive water-proof coat or "kamlayka" and boots or tarbosars, which are made from seal and sea-lion intestines. In the poverty-smitten stations of Akoon and Avatanak the early bird-skin "parkas" will probably be most commonly worn ; (but it is because these natives are so miserably poor in furs that they do so). They get from the trader's store at every village a full assortment of our own shop-made clothes, and, on Sunday in especial, many shiny broadcloth suits will be displayed by the luckier hunters. The women are all attired in cotton dresses and gowns, made up pretty closely in imitation of the prevailing fashions among our own people. They wear the boots and shoes which are regularly brought up from San Francisco. But whenever they go out fox-trapping, or enter their bidarkas, they wear the "tarbosar" or water-proof boot of primitive use—the uppers to it are made from the intestines or the gullets of marine mammalia, and it is soled with the tough flipper palms of a sea-lion.

They have the same weakness for our conventional high stove-pipe hats which we display ; but the prevalence of those boisterous gales and winds peculiar to these latitudes prevents that use of the cherished "beaver" that they otherwise would make of it. Instead, they universally wear low-crowned, leather-peaked caps, to which they love to add a gay red-ribbon band, suggested most likely by the recollection which they have of that gorgeous regalia of the Russian army and naval officers, who were wont to appear in full dress very often when among them in olden time.

The Aleutian men dress very plainly, young and old alike, little or no attention being given by them to details of color or ornamentation, as is the common usage and practice of most semi-civilized races ; but they do lavish a great deal of care and skill in the decoration of their antique "kamlaykas," "tarbosars," and their bidarkas : the seams of these garments and the boats are frequently embellished with gay tufts of gaily colored sea-bird feathers and lines of goose-quill embroidery.

True feminine desire for all the bright ribbons and cheap jewelry that a trader spreads before her consumes the heart of the Aleutian woman, especially if she be young and admired by her people. The women are, therefore, only limited by their means, when it comes to bedecking themselves with all of these trinkets and gewgaws of the kind which the artful trader exhibits for that purpose. They braid their hair up in two queues usually and let them hang down behind upon their backs. They never wear bonnets, or hats, for that matter ; but as they go to church or from hut to hut they tie cotton handkerchiefs over their heads. When hasty little errands out of doors, or sudden gossiping trips are undertaken, a shawl is thrown over the woman's head and held there, with the gathered ends together, under her chin by one hand. The shawls are of bright colors, and supply the place of woollen garments, though ready-made cloaks and dolmans are not uncommon at those points where the sea-otter-hunting harvest is the best : her skirts, overskirts, waists, and stockings are all of cotton.

As these people have really but one idea and no variation of occupation, they all live alike, in the same general manner. The difference between the families is only that of relative cleanliness and thrift. The most important and serious business of their shore-life is that embodied in the construction and repair of their huts, or barrabkies. If it is well built it makes a warm, dry shelter, and answers every requirement of a comfortable domicile. An excavation is made in the earth on the spot selected in the village site, ten or twelve feet square, and three or four feet deep. A wooden frame and lining is then put into this sub-cellar, and the excavated earth is then thrown back against and over it, with an outer wall of carefully-cut sod and boggy peat, being laid up two and three feet thick, sloping down to which is a well-thatched roof of grass and sedge, that abounds everywhere on the sandy margins of the sea-shore. Some of these huts are made very much larger than this pattern just defined, having regularly spread wings, like a Maltese cross, on the floor. The entrance to the barrabkie is usually through a low doorway that is made to a small annex or storm hallway, also built of sod and peat. This shields another little door, which opens into the living-room that the architect steps down into as he enters. A single window is put at the opposite end of the room from the door, in which a small glazed sash is usually employed. The floor is either covered with boards which the native has purchased from the trader, or else it is the hard-trodden earth itself, upon which the women strew grass and spread mats of the same texture.

A diminutive cast-iron stove is now very generally used by the Aleutes. It commonly stands right in the centre of the room, and upon it the cooking can be done, instead of being driven to the hallway fireplace, or " povarnik," of the olden time, when the smoke then stifled them from the burning of that fat of seals, fish and birds, which was used very largely for fuel. Therefore, they were obliged to stew and broil on a special fireplace constructed outside of the living-room. A great many old-style "peechka" stoves of the Russians are still in use, but no new ones are being made any more, since the introduction of our little iron stoves. This living-room of the hut is usually curtained or partitioned into two sections, one of which is the bedchamber, or "spalniah." They have a great variety of beds and bedsteads, or bunks rather. They are proud of a well-stuffed couch of feathers, and take more real, solid comfort in sleeping thereon than in anything else that transpires of an enjoyable nature in their lives. The dealers sell a series of the most gaudily printed spreads for these beds, and sometimes you will be much surprised to see a white counterpane and fluted pillow-shams spread over an Aleutian couch. Those beds are always raised well up from the floor, and sometimes a curtain is specially hung around them—a borrowed Russian idea, unquestionably. A rude table, two or three empty cracker-boxes from the trader's store for chairs, and a rough bench or two, is about all the furniture ever seen in a barrabkie. The table-ware and household utensils do not require a large cupboard for their reception. Cups and saucers of white crockery, highly decorated in flaring blue and red floral designs, plates to match, a few pewter teaspoons, will usually be found in sufficient quantity for the daily use of the family ; and these are loaned out to a neighbor also, on occasions of festivity, when an entire circle of chosen friends join under the roof of some one barrabora in tea-drinking and " praznik " feasting.

The traders say that recently a great desire has come upon the natives to possess granite ware cooking utensils and drinking cups, or those porcelain or silicon-plated iron vessels which we designate by that name ; they do not require washing, and can be easily wiped out and never rust. Tin-ware is at a great discount among them— it rusts. The odor of coal-oil will be noticed among many others in the barraboras of the Aleutians and Kadiaks in these days, for the general use of this fluid has been established. The glass lamps and the smell suggestive of that illuminant can be plainly detected by any stranger who goes into a village up there now, in spite of the fishy and other indigenous strong aromas, which are in themselves equally odious and penetrating. However, an old Aleutian fogy will occasionally insist upon using a primitive stone lamp, with a wicking of moss or strips of cotton cloth.

A marked fondness for pictures, old engravings, chromos, in

fact anything that goes in the line of caricature or illustration, is manifested by the Aleutes. They paste all sorts of scraps from newspapers, magazines, and theatrical posters, which the traders give them, upon the walls of the barrabora. The Russians early took notice of this trait, and the priests of the Greek Church made good use of it by distributing richly-colored and gilded portraits of holy men and women, the Imperial family, and mythological church groups.

As the Aleut, his wife and children, and a relative or two, perhaps, are living in the barrabora, he enjoys a warm and comfortable shelter as long as he keeps it in good repair. He does not place what he has of surplus supply in a cellar—such fish, fowl, or meat as he may have in excess of immediate consumption is hung up outside of his door on a wooden frame, or "laabaas." Here it is beyond the reach of dogs, and is quite secure, inasmuch as he lives in no danger or dread of theft from the hands of his neighbors.

He is a fish-eater, like a vast majority of the rest of native Alaskans. He has cod, halibut, salmon, trout, and herrings in overflowing abundance, and all swim close to his door. He hooks and nets his piscine food-supply all the year round as it rotates with the seasons. He varies this steady diet with all the tea, sugar, and hard bread, or flour, that he can purchase from the trader's store ; some other little articles in the grocery line, such as canned California fruits, are especially agreeable to his palate. These natives call on the trader for biscuits, or sea-crackers, not because they like this hard bread best, but on account of the scarcity of fuel wherewith to properly bake up flour.

While fish is the staff of Aleutian daily life, yet nature has vouchsafed many simple luxuries to those people : these are sea-urchins, or echinoderms, and the eggs and flesh of the several species of water-fowl peculiar to and abundant in such latitudes. Then, in August and September, the valleys, hillsides, and margins of the sea are resorted to by the natives for the huckleberries, the "moroshkies," the crowberries, and giant umbelliferous stalks of the *Archangelica*, found ripe and ripening there. The Aleutian huckleberries are much better than those of the Sitkan district, and are really the only good indigenous fruit, according to the evidence of our palates.

Another peculiarity of an Aleutian village, which strikes a stranger's eye, is the irregular but frequent coming and going of

OONALASHKAN NATIVES COD-FISHING

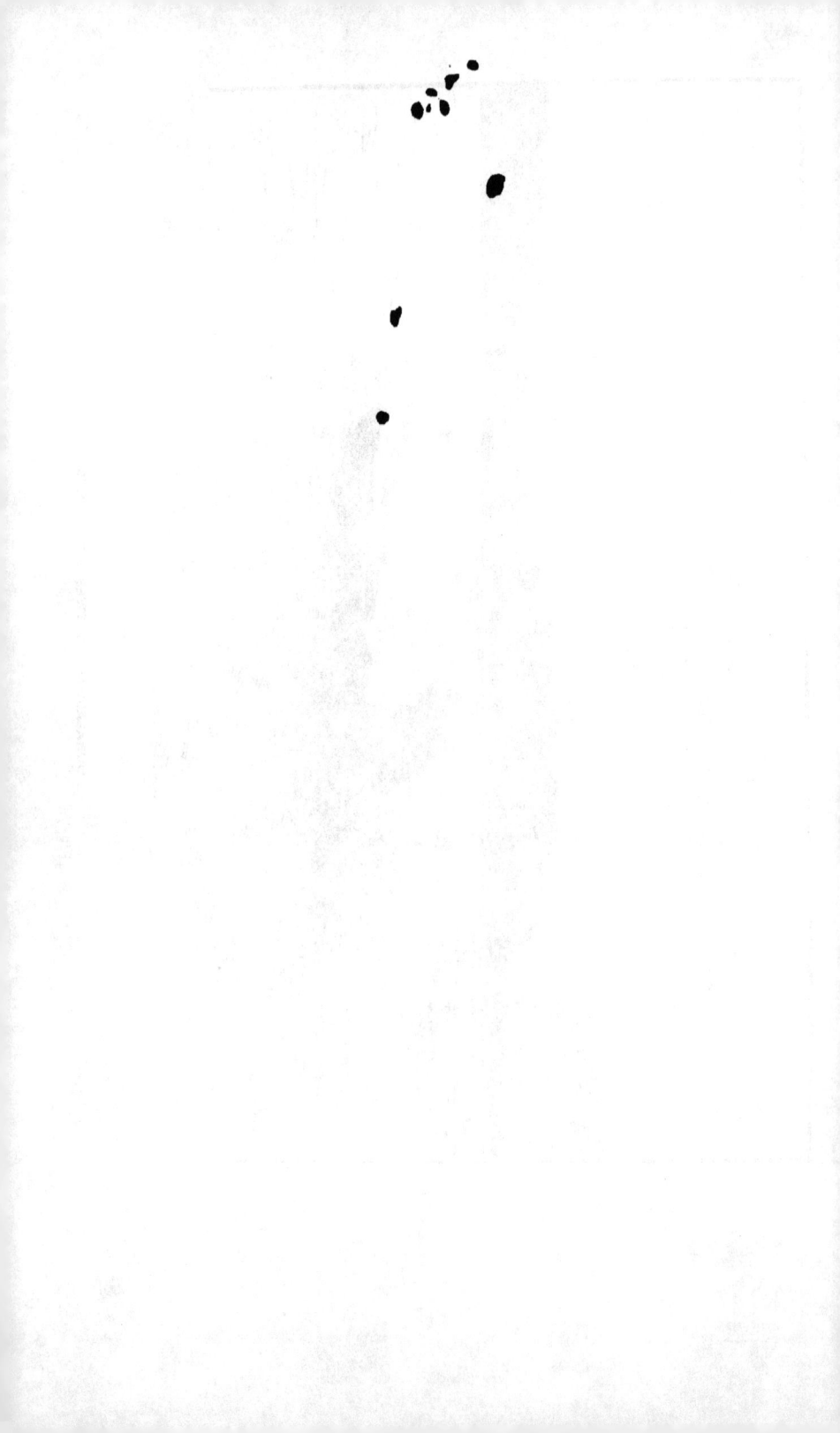

a number of old women, and younger ones, to and from the mountains; they always return with a burden of what appears to be
coarse grass upon their backs, in such huge bundles that the bear-
ers are quite hidden from view. These females, while not literally
hewers of wood, are really working as hard. They are gathering
the only natural resource which is afforded them for fuel. When
long and tedious journeys along the coast fail to reward a search
for drift-logs, which are found here and there in scant number at
the best, then the women repair to those spots on the mountain
sides where the slender strawberry-like runners of the crowberry*
have grown and intergrown into thick masses. These they pull
from the earth, as we would gather dried grasses. A large bundle
is made for each woman in the party, and then, assisting each other
to load up, they stagger down the hillside trails, under these heavy
burdens, back to their respective barraboras. This "sheeksa" is
then air-dried, or weathered several weeks, so as to get it
ready and fit for use in those odd Russian ovens or "peechka"
stoves. It is twisted into short wisps, two or three of which at one
time are ignited, and thrust as they blaze, into the oven; then the
door of the peechka is closed tightly and promptly. This makes a
hot fire for a few moments; every particle of the heat is absorbed
by the thick, brick walls of. the oven, so that, as it radiates slowly,
the small apartment within the earthen walls of the barrabora is
kept at a tropical temperature, for several hours at a time, without
a renewal of this fire. To-day, however, at Oonalashka, and at three
or four other central sea-otter villages, the natives are buying cord-
wood and coal from the traders. The wood is brought from
Kadiak, while the coal comes up as ballast from San Francisco in
the traders' vessels.

Housed and fed in this manner, the entire Aleutian population
have been, and are living; as their children grow up and inherit
the parental homes, or branch out, after marrying, to erect barraboras of their own, they repeat the same methods of their ancestry.
In a normal condition the Aleut is a quiet, peaceful parent, affectionate but yet not demonstrative; he is kind to his wife and
imposes no real burden upon her which he does not fully share

---

* *Empetrum nigrum.* The fruit is a small black berry very much like
that borne upon those hedges of an English privet, which grows in our garden
here at home.

himself. The children grow up without harsh discipline ; still they are not the recipients of marked attention. The family life, when the head or the hunter is at home, is one of very simple routine ; he is in bed most of the time, striving to balance that account of the very many sleepless nights he has passed in his bidarka scouring sea-otter reefs during his recent three months' trip, to Saanak or the Chernaboors. The others rise at broad day-light, light their blubber-fire in the outside kitchen, and prepare a slight breakfast of crackers, tea, and boiled fresh fish. This meal is carried into the living-room, where the "pecchka" has been started up, so as to thoroughly warm that apartment. If this native is the possessor of a little iron stove, of our own make, then all heating and cooking is done on the one fire made in it and the smoke of that burning fat and oil with which so much of their fuel of drift-wood and sheeksa is mixed, goes up the pipe and leaves no annoying trace behind. Between the members of the household there is never much conversation—the topics are few, indeed, beyond the ordinary routine of irregular meals, and the desultory rising and retiring of a family. This monotony of their lives is very much enlivened by exercises of the church, to which they are constantly going and coming from. But when they meet in a neighboring barrabkie, or receive friends in their own, then tongues are loosened, and conversation flows freely, especially over cups of tea between the old men and women ; the latter are incessant talkers under such genial encouragement.

Although the Aleut does not give you, at first, the least idea that he has ever had any severe training of a heroic order, yet it is a fact that most of the young men, ere they become recognized hunters, had to "win their spurs," as it were. The old men always impress upon the native youth that great importance of strictly observing the customs of their forefathers in conducting the chase, and that neglect in this respect will surely bring upon them disaster and punishment ; therefore the young men are encouraged to go to sea in gales of wind, and make difficult landings with their bidarkas at surf-washed places. Before the advent of Russian priests, every village had one or two old men at least, who considered it their especial business to educate the children ; thereupon, in the morning or the evening when all were at home, these aged teachers would seat themselves in the centre of one of the largest village yourts or "oolagamuh :" the young folks surrounded them, and listened attentively to what they said—sometimes failing memory would cause the old preceptors to repeat over and over again the same advice or legend in the course of a lecture. The respect of the children, however, never allowed or occasioned an interruption of such a senile oration.

But to-day their education, in so far as the strict sense of the term goes, is received from the priests and deacons of the Greek Church. They seem to have abandoned all the shamanism, the mummery and savagery of their primitive lives eagerly and willingly for those practices and precepts of the Christian faith; in this strange accord the Kadiakers also joined. No recourse to violent measures was ever resorted to by the Russian missionaries, who were always met more than half-way by these singular heathen. An Aleut is the better Christian when fairly compared with the Kaning—the latter is not half as sincere or faithful.

Stepan Glottov, in 1759, wintered, first of all white men, at Oomnak Island, and he lived there then in perfect peace and quiet with the natives; so amicable were his relations with these people, that he persuaded their chief to be baptized, and to allow a little son to go with Glottov to Kamchatka, where the youth lived several years, then returned, well versed in the Russian language, and assumed the title of supreme chief of the Aleutians; this is the earliest record made of the conversion of these people. In 1795 the first priest or missionary came among them; and never, from that time to the present moment, has a representative of that church ever been treated otherwise than well by these islanders.

The Aleutian brain has streaks of genuine philosophy and a keen sense of humor. A priest once reproached an aged sire for allowing a worthless son to worry and vex the household. "Why, Ivan," said he, "do you, who are so good, and Natalie, your wife, also most excellent, permit this rude child to so deport himself?" "Ah, father," replied the old man with great emotion, "not out of every sweet root grows a sweet plant!"

This inherent religious disposition of the Aleut is the reason why we find a Greek church or a chapel in every little hamlet where his people live. The exclusion of all other sects, however, is natural, since the character of the ornate service and frequent "prazniks," or festivals of that chosen denomination, suits him best. The Greek Catholic Bishop of the Alaskan diocese now resides in Oonalashka. He used to make Sitka his headquarters, but the depopulation of the whites there after the transfer of the country made that spot too lonely for him, and he soon removed to San Francisco. A few years ago a final transfer was made to Illoolook. As far as possible the natives support their own respective chapels, erect the church structures, keep them in repair, and make an annual contribution sufficient to support a reader, or "deacon," so that the order of daily services shall be constantly in operation. When a community is too poor, however, to do this, then the bishop has money supplied to him by the Russian Home Church Fund, which he uses to maintain the proper conduct of those chapels situated at impecunious settlements. Of course these outlying and far-distant hamlets of the Aleutian archipelago are unable to secure and pay, each one, for the services of a regularly ordained resident priest. Therefore a parish priest, either from Oonalashka, Belcovsky, Sitka, or even San Francisco, is in the habit of making a tour of the entire Alaskan circuit once in every year or two, so as to administer the higher offices of the church, such as baptism, marriage, etc.

Most amusing is that intense outward piety of these grimy people—they greet you with a blessing and a prayer for your good health in the same breath, and they part from you murmuring a benediction. They never sit down to their rude meals without asking the blessing of God ; never enter a neighbor's house without crossing themselves at the threshold ; and in most of the barraboras a little image-picture of a patron saint will be found in one corner, high up on a shelf, to which the face of every member of the family is always turned when they rise and retire—the head bowed and the cross sign made before this "eikon," in humility and silence. These people also carry the precepts and phraseology of the church upon their lips, constantly repeating them during holy weeks and pious festivals.

The fact that among all the savage races found on the northwest coast by Christian pioneers and teachers, the Aleutians are the only practical converts to Christianity, goes far, in my opinion, to set them apart as very differently constituted in mind and disposition from our Indians and our Eskimo of Alaska. To the latter, however, they seem to be intimately allied, though they do not mingle in the slightest degree. They adopted the Christian faith with very little opposition, readily exchanging their barbarous customs and wild superstitions for the rites of the Greek Catholic Church and its more refined myths and legends.

At the time of their first discovery, they were living as savages in every sense of the word, bold and hardy, throughout the Aleutian chain, but now they respond, on these islands, to all outward signs of Christianity, as sincerely as our own church-going people. The question as to the derivation of those natives is still a mooted one among ethnologists, for in all points of personal bearing, intelligence, character, as well as physical structure, they seem to form a perfect link of gradation between the Japanese and Eskimo, notwithstanding their traditions and their language are entirely distinct and peculiar to themselves; not one word or numeral of their nomenclature resembles the dialect of either. They claim, however, to have come first to the Aleutian Islands from a " big land in the westward," and that when they came there first they found the land uninhabited, and that they did not meet with any people, until their ancestors had pushed on to the eastward as far as the peninsula and Kadiak. Confirmatory of this legend, or rather highly suggestive of it, is the fact that repeated instances have occurred within our day-where Japanese junks have been, in the stress of hurricanes and typhoons, dismantled, and have drifted clear over and on to the reefs and coasts of the Aleutian Islands. Only a short time ago, in the summer of 1871, such a craft was so stranded, helpless and at the mercy of the sea, upon the rocky coast of Adak Island, in this chain; the few surviving sailors, Japanese, five in number, were rescued by a party of Aleutian sea-otter hunters, who took care of them until the vessel of a trader carried them back, by way of Oonalashka, to San Francisco, and thence they returned to their native land.

A number of the males in every Aleutian village will be found who can read and write with the Russian alphabet. This education they get in the line of church exercises, inasmuch as they are all conducted in the Russian language, though the responses for the congregation usually are made by Aleutian accents. An Aleut grammar and phonetic alphabet, adapted to the expression of the Russian language, is used in all of these hamlets. It was prepared by that remarkable man, Veniaminov, in 1831 : a large number of the books were printed, and they have been in use ever since. The young men and boys are taught as they grow up, by the church deacon usually, to read, first in the Aleut dialect, then in the Russian. The traders take advantage of this understanding among these people, and facilitate their bartering very materially. They give every hunter a pass, or grocer's book, in which he keeps a regular account, charging what he may need, in advance of payment, so enabling his family to get what it requires during his long absence on the hunting-grounds. In short, that book is a regular letter of credit at the store, and the traders have found it the best way of influencing the natives in their favor, and also of aiding the superior hunters.

This method of credit has developed, and made manifest the truth of a strong statement in which Veniaminov declared his belief that these people were really honest at heart, totally unlike all other savages in Alaska, or elsewhere on the American continent, for that matter. Many of the hunters, when they are about to depart for a long four or five months' sea-otter chase, and consequent absence for such length of time from home, go to the trader and tell him to restrict their wives from overdrawing a certain pecuniary limit, which they fix of their own idea as to what they can afford. This action, however, is the purpose of true honesty only, for those same hunters, when they get back, after first religiously settling every indebtedness in full, make at once a heavy draft upon any surplus that they may have, going so far in the line of extravagance and singular improvidence, in some instances, as to purchase, on the spur of the moment, music-boxes worth two hundred dollars each, or whole bolts of silk and costly packages of handkerchiefs, neckties, and white linen, and many other things of a like nature, wholly unwarranted by the means of the hunter, or of any real service to him or his family.

The church "prazniks," or festivals, are very quiet affairs, but when the Aleut determines to celebrate his birthday or "eman nimik," he goes about it in full resolution to have a stirring and vociferous time. Therefore he brews a potential beer by putting a quantity of sugar, flour, rice, and dried apples (if he can get the latter) into a ten or twenty-gallon barrel, which is filled with water. He sends invitations out to his friends so dated as to bring them to the barrabkie when a right degree of fermentation in the kvass-barrel shall have arrived; sometimes the odor of that barrel itself is sufficient to gather them in all on time. Some one of the natives who is famous for natural and cultivated skill in playing the accordion or concertina, is given the post of honor and the best of the beer; he or she, as the case may be, soon starts the most hilarious dancing, because Aleutes are exceedingly fond of this amusement, especially so when stimulated by beer. If the apartment is large enough, the figures of an old Russian quadrille are gone through with, accompanied by indescribable grimaces and grotesque side-shuffles of the dancers, the old women and young men being the most demonstrative. Usually, however, a single waltzing couple has the floor at one time, whirling around with the liveliest hop-waltz steps, and as it settles down out of breath, a fresh pair springs up from the waiting and watching circle. The guests rapidly pass from their normal sedateness into the varying stages that rotate between slight and intense drunkenness.

These kvass orgies, on such occasions, are the only exhibitions of disorder that the people of the Aleutian Islands and Kadiak ever afford. At Belcovsky, and at every point where the sea-otter industry is most remunerative to the native hunter, there you will find the greatest misery, due wholly to those beery birthday celebrations as sketched above.

Some traders often give entertainments to the natives, in which they wisely offer plenty of strong tea, with white sugar-lumps, and nothing else ; these parties are quite reputable and highly enjoyed by all concerned. The floor of the warehouse, or the living-rooms of the trader himself, are cleared, and this allows ample space for a full-figured cotillon or quadrille, or a dozen or two of dancing couples. The ball-room of the chief trading-firm at Oonalashka is a very animated and extensive prospect when an evening-party of this sort is in fine motion. The familiar strains of "Pinafore," the "Lancers," "John Brown," and "Marching through Georgia," rise in piercing strength from the vigorous men and women who are squeezing the accordions, and every now and then a few of the young Aleutes break out into a short singing refrain, using English words to suit the music, as they caper in the high-tide of this festivity. It is the young men, however, only, who thus vocalize ; the women, when sober, old and young, are always silent, with downcast eyes, and are very abject in demeanor.

The great feminine solace in a well-to-do native hut is recourse to a concertina or accordion, as the case may be. These instruments are especially adapted to the people. Their plaintive, slow measure, when fingered in response to native tunes and old Slavonian ballads, always rise upon the air in every Aleutian hamlet, from early morning until far into night. An appreciation

of good music is keen : many of the women can easily pick up
strains from our own operas, and repeat them correctly after listen-
ing a short while to the trader or his wife play and sing. They
are most pleased with sad, wailing tunes, such as "Lorena," the
"Old Cabin Home," and the like.

Thus we note those salient characteristics of Aleutians, who
are the most interesting and praiseworthy inhabitants of Alaska.
There are not a great many of them, however, when contrasted
numerically with the Indians and the Eskimo of this region ; but
they come closer, far nearer to us in good fellowship and human
sympathy. We turn, therefore, from them again to resume our
contemplation of the country in which they live. The sun is burning through a gray-blue bank of sea-swept fog, ever and anon
shining down brilliantly upon the beautiful, vividly green mountains, and glancing from the clear waters of Oonalashka's harbor.
It tempts us to walk, to stroll, when the trader tells us that we can
easily cross over to Beaver Bay, where Captain Cook anchored and
refitted in 1778. So we go, and a patient, good-humored native
trots ahead to keep us on the road and bring us back safely, lest
the fog descend and shut all in darkness which is now so light and
bright. A narrow foot-trail that is deeply worn by the pigeon-toed
Aléutes into moss and sphagnum, or fairly choked by rank-
growing grasses and annuals in low warm spots, winds around and
over the divide between Oonalashka village and Borka. As we reach
the rippling, rocky strand of Beaver Bay, a cascade arrests our attention on account of its exceeding beauty. Tumbling down from
the brow of a lofty bluff of brown and reddish rocks, a rivulet falls
in a line of snowy spray, which reflects prismatic colors from the
rocks and sunlight as it drops into the cold embrace of the sea.
While we, resting on the grassy margin of the beach, enjoy this
charming picture, our native turns his face to the bay, and he
points out to us the pebbly shore where Captain Cook "hove
down" his vessel, more than a century ago, and then scraped those
barnacles and sea-weed growths from that ship's bottom. Here the
English discoverer first came in contact with the natives of Oonalashka, and there are people over on Spirkin or "Borka" Island,
just across the bay from us, who will recite the legend of this early
visit of that Englishman with great earnestness, circumspec-
tion, and detail, so faithfully has the story been transmitted from
father to son. Their own name of Samahgaanooda was changed

voluntarily to English Bay, or "Angliceski Bookhta," by which designation they themselves call the harbor to-day.

A broad expanse of this bay lies directly between us on the north side and the village of Borka, which is perched on a narrow bench-level shelf of an island that rises bold and abruptly, high from the sea. This hamlet is the most remarkable native settlement in all Alaska with respect to a strange and unwonted cleanliness which is exhibited in this community of one hundred and forty Aleutes, who are living here to-day in twenty-eight frame houses, barraboras, and a chapel. What makes it still more remarkable is the fact that these people are in close communication with their kindred of Oonalashka, who are distant only a few hours' journey by canoe and portage, and who are not especially cleanly to the slightest noteworthy degree. Those people of Borka are living in the cleanest and neatest of domiciles. They are living so without an exceptional instance, every hut being as tidy and as orderly as its neighbor. They have large windows in the small frame houses and barraboras, scrub and sand the floors, and keep their simple furniture, their beds, and window-panes polished and bright. Glass tumblers, earthen pots, and wooden firkins filled with transplanted wild-flowers stand on the tables and deep window-sills to bloom fresh and sweet all the year round. A modest, unassuming old Russian Creole trader, who has lived there all his life, and who was living recently, is credited with this influence for the better with the natives. Certainly he is the only one who has ever succeeded in working such a revolution in the slovenly, untidy household habits of these amiable but shiftless people.

As we retrace our steps to Oonalashka village we become fully impressed with the size of this island. It bears so many mountain spurs, with a singularly rugged, cut-up coast, in which the deep indentations or gulf-fiörds nearly sever the island in twain as they run in to almost meet from the north and south sides. Beautiful mats of wild poppies are nodding their yellow heads as the gusts sweep over them on the hillsides, and a rank, rich growth of tall grasses by the creek-margins and the sea-shore in sheltered places shimmers and sways like so many fields of uncut green grain do at home. Vegetation everywhere, except on the summits of the highest peaks and ridges and the mural faces of the bluffs! Even there some tiny lichens grow, however, and give rich tones of golden ochre and purplish-bronzed reflections from the cold, moist rocks, whereto they cling.

12

We pause in that little cemetery, just outside of the village of Illoolook. It is on a small knoll, under higher hills that rear themselves over it. Its disorder and neglect is a fair reminder of what we see in most of our own rural graveyards. The practice of all these natives is to inter by digging a shallow grave. The body is prepared in its best clothes, and coffined in a plain wooden box. A small mound and a larger or smaller wooden Greek cross is the only monument. Tiny oil-portraits of their patron saints, painted on tin or sheet-iron, especially made for these purposes, and furnished by the Church, are tacked to the crosses, with now and then a rude Russian inscription carved or painted thereon in addition. During certain periods of the summer, when the weather is pleasant, little squads of relatives will come out here from the village and pass a whole day in tea-drinking and renovating the crosses, sitting on the mounds as they chat, work, and boil their samovar. The Illoolook church-bells ring—they arouse us to resume the walk thus interrupted in this small city of Aleutian dead. As we enter the town, we see the occupants of turfy barraboras and frame cottages hastening from every quarter and trooping to the door of a yellow-walled and red-roofed house of worship. Perched on that three-barred cross which crowns the cupola of this chapel are half a dozen big black ravens, all croaking most lugubriously, as the clanging chimes peal out below them. That is their favorite roosting-place. The natives take no notice of those ill-omened birds, which as feathered scavengers, hop around the barraboras in perfect security, since no one ever disturbs them, unless it be some graceless trader who is anxious to test the killing power of a new shotgun. They breed in high chinks of the bluffs, and find abundant food cast upon the beaches by the sea. A few domestic fowls, some with broods of newly-hatched chicks, are running about or scratching around the place. The priest's shaggy little bull and cow stand in front of a small stable or "scoatnik," lazily chewing their cud. There is no other live-stock in the hamlet, except a few dogs and cats; not a great many of the latter, however.

West of this Island of Oonalashka is a narrow-lined stretch of more than eight hundred miles of rapidly-succeeding islets and islands, until the extreme limit of the Alaskan border is reached at Attoo. In all this dreary wilderness of land and water only three small human settlements are to be found to-day, with a population of less than five hundred natives and six or seven white men.

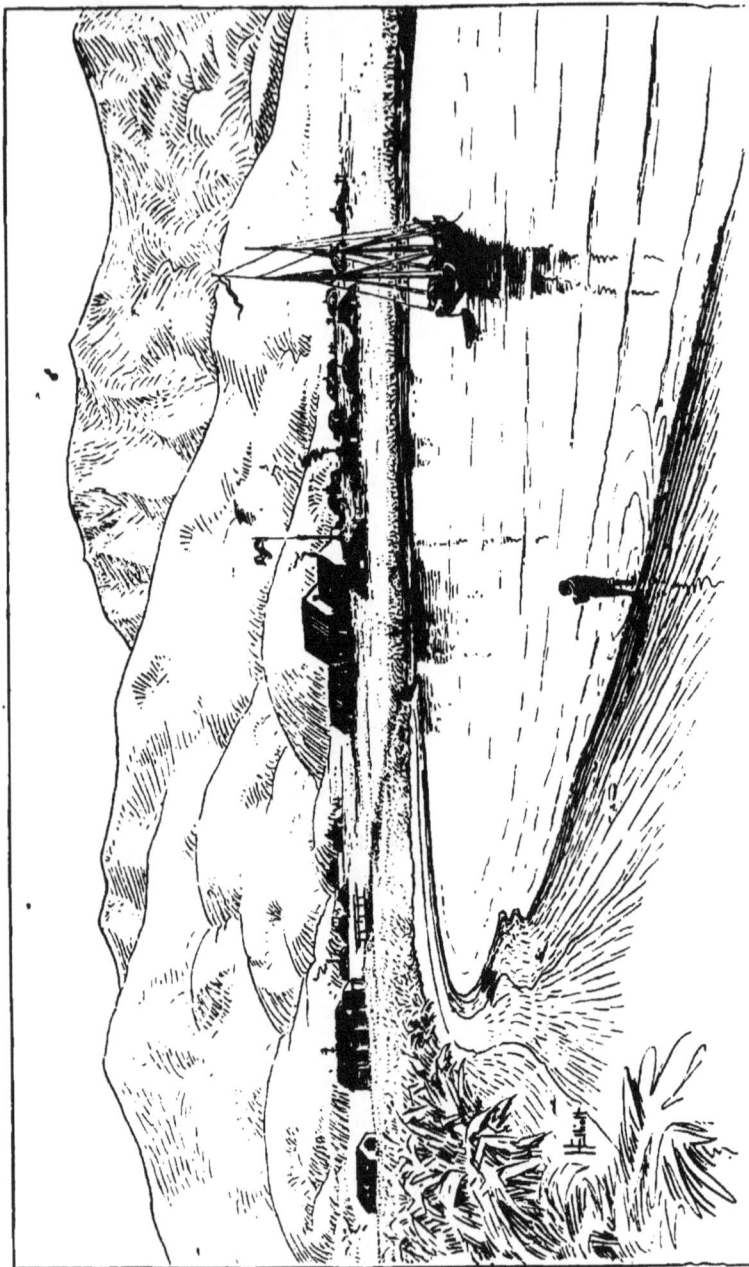

THE VILLAGE AND HARBOR OF ATTOO

Attoo, Atkha, and Oomnak are the only villages, the last closely adjoins Oonalashka, on a large island of the same name.

Attoo is the extreme western town which is or can be located on the North American continent. It is the first land made and discovered by the Russians, as they became acquainted with the Aleutian chain. Michael Novodiskov, a sailor who had survived Bering and the wreck of the *St. Peter* in 1741, took command of a small shallop in 1745, and sailed from Lower Kamchatka. He reached Attoo, and also landed on its sister island of Aggatoo, in the same season. The Aleutes were then numerous, bold, and richly supplied with sea-otter skins. Now, nothing but the ruined sites of once populous villages remain behind to attest the truth of that early Russian narrative; and the descendants, who number but a little more than one hundred souls, are living in a small hamlet that nestles in the shelter of that beautiful harbor on the north side of Attoo Island, at the rear of which abrupt hills and high mountains suddenly rise. A sand-beach before the village is fringed by a most luxurious growth of rank grass, that wild wheat of the north, the tasselled seed-plumes of which are waving as high as the waists, and even the heads of the natives themselves.

Sea-otters have been virtually exterminated or driven from the coast here, so that the residents of Attoo are, in worldly goods, poor indeed; and a small trader's store is stationed here, more for the sake of charity than of commercial gain. But they have an abundance of sea-lion meat, of eggs and water-fowl; a profusion of fish—cod, halibut, algæ mackerel, and a few salmon. They have a liberal supply of drift-wood landed by currents upon the shores of this and the contiguous rugged islets. Several times during the last ten years have traders endeavored to coax these inhabitants to abandon Attoo and go with them to better situations for sea-otter hunting. But, although pinched by poverty, yet so strongly attached are they to this lonely island of their birth, that they have obstinately declined. Though they are poor, yet the contrast between their cheerful, healthy faces and those debauched countenances which we observed at the wealthy villages of Protassov and Belcovsky is a delightful one, and preaches an eloquent sermon in its own reflection. Naturally the people of Attoo do not enjoy much sugar, tea, cloth, and other little articles which they have learned to covet from the trader's store; so we find them living nearer the primitive style of Aleutian ancestry than elsewhere in the archipelago, being dressed largely in tanned seal and bird-skins, of the fashion made and worn by their forefathers who welcomed Novodiskov long, long ago.

The necessity of doing something in order to gain from the trader a few of the simplest articles, such as the natural resources of Attoo utterly failed to supply, has driven the natives to the care and conservation of blue foxes, which they introduced here many years past, and of which they secure, in traps, two or three hundred every season. The common red fox * of the whole Aleutian chain became extinct here in prior time; so, taking advantage of this fact, those blue foxes, so abundant and so valuable on the Seal Islands, were imported, and have ranged without deterioration, since ice-floes never bridge the straits that isolate this island from the nearest adjacent land, and upon which the common breed might cross over to ruin the quality of the fur of that transported *Vulpes lagopus.* They have also domesticated the wild goose, and rear flocks of them around their barraboras, being the only people in Alaska who have ever done so.

It hardly seems credible, at first thought, but the village of Attoo makes San Francisco practically the half-way town as we go from Calais, Me., to it, our westernmost settlement! It is really but slightly short of being just midway, since Attoo stands almost three thousand miles west of the Golden Gate. † A strict geographical centre of the American Union is that point at sea forty miles off the Columbia River mouth, on the coast of Oregon.

The nearest neighbors of the Attoo villagers are not of their own kith and kin—they are the Atkhan and Kamchadale Creoles and natives of the Russian Seal Islands, some two hundred miles

---

* The only fur-bearing animal found in every section of Alaska is the red fox (*Vulpes fulvus*). From Point Barrow to the southern boundary, and from the British line to the Island of Attoo, this brute is omnipresent. It varies greatly in size and quality of fur, from the handsome specimens of Nooshagak down to the diminutive yellow-tinged creatures that ramble furtively over the Aleutian Islands.

† "The distance in statute miles between San Francisco and a point due south of Attoo, measured on the parallel of San Francisco, is 2,943.1 miles. The distance east from Attoo of a point due north of San Francisco, measured on the parallel of Attoo, is 2,214.5 miles. The amount of westing made in sailing from San Francisco to Attoo, on a great circle, is very nearly 2,582.5 miles."—(Henry Gannett, Geographer U. S. Geological Survey: letter to author.)

west ; but on our side they are separated by more than four hundred and thirty miles of stormy water from the first inhabited island, which is Atkha, where a much larger and a much more fortunately situated settlement exists on its east coast, at Nazan Bay. Here is a community of over two hundred and thirty souls, being all the people gathered together who previously lived in small scattered hamlets on the many large and small islands between Atkha and Attoo. They secure a comparatively good number of sea-otters, and are relatively well-to-do, being able to excite and sustain much activity in the trader's store.

General agreement among those who have visited the Atkhans, as traders and agents of the Government, is that these natives are the finest body of sea-otter hunters in all respects known to the business. They make long journeys from their homes, carried to the outlying islands of Semeisopochnoi, Amchitka, Tanaga, Kanaga, Adahk and Nitalikh, Siguam and Amookhta, some of them far distant, on which they establish camps and search the reefs and rocks awash, as they learn by experience where the chosen haunts of the shy sea-otter are. Here they remain engaged in the chase over extended periods of months at a time, when, in accord with a pre-concerted date arranged with the traders, those schooners which carried them out from Atkha, return, pick them up, and take them back. Then the trader's store is made a grand rendezvous for the village ; the hunters tally their skins, settle their debts, make their donations to the church, and then promptly invest their surplus in every imaginable purchase which the goods displayed will warrant.

The women of Atkha employ long intervals, in which their husbands and sons are absent, by making the most beautifully woven grass baskets and mats. The finest samples of this weaving ever produced by a savage or semi-civilized people are those which come from Atkha. The girls and women gather grasses at the proper season, and prepare them with exceeding care for their primitive methods of weaving ; and they spare no amount of labor and pains in the execution of their designs, which are now almost entirely those suggested by the traders, such as fancy sewing-baskets, cigar-holders, table-mats, and special forms that are eagerly accepted in trade, for they find a ready sale in San Francisco.

A peculiar and valuable food-fish is found in the Atkhan waters which has been attracting a great deal of attention as a substitute for the mackerel of our east coast, inasmuch as there is no such

fish found on the Alaskan coasts. Among the sea-weed that floats
in immense rafts everywhere throughout the Aleutian passes, the
"yellow-fish," or "Atkha mackerel," * is very abundant ; it is also
plentiful' off the Shoomagin Islands. It is a good substitute for
the real mackerel,† resembling it in taste after salting, as well as in
size and movements.

During early days of Russian order and control, the people of
Atkha lived altogether on the north side of the island, and it was
then the grand central depot of the old Russian American Company. A chief factor was in charge, who had exclusive jurisdiction
over all that country embraced in the Kurile archipelago, and the
Commander group of Kamchatka, and the Aleutian chain as far
east as Oomnak. It was a very important place then, and this territory of its jurisdiction was styled the "Atkhan Division." But
within the last ten or twelve years, fish and drift-wood became very
scarce on the Bering Sea coast, so the inhabitants made a sweeping
removal of everything from the ancient site on Korovinsky Bay to
that of Nazan, where the little hamlet now stands, overtopped by
lofty peaks and hills on every side, except where it looks out over
the straits to the bold headlands of Seguam. So thorough were
they in this " nova-sailnah," that they even disinterred the remains
of their first priest and re-buried them in front of their new chapel
—a delightful exhibition of fond memory and respect where we
might, perhaps, have least thought to have found them manifested.

The curious Island of Amlia shuts out the heavy swell of the
Pacific Ocean from Nazan Harbor, and gives that little bay great
peace and protection. This island is thirty miles in length, and
nowhere has it a breadth of over four miles ; most of its entire
extent is only some two miles from ocean to ocean. It consists of
a string of sharp, conical peaks, which once were active volcanoes,
but now cold and silent as the tomb. So abruptly do they rise
from the oceans which they divide, that there is but one small spot
on the south side where a vessel can lie at anchor and effect a
landing.

Atkha is a large island, and it has very slight resemblance to
that of Oonalashka in shape ; its indented fiörds are, however, less
deep and not near so commodious and accessible. The snowy,
smoking crater of Korovinsky Sopka stands like a grim sentinel

---

overlooking the north end of the island, a sheer five thousand feet above the sea-beach at its feet. A few miles to the south another rises to almost as great an elevation, from the flanks of which a number of hot springs pour out a steady boiling flood; then, at the northeast extremity, and handsomely visible from the village, is a silent, snowy crater which they call Sarichev. Korovinsky is the only disturber of the peace that rightfully belongs to Atkha. It is constantly emitting smoke and ashes, while earthquakes and subterranean noises are felt and heard all over the island at frequent though irregular periods during the entire year. In the ravines and cañons of this volcano and its satellites are the only glaciers which the geologist has ever been able to find on any of these peaked, lofty islands west of Oonalashka, though a hundred eternal snow-clad summits and a thousand snow-filled gorges are easily discerned. The natives here also describe a series of mud-volcanoes, or "mud-pots," that exist on the island, in which this stuff is constantly boiling up with all the colors of the rainbow, about as they seethe and puff in the Geyser Basin of the Yellowstone Park.

There are a dozen or so small, mountainous, uninhabited islands between Atkha and the larger island of Adakh in the west. Very little is known of them, since they endanger life if a landing is made. The most imposing one is Sitkhin, a round, mountainous, lofty mass which culminates in a snow-covered peak over five thousand feet in height. The crater is dead, however, and no sign of ancient volcanic energy is now displayed, beyond the emission of hot springs from fissures in its rocky flanks. Adakh itself is quite a large island, rough and hilly to an excessive degree. A grand cone, which rises up directly in the centre high above all the rest, is called the "white crater." It is also a dead volcano like Sitkhin; but steamy vapors from outpouring hot waters rise in many valleys and from the uplands. A succession of volcanic peaks reared from the sea, a few of them still smoking and muttering, constitute the islands of Tanaga and Kanaga in the vicinity of Adakh. No place is feasible for a boat to land on either of these wild islets, except on the west shore of Tanaga in Slava Rossia Bay.

A single immense peak, rising all by itself, solitary and alone, from the girdle of surf that encircles it—a band of foaming breakers eighteen miles in circumference, is the islet of Goreloi. It is a formidable rival of the majestic volcano of Shishaldin, on Oonimak. Though nearly as high, yet it is not so symmetrical a cone. Wreaths and solid banks of fog are pressed against its volcanic sides, and hang around its glittering white head, so that the full impression of its grandeur cannot strike us as we gaze at its defiant presence, where, unsupported, it alone beats back the swell of a vast ocean.

That cluster of islands which stand between Goreloi and Attoo is an aggregate of cold volcanic peaks—Amchitka and Kyska being the largest—the Seven Peaks, or Semiseisopochnoi, smoking a little, all the rest entirely quiet. They offer no hospitality to a traveller, and the natives have done wisely in abandoning these savage island-solitudes to reside at Nazan Bay, where the country has a most genial aspect, and many stretches of warm sand-dune tracts are found, upon which vegetation springs into luxuriant life. Here, also, quite a herd of Kamchatkan cattle were cared for when the Russian régime was in vogue. This stock-raising effort was not a practical success, however, and the last of the bovine race disappeared very shortly after the country changed ownership. Goats were also introduced here, as well as elsewhere throughout the fur-trading posts of the old company in Alaska ; but the morbid propensity of those pugnacious little animals to feed upon the grasses which grow over roofs of the barraboras, and thus break in and otherwise damage such earthen tenements, made them so unpopular that their propagation was energetically and successfully discouraged by the suffering Aleutes.

Two hundred miles of uninhabited waste extends between the natives of Atkha and their nearest neighbors, the villagers of Nikolsky, who live in a small, sheltered bight of the southwest shore of Oomnak. This is one of the largest islands of the whole Aleutian group, very mountainous, with three commanding, overlooking peaks that are most imposing in their rugged elevation. Several large lakes nestle in their hilly arms, and feed salmon rivulets that rush in giddy rapids and cascades down to the ocean. Everything grows at Oomnak which we have noticed on its sister island of Oonalashka, except the willow ; while cross and red foxes are much more abundant here than at any other place in the whole archipelago. A great many hot springs boil up on the north side, and only as recently as 1878 a decided volcanic shock was experienced, which resulted in the upheaval of a small mud-crater between the village and Toolieskoi Sopka, a huge fire-mountain of the middle interior. Subterranean noises and tremors of the earth are chronic phenomena here, but the natives pay no attention to them. They complain, however, of inability to find fish where they usually found them in abundance prior to these earthquakes. Redoubled attention, however, is paid to the salmon when they run, and thus the deficiency is made up.

Before the coming of the Russians, Oomnak was one of the most populous islands ; then there were over twenty villages, some of them quite large. One was so big that "the inhabitants of it were able to eat the carcass of an enormous whale in a single day ! " The most stubborn and independent spirit displayed by the Aleutes prior to their subjugation was exhibited by the inhabitants of this

An Aleutian Mummy.

[*Unrolled from its cerements.*]

island. The four or five thousand hardy savages which the promishlyniks met here in 1757–59 have dwindled to a microscopic number of less than one hundred and thirty souls, who reside at Nikolsky to-day. They enjoy a somewhat better climate, for a good deal less snow falls here than at Oonalashka, and the small vegetable-garden does much better than elsewhere, except at Attoo. They raise domestic fowls, and have a very fair sea-otter catch every winter, when they scour the south coast, and reside for months at Samalga, hunting that animal. Furious gales which prevail during certain seasons drive kahlans out upon the south beach, there to rest from the pelting of storms: then they are speedily apprehended and clubbed by the watchful Oomnak hunter.

That curious group, the "Cheetiery Sopochnie," or Islands of the Four Mountains, stands right across the straits, opposite Oomnak. From Kaygamilak, which lies nearest, eleven mummies were taken as they were found in a warm cave on the northeast side of this island. These bodies were placed there in 1724, or some twenty-five or thirty years before the Russians first appeared. The mummies * were in fine preservation, and were the remains of a noted chief and his family, who in that time ruled with an iron hand over a large number of his people. The Island of Kayamil is a mere vol: canic series of fire-chimneys, the walls of which are not yet cool. The southeast shore in olden times was the site of several large settlements, where the people lived well upon an abundance of sea-lions, hair-seals, and water-fowl, which still repair to its borders. Now that it is desolate and uninhabitable, large flocks of tundra geese spend the summers here, as they shed their feathers and rear their young, not a fox to vex or destroy them having been left by those prehistoric Aleutian hunters.

But on Tahnak, which is the largest of the group, plenty of red foxes are reported. The loftiest summits are also on this one of the four islets, and on the south side once lived a race of the most warlike and ferocious of all Aleutes. They were destroyed to a man by Glottov, and their few descendants have long since been merged with those of Oomnak, where they now live. Several small, high, bluffy islands stand around Tahnak, and between it and its sister, Oonaska, which is nearly as large, equally rugged and precipitous. Amootoyon is a quite small islet, and completes the quartet of "Cheetiery Sopochnie."

A most interesting volcanic phenomenon of recent record is afforded by the study of that small Bogaslov islet which now stands hot and smoking twenty miles north of Oomnak, and which, two years ago, raised a great commotion by firing up anew. In the autumn of 1796 the natives of Oomnak and Oonalashka were startled by a series of loud reports like parks of artillery, followed

---

* These specimens were procured at the urgent request of the author, who induced a trader to make the attempt, September 22, 1874. They were presented by the Alaska Commercial Company of San Francisco to the Smithsonian Institution.

by tremblings of the earth upon which they stood. Then a dense dark cloud, full of gas and ashes, came down upon them from Bering Sea, swept by a northerly wind, and it hung over their astonished heads for a week or ten days, accompanied by earth-quakes and subterranean thunder ; then when an interval of clear-ing occurred by a change of winds, they saw distinctly to the northward a bright light burning over the sea. The boldest launched their bidarkas, and, after a close inspection, saw that a small island had been elevated about one hundred feet above the level of the surrounding waters ; that it had been forced up from some fissure of the bottom to the sea, and was still rising, while liquid streams of lava and scoriæ made it impossible for them to land. This plutonic action did not cease here until 1825, when it left above the green waters of Bering Sea an isolated oval peak with a serrated crest, almost inaccessible, some two hundred and eighty feet high, and two or three miles in circumference. The Russians landed here then for the first time, and the rocks were so hot that they passed but a few moments ashore. It has, however, cooled off enough now to be occupied by large herds of sea-lions, and is re-sorted to by flocks of sea-fowl. In this fashion of the making of Bogaslov was our vast chain of the Aleutian archipelago cast up from that line of least resistance in the earth's crust which is now marked by the position of these islands, as they alternately face the billows of the immense wastes of the Pacific, and those storm-tossed waves of the shoal sea of Bering.

CHAPTER IX.

WONDERFUL SEAL ISLANDS.

> When they the approaching time perceive,
> They flee the deep, and watery pastures leave ;
> On the dry ground, far from the swelling tide,
> Bring forth their young, and on the shores abide
> Till twice six times they see the eastern gleams
> Brighten the hills and tremble on the streams.
> The thirteenth morn, soon as the early dawn
> Hangs out its crimson folds or spreads its lawn,
> No more the fields and lofty coverts please,
> Each hugs her own and hastes to rolling seas.
> —OVID.

THE story of the gloomy grandeur of Alaskan scenery and the wild existence of its inhabitants is not half told until that picture of what we observe on the Pribylov Islands of Bering Sea is graphically drawn. Here is annually presented one of the most mar-

vellous exhibitions of massed animal-life that is known to man,
civilized or savage ; here is exhibited the perfect working of an
anomalous industry, conducted without a parallel in the history of
human enterprise, and of immense pecuniary and biological value.
In treating this subject the writer has trusted to nothing save
what he himself has seen, for, until these life-studies were made by
him, no succinct and consecutive history of the lives and movements of these animals had been published by any man. Fanciful
yarns, woven by the ingenuity of whaling captains, in which the
truth was easily blended with that which was not true, and short
paragraphs penned hastily by naturalists of more or less repute,
formed the knowledge that we had. Best of all was the old diary
of Steller, who, while suffering bodily tortures, the legacy of gan-
grene and scurvy, when wrecked with Bering on the Commander
Islands, showed the nerve, the interest, and the energy of a true
naturalist. He daily crept, with aching bones and watery eyes,
over the boulders and mossy flats of Bering Island to catch glimpses
of those strange animals which abode there then as they abide to-
day. Considering the physical difficulties that environed Steller,
the notes made by him on the sea-bears of the North Pacific are
remarkably good ; but, as I have said, they fall so far from giving
a fair and adequate idea of what these immense herds are and do
as to be absolutely valueless for the present hour. Shortly after
Steller's time great activity sprang up in the South Atlantic and
Pacific over the capture and sale of fur-seal skins taken in those
localities. It is extraordinary that, though whole fleets of American, English, French, Dutch, and Portuguese vessels engaged during a period of protracted enterprise of over eighty years in length
in the business of repairing to the numerous rookeries of the Antarctic, returning annually laden with enormous cargoes of fur-seal
skins, yet, as above mentioned, hardly a definite line of record has
been made in regard to the whole transaction, involving, as it did,
so much labor and so much capital.

The fact is, that the acquisition of these pelagic peltries had engaged thousands of men, and that millions of dollars had been employed in capturing, dressing, and selling fur-seal skins during the
hundred years just passed by ; nevertheless, from the time of Steller, away back as far as 1751, up to the beginning of the last decade, the scientific world actually knew nothing definite in regard
to the life history of this valuable animal. The truth connected

with the life of the fur-seal, as it herds in countless myriads on the Pribylov Islands of Alaska, is far stranger than fiction. Perhaps the existing ignorance has been caused by confounding the hair-seal, *Phoca vitulina*, and its kind, with the creature now under discussion. Two animals, more dissimilar in their individuality and method of living, can, however, hardly be imagined, although they belong to the same group, and live apparently upon the same food.

The following notes, surveys, and hypotheses herewith presented are founded upon the writer's personal observations in the seal-rookeries of St. Paul and St. George, during the seasons of 1872 to 1874 inclusive, supplemented by his confirmatory inspection made in 1876. They were obtained during long days and nights of consecutive observation, from the beginning to the close of each seal-season, and cover, by actual surveys, the entire ground occupied by these animals.

During the progress of heated controversies that took place pending the negotiation which ended in the acquisition of Alaska by our Government, frequent references were made to the fur-seal. Strange to say, this animal was so vaguely known at that time, even to scientific men, that it was almost without representation in any of the best zoölogical collections of the world ; even the Smithsonian Institution did not possess a perfect skin and skeleton. The writer, then as now, an associate and collaborator of that establishment, had his curiosity very much excited by these stories; and in March, 1872, he was, by the joint action of Professor Baird and the Secretary of the Treasury, enabled to visit the Pribylov Islands for the purpose of studying the life and habits of these animals.*

All writers on the subject of Alaskan exploration and enterprise agree as to the cause of the discovery of the Pribylov Islands in the last century. It was due to the feverish anxiety of a handful of

---

* It was with peculiar pleasure that the writer undertook, at the suggestion of Professor Baird, who is the honored and beloved secretary of the Smithsonian Institution, the task of examining into and reporting upon this subject ; and it is also gratifying to add, that the statements of fact and the hypotheses evolved therefrom by him in 1874 have, up to the present time, been verified by an inflexible sequence of events on the ground itself. The concurrent testimony of the numerous agents of the Treasury Department and the Government generally, who have trodden in his footsteps, amply testifies to their stability.

Russian fur-gatherers, who desired to find new fields of gain when they had exhausted those last uncovered. Altasov and his band of Russians, Tartars, and Cossacks arrived at Kamtchatka toward the end of the seventeenth century, and they were the first discoverers of the beautiful, costly fur of the sea-otter. The animal bearing this pelage abounded then on that coast, but by the middle of the eighteenth century they and those who came after them had entirely extirpated it from that country. Then the survivors of Bering's second voyage of observation, in 1741–42, and Tschericov brought back an enormous number of skins from Bering Island ; then Michael Novodiskov discovered Attoo and the contiguous islands in 1745 ; Paicov came after him, and opened out the Fox Islands, in the same chain, during 1759 ; then succeeded Stepan Glottov, of infamous memory, who determined Kadiak in 1763 ; the peninsula of Alaska was discovered by Krenitsin in 1768. During these long years, from the discovery of Attoo until the last date mentioned above, a great many Russian companies fitted out at the mouth of the Amoor River and in the Okotsk Sea ; they prospected therefrom this whole Aleutian archipelago in search of the sea-otter. There were, perhaps, twenty-five or thirty different companies, with quite a fleet of small vessels ; and so energetic and thorough were they in their search and capture of the sea-otter that as early as 1772 and 1774 the catch in that group had dwindled from thousands and tens of thousands at first to hundreds and tens of hundreds at last. As all men do when they find that that which they are engaged in is failing them, a change of search and inquiry was in order ; and, then the fur-seal, which had been noted, but not valued much, every year as it went north in the spring through the passes and channels of the Aleutian chain, then going back south again in the fall, became the source of much speculation as to where it spent its time on land and how it bred. No one had ever known of its stopping one solitary hour on a single rock or beach throughout all Alaska or the northwest coast. The natives, when questioned, expressed themselves as entirely ignorant, though they believed, as they believe in many things of which they have no knowledge, that these seals repaired to some unknown land in the north every summer and left it every winter. They also reasoned then, that when they left the unknown land to the north in the fall, and went south into the North Pacific, they travelled to some other strange island or continent there, upon which in turn to spend the winter. Naturally

the Russians preferred to look for the supposed winter resting-
places of the fur-seal, and forthwith a hundred schooners and shallops sailed into storm and fog, to the northward occasionally, and
always to the southward, in search of this rumored breeding-ground.
Indeed, if the record can be credited, the whole bent of this
Russian attention and search for the fur-seal islands was devoted
to that region south of the Aleutian Islands, between Japan and
Oregon.

Hence it was not until 1786, after more than eighteen years of
unremitting search by hardy navigators, that the Pribylov Islands
were discovered. It seems that a rugged Muscovitic "stoorman,"
or ship's "mate," Gerassim Pribylov by name, serving under the
direction and in the pay of one of the many companies engaged in
the fur-business at that time, was much moved and exercised in his
mind by the revelations of an old Aleutian shaman at Oonalashka,
who pretended to recite a legend of the natives, wherein he declared that certain islands in Bering Sea had long been known to
the Aleutes.

Pribylov * commanded a small sloop, the *St. George*, which
he employed for three successive years in constant, though fruitless, explorations to the northward of Oonalashka and Oonimak,
ranging over the whole of Bering Sea from the straits above. His
ill-success does not now seem strange as we understand the currents, the winds, and fogs of those waters. Why, only recently the
writer himself has been on one of the best-manned vessels that ever

---

* Pribylov, the discoverer of the Seal Islands, was a native of "old Russia." His father was one of the surviving sailors of the *St. Peter*, which
was wrecked, with Bering in command, November 4, 1741, on Bering Island.
The only reference which I can find to him is the vague incidental expressions,
used here and there throughout an extended series of lengthy Russian letters
published by Techmainov, as illustrative of the condition of affairs in regard
to the Russian American Company. Pribylov was, when cruising in 1783–86
for the rumored seal-grounds, merely the first mate of the sloop *St. George*.
The captain and part owner was one M. Subov, who was a member of a trading association then well organized in Alaska, and widely known as the "*Lay-
buidev Lastochin*" Company. It does not appear that Pribylov took any part
in the business of sealing other than that of remaining in charge of the company's vessels. He died while in discharge of these duties at Sitka, March,
1796, on his ship *The Three Saints*.

Pribylov himself called these islands of his discovery after Subov; but
the Russians then, and soon after unanimously, indicated the group by its
present well-deserved title, "*Ostrovie Pribylova*."

sailed from any port, provided with good charts and equipped with
all the marine machinery known to navigation, and that vessel has
hovered for nine successive days off the north point and around
St. Paul Island, sometimes almost on the reef, and never more
than ten miles away, without actually knowing where the island
was! So Pribylov did well, considering, when at the beginning of
the third summer's tedious search, in July, 1786, his old sloop ran
up against the walls of Tolstoi Mees, at St. George, and, though
the fog was so thick that he could see scarce the length of his vessel, his ears were regaled by the sweet music of seal-rookeries
wafted out to him on the heavy air. He knew then that he had
found the object of his search, and he at once took possession of
the island in the Russian name and that of his craft.

His secret could not long be kept. He had left some of his
men behind him to hold the island, and when he returned to Oonalashka they were gone; and ere the next season fairly opened, a
dozen vessels were watching him and trimming in his wake. Of
course, they all found the island, and in that year, July, 1787, the
sailors of Pribylov, on St. George, while climbing the bluffs and
straining their eyes for a relief-ship, descried the low coast and
scattered cones of St. Paul, thirty-six miles to the northwest of
them. When they landed at St. George, not a sign or a vestige of
human habitation was found thereon ; but during the succeeding
year, when they crossed over to St. Paul and took possession of
it, in turn, they were surprised at finding on the south coast of
that island, at a point now known as English Bay, the remains of a
recent fire. There were charred embers of driftwood and places
where grass had been scorched ; there was a pipe and a brass knife-
handle, which, I regret to say, have long passed beyond the cognizance of any ethnologist. This much appears in the Russian
records.

But, if we can believe the Aleutes in what they relate, the islands
were known to them long before they were visited by the Russians.
They knew and called them "Ateek," after having heard about
them. The legend of these people was as follows :

Eegad-dah-geek, a son of an Oonimak chief of the name of Ah-
kak-nee-kak, was taken out to sea in a bidarka by a storm, the
wind blowing strong from the south. He could not get back to
the beach, nor could he make any other landing, and was obliged
to run before the wind three or four days, when he brought up
13

on St. Paul Island, north from the land which he had been compelled to leave. Here he remained until autumn, and became acquainted with the hunting of different animals. Elegant weather
one day setting in, he saw the peaks of Oonimak. He then resolved to put to sea, and return to receive the thanks of his people
there, and after three or four days of travelling he arrived at Oonimak with "many otter tails and snouts." *

The Pribylov Islands lie in the heart of Bering Sea, and are
among the most insignificant landmarks known to that ocean.
They are situated one hundred and ninety-two miles north of Oonalashka, two hundred miles south of St. Matthews, and about the
same distance westward of Cape Newenham on the mainland.

The islands of St. George and St. Paul are from twenty-seven
to thirty miles apart, St. George lying southeastward of St. Paul.
They are far enough south to be beyond the reach of permanent
ice-floes, upon which polar bears would have made their way to the
islands, though a few of these animals were doubtless always present. They were also distant enough from the inhabited Aleutian
districts and the coast of the mainland to have remained unknown
to savage men. Hence they afforded the fur-seal the happiest
shelter and isolation, for their position seems to be such as to
surround and envelop them with fog-banks that fairly shut out the
sun nine days in every ten during the summer and breeding-season.

In this location ocean-currents from the great Pacific, warmer
than the normal temperature of this latitude, trending up from
southward, ebb and flow around the islands as they pass, giving rise
during the summer and early autumn to constant, dense, humid
fog and drizzling mists, which hang in heavy banks over the ground
and the sea-line, seldom dissolving away to indicate a pleasant day.
By the middle or end of October strong, cold winds, refrigerated
on the Siberian steppes, sweep down over the islands, carrying off

---

* Veniaminov says that he does feel inclined to believe this story, as the
peaks of Oonimak can be seen occasionally from St. Paul. I have no hesitation in saying that they were never observed by any mortal eye from the
Pribylov group. The wide expanse of water between these points, and the
thick, foggy air of Bering Sea, especially so at the season mentioned in this
story above, will always make the mountains of Oonimak invisible to the eye
from Saint Paul Island. A *mirage* is almost an impossibility. It may have
been much more probable if the date was a winter one.—Veniaminov: Zapieskie ob Oonalashkenskaho Otdayla, etc., 1842.

the moisture and clearing up the air. By the end of January, or early in February, they usually bring, by their steady pressure, from the north and northwest, great fields of broken ice, sludgy floes, with nothing in them approximating or approaching glacial ice. They are not very heavy or thick, but as the wind blows they compactly cover the whole surface of the sea, completely shutting in the land, and for months at a time hush the wonted roar of the surf. In the exceptionally cold seasons that succeed each other up there every four or five years, for periods of three and even four months—from December to May, and sometimes into June—the islands will be completely environed and ice-bound. On the other hand, in about the same rotation, occur exceptionally mild winters. Not even the sight of an ice-blink is recorded then during the whole winter, and there is very little skating on the shallow lakes and lagoons peculiar to St. Paul and St. George. This, however, is not often the case.

The breaking up of winter-weather and the precipitation of summer (for there is no real spring or autumn in these latitudes), usually commences about the first week in April. The ice begins to leave or dissolve at that time, or a little later, so that by the 1st or 5th of May, the beaches and rocky sea-margin beneath the mural precipices are generally clear and free from ice and snow, although the latter occasionally lies, until the end of July or the middle of August, in gullies and on leeward hill-slopes, where it has drifted during the winter. Fog, thick and heavy, rolls up from the sea, and closes over the land about the end of May. This, the habitual sign of summer, holds on steadily to the middle or end of October again.

The periods of change in climate are exceedingly irregular during the autumn and spring, so-called, but in summer a cool, moist, shady gray fog is constantly present. To this certainty of favored climate, coupled with the perfect isolation and an exceeding fitness of the ground, is due, without doubt, that preference manifested by the warm-blooded animals which come here every year, in thousands and hundreds of thousands to breed, to the practical exclusion of all other ground.*

I simply remark here, that the winter which I passed upon St.

---

* A large amount of information in regard to the climate of these islands has been collected and recorded by the signal service, United States Army, and similar observations are still continued by the agents of the Alaska Commercial Company.

Paul Island (1872-73) was one of great severity, and, according to the natives, such as is very seldom experienced. Cold as it was, however, the lowest marking of the thermometer was only 12° Fahr. below zero, and that lasted but a few hours during a single day in February, while the mean of that month was 18° above. I found that March was the coldest month. Then the mean was 12° above, and I have since learned that March continues to be the meanest month of the year. The lowest average of a usual winter ranges from 22° to 26° above zero; but these quiet figures are simply inadequate to impress the reader with this exceeding discomfort of a winter in that locality. It is the wind that tortures and cripples out-door exercise there, as it does on all the sea-coasts and islands of Alaska. It is blowing, blowing, from every point of the compass at all times; it is an everlasting succession of furious gales, laden with snow and sleety spiculæ, whirling in great drifts to-day, while to-morrow the wind will blow from a quarter directly opposite, and reverse its drift-building action of the day preceding.

Without being cold enough to suffer, one is literally confined and chained to his room from December until April by this Æolian tension. I remember very well that, during the winter of 1872-73, I was watching, with all the impatience which a man in full health and tired of confinement can possess, every opportunity to seize upon quiet intervals between the storms, in which I could make short trips along those tracks over which I was habituated to walk during the summer; but in all that hyemal season I got out but three times, and then only by the exertion of great physical energy. On a day in March, for example, the velocity of the wind at St. Paul, recorded by one of the signal-service anemometers, was at the rate of eighty-eight miles per hour, with as low a temperature as —4°! This particular wind-storm, with snow, blew at such a velocity for six days without an hour's cessation, while the natives passed from house to house crawling on all fours. No man could stand up against it, and no man wanted to. At a much higher temperature —say at 15° or 16° above zero—with the wind blowing only twenty or twenty-five miles an hour, it is necessary, when journeying, to be most thoroughly wrapped up so as to guard against freezing.

As I have said, there are here virtually but two seasons—winter and summer. To the former belongs November and the following months up to the end of April, with a mean temperature of 20° to 28°; while the transition of summer is but a very slight elevation of that temperature, not more than 15° or 20°. Of the summer months, July, perhaps, is the warmest, with an average temperature between 46° and 50° in ordinary seasons. When the sun breaks out through the fog, and bathes the dripping, water-soaked hills and flats of the island in its hot flood of light, I have known the thermometer to rise to 60° and 64° in the shade, while the natives crawled out of that fervent and unwonted heat, anathematizing its brilliancy and potency. Sunshine does them no good ; for, like the seals, they seem under its influence to swell up at the neck. A little of it suffices handsomely for both Aleutes and pinnipedia, to whom the ordinary atmosphere is much more agreeable.

It is astonishing how rapidly snow melts here. This is due, probably, to the saline character of the air, for when the temperature is only a single degree above freezing, and after several successive days in April or May, at 34° and 36°, grass begins to grow, even if it be under melting drifts, and the frost has penetrated the ground many feet below. I have said that this humidity and fog, so strongly and peculiarly characteristic of the Pribylov group, was due to the warmer ocean-currents setting up from the coast of Japan, trending to the Arctic through Bering's Straits, and deflected to the southward into the North Pacific, laving, as it flows, the numerous passes and channels of the great Aleutian chain ; but I do not think, nor do I wish to be understood as saying, that my observation in this respect warrants any conclusion as to so large a Gulf Stream flowing to the north, such as mariners and hydrographers recognize upon the Atlantic coast. I do not believe that there is anything of the kind equal to it in Bering Sea. I believe, however, that there is a steady set up to the northward from southward around the Seal Islands, which is continued through Bering's Straits, and drifts steadily off up to the northeast, until it is lost beyond Point Barrow. That this pelagic circulation exists, is clearly proven by the logs of the whalers, who, from 1845 to 1856, literally filled the air over those waters with the smoke of their "try-fires," and ploughed every square rod of that superficial marine area with their adventurous keels. While no two, perhaps, of those old whaling captains living to-day will agree as to the exact course of tides,* for Alaskan tides do not seem to obey any law, they all

---

* The rise and fall of tide at the Seal Islands I carefully watched one whole season at St. Paul. The irregularity, however, of ebb and flow is the

affirm the existence of a steady current, passing up from the south to the northeast, through Bering's Straits. The flow is not rapid, and is doubtless checked at times, for short intervals, by other causes, which need not be discussed here. It is certain, however, that there is warm water enough, abnormal to the latitude, for the evolution of those characteristic fog-banks, which almost discomfited Pribylov, at the time of his discovery of the islands, nearly one hundred years ago, and which have remained ever since.

Without this fog the fur-seal would never have rested there as he has done ; but when he came on his voyage of discovery, ages ago, up from the rocky coasts of Patagonia mayhap, had he not found this cool, moist temperature of St. Paul and St. George, he would have kept on, completed the circuit, and returned to those congenial antipodes of his birth.

Speaking of the stormy weather brings to my mind the beautiful, varied, and impressive nephelogical display in the heavens overhead here during October and November. I may say, without exaggeration, that the cloud-effects which I have witnessed from the bluffs of this little island, at this season of the year, surpass anything that I had ever seen before. Perhaps the mighty masses of cumuli, deriving their origin from warm exhalations out of the sea, and swelled and whirled with such rapidity, in spite of their appearance of solidity, across the horizon, owe their striking brilliancy of color and prismatic tones to that low declination of the sun due to the latitude. Whatever the cause may be, and this is not the place to discuss it, certainly no other spot on earth can boast of a more striking and brilliant cloud-display. In the season of 1865-66, when I was encamped on this same parallel of latitude in the mountains eastward of Sitka and the interior, I was particularly attracted by an exceeding brilliancy, persistency, and activity of the aurora ; but here on St. Paul, though I eagerly looked for its dancing light, it seldom appeared ; and when it did it was a sad disappointment, the exhibition always being insignificant as compared in my mind with its flashing of my previous experience. A quaint old writer, a hundred years ago, was describing Norway and its peo-

---

most prominent feature of the matter. The highest rise in the spring-tides was a trifle over four feet, while that of the neap-tides not much over two. Owing to the nature of the case, it is impossible to prepare a tidal calendar for Alaska, above the Aleutian Islands, which will even faintly foreshadow a correct registration in advance.

ple : he advanced what he considered a very plausible theory for the cause of the aurora; he cited an ancient sage, who believed that the change of winds threw the saline particles of the sea high into the air, and then by aërial friction, " fermentation " took place, and the light was evolved! I am sure that the saline particles of Bering Sea were whirled into the air during the whole of that winter of my residence there, but no "fermentation" occurred, evidently, since rarely indeed did the aurora greet my eyes. In the summer season there is considerable lightning ; you will see it streak its zigzag path mornings, evenings, and even noondays, but from the dark clouds and their swelling masses upon which it is portrayed no sound returns—a *fulmen brutum*, in fact. I remember hearing but one clap of thunder while in that country. If I recollect aright, and my Russian served me well, one of the old natives told me that it was no mystery, this light of the aurora, for, said he, " we all believe that there are fire-mountains away up toward the north, and what we see comes from their burning throats, mirrored back on the heavens."

The formation of these islands, St. Paul and St. George, was recent, geologically speaking, and directly due to volcanic agency, which lifted them abruptly, though gradually, from the sea-bed. Little spouting craters then actively poured out cinders and other volcanic breccia upon the table-bed of basalt, depositing below as well as above the water's level as they rose ; and subsequently finishing their work of construction through the agency of these spout-holes or craters, from which water-puddled ashes and tufa were thrown. Soon after the elevation and deposit of the igneous matter, all active volcanic action must have ceased, though a few half-smothered outbursts seem to have occurred very recently indeed, for on Bobroyia or Otter Island, six miles southward of St. Paul, is the fresh, clearly blown-out throat, with the fire-scorched and smoked, smooth, sharp-cut, funnel-like walls of a crater. This is the only place on the Seal Islands where there are any evidences of recent discharges from the crater of a volcano.

Since the period of the upheaval of the group under discussion, the sea has done much to modify and even enlarge the most important one, St. Paul, while the others, St. George and Otter, being lifted abruptly above the power of water and ice to carry and deposit sand, soil, and boulders, are but little changed from the condition of their first appearance.

The Russians tell a somewhat strange story in connection with Pribylov's landing. They say that both the islands were at first without vegetation, save St. Paul, where there was a small "talneek," or willow, creeping along on the ground; and that on St. George nothing grew, not even grass, except on the place where the carcasses of dead animals rotted. Then, in the course of time, both islands became covered with grass, a great part of it being of the sedge kind, *Elymus*. This record of Veniaminov, however, is scarcely credible; there are few, surely, who will not question the opinion that the seals antedated the vegetation, for, according to his own statements, these creatures were there then in the same immense numbers that we find them to-day. The vegetation on these islands, such as it is, is fresh and luxuriant during the growing season of June and July and early August, but the beauty and economic value of trees and shrubbery, of cereals and vegetables, are denied to them by climatic conditions. Still I am strongly inclined to believe that, should some of those hardy shrubs and spruce trees indigenous at Sitka or Kadiak, be transplanted properly to any of the southern hill-slopes of St. Paul most favored by soil, drainage, and bluffs, for shelter from saline gales, they might grow, though I know that, owing to the lack of sunlight, they would never mature their seed. There is, however, during the summer, a beautiful spread of grasses, of flowering annuals, biennials, and perennials, of gayly-colored lichens and crinkled mosses,* which have always afforded me great delight whenever I have pressed my way over the moors and up the hillsides of the rookeries.

There are ten or twelve species of grasses of every variety, from close, curly, compact mats to tall stalks—tussocks of the wild wheat, *Elymus arenaria*, standing in favorable seasons waist high—the "wheat of the north"—together with over one hundred varieties of annuals, perennials, sphagnum, cryptogamic plants, etc., all flourishing in their respective positions, and covering nearly every point of rock, tufa, cement, and sand that a plant can grow upon, with a living coat of the greenest of all greens—for there is not sunlight enough there to ripen any perceptible tinge of ochre-yellow into it—so green that it gives a deep blue tint to gray noonday shadows, contrasting pleasantly with the varying russets, reds,

---

* The mosses at Kamminista, St. Paul, are the finest examples of their kind on the islands ; they are very perfect, and many species are beautiful.

THE NORTH SHORE OF SAINT GEORGE

View of the Coast looking East from the North Rookery over to the Village and the Landing

lemon-yellows, and grays of the lichen-covered rocks, and the
brownish-purple of the wild wheat on the .sand-dune tracts in
autumn, together, also, with innumerable blue, yellow, pink, and
white phænogamous blossoms, everywhere interspersed over the
grassy uplands and sandy flats. Occasionally, on looking into the
thickest masses of verdure, our common wild violet will be found,
while the phloxes are especially bright and brilliant here. The
flowers of one species of gentian, *Gentiana verna* are very marked
in their beauty ; also those of a nasturtium, and a creeping pea-vine
on the sand-dunes. The blossom of a species of the pulse family
is the only one here that emits a positive, rich perfume ; the
others are more suggestive of that quality than expressive. The
most striking plant in all of a long list is the *Archangelica offici-
nalis*, with its tall seed-stalks and broad leaves, which grows first
in spring and keeps green latest in the fall. The luxuriant rhu-
barb-like stems of this umbellifer, after they have made their rapid
growth in June, are eagerly sought for by the natives, who pull
them and crunch them between their teeth with all the relish that
we experience in eating celery. The exhibition of ferns at Kamminista, St. Paul, during the summer of 1872, surpassed anything
that I ever saw : I recall with vivid detail the exceedingly fine display made by these luxuriant and waving fronds, as they reared
themselves above the rough interstices of that rocky ridge. From
the fern roots, and those of the gentian, the natives here draw their
entire stock of vegetable medicines. This floral display on St. Paul
is very much more extensive and conspicuous than that on St.
George, owing to the absence of any noteworthy extent of warm
sand-dune country on the latter island.

When an unusually warm summer passes over the Pribylov
group, followed by an open fall and a mild winter, the elymus
ripens its seed, and stands like fields of uncut grain in many places
along the north shore of St. Paul and around the village, the snow
not falling enough to entirely obliterate it ; but it is seldom allowed
to flourish to that extent. By the end of August and the first week
of September of normal seasons, the small edible berries of *Empetrum nigrum* and *Rubus chamœmorus* are ripe. They are found in
considerable quantities, especially at "Zapadnie," on both islands,
and, as everywhere else throughout circumpolar latitudes, the
former is small, watery, and dark, about the size of an English or
black currant ; the other resembles an unripe and partially decayed

raspberry. They are, however, keenly relished by the natives, and even by American residents, being the only fruit growing upon the islands. Perhaps no one plant that flowers on the Seal Islands is more conspicuous and abundant than is the *Saxifraga oppositofo-lia ;* it grows over all localities, rank and tall in rich locations, to stems scarcely one inch high on the thin, poor soil of hill summits and sides, densely cespitose, with leaves all imbricated in four rows ; and flowers almost sessile. I think that at least ten well-defined species of the order *Saxifragaceæ* exist on the Pribylov group. The *Ranunculaceæ* are not so numerous ; but, still, a buttercup grow-ing in every low slope where you may chance to wander is always a pleasant reminder of pastures at home ; and, also, a suggestion of the farm is constantly made by the luxuriant inflorescence of the wild mustard (*Cruciferæ*). The chickweeds (*Caryophyllaceæ*) are well represented, and also the familiar yellow dandelion, *Taraxacum palustre.* Many lichens (*Lichenes*) and soft mosses (*Musci*) are in their greatest exuberance, variety, and beauty here ; and myriads of golden poppies (*Papaveraceæ*) are nodding their graceful heads in the sweeping of the winds—the first flowers to bloom, and the last to fade.

The chief economic value rendered by the botany of the Pribylov Islands to the natives is an abundance of the basket-making rushes (*Juncaceæ*), which the old "barbies" gather in the margins of many of the lakes and pools.

The only suggestion of a tree* found growing on the Pribylov

---

* That spruce-trees can be made to live transplanted from indigenous lo-calities to the barren slopes of the Aleutian Islands, has been demonstrated ; but in living these trees do nothing else, and scarcely grow to any appreciable degree. A few spruce were transferred to Oonalashka when Veniaminov was at work there in 1830-35. They are still standing and keep green, but the change which such a long lapse of time should produce by growth has been as difficult to determine as it is to find evidence of increased altitude to the mountains around them since these Sitkan trees were planted with pious hope at their feet fifty years ago. Though I can readily understand why the sal-mon-berries of Oonalashka should not do well on the Seal Islands (though I think they would at the Garden Cove of St. George), yet I believe that the huckleberries of that section would thrive at many places if carefully transplanted to these localities : the southern slopes of Cemetery Ridge at Zapadnie ; the southern slopes of Telegraph Hill, and eastern fall of Tolstoi peninsula down to the shore of the lagoon. They might also do well set out at picked places around the Big Lake and on Northeast Point, around the little lake thereon. If these bushes really throve here, they would be the means

group is the hardy "talneek" or creeping willow; there are three
species of the genus *Salix* found here, viz., *reticulata, polaris,* and
arctica ; the first named is the most common and of largest growth;
it progresses exactly as a cucumber-vine does in our gardens; as
soon as it has made from the seed a growth of six inches or a foot
upright from the soil, then it droops over and crawls along prostrate upon the earth, rocks, and sphagnum; some of the largest talneek trunks will measure eight or ten feet in decumbent length
along the ground, and are as large around the stump as an average
wrist of man. The usual size, however, is very, very much less;
while the stems of *polaris* and *arctica* scarcely ever reach the diameter of a pencil case, or the procumbent length of two feet.

Although *Rubus chamæmorus* is a tree-shrub, and is found here
very commonly distributed, yet it grows such a slender diminutive
bush, that it gives no thought whatever of its being anything of
the sort. Herbs, grasses, and ferns tower above it on all sides.

The fungoid growths on the Pribylov Islands are abundant and
varied, especially in and around the vicinity of the rookeries and
the killing-grounds. On the west slope of the Black bluffs at St.
Paul the mushroom, *Agaricus campestris,* was gathered in the season of 1872 by the natives, and eaten by one or two families in the
village, who had learned to cook them nicely from the Russians.
These Seal Island mushrooms have deeper tones of pink and purple-
red in their gills than do those of my gathering in the States. I
kicked over many large spherical "puff-balls" (*Lycoperda*) in my
tundra walks; myriads of smaller ones (*Lycoperdon cinercum ?*) cover
patches near the spots where carcasses have long since rotted, together with a pale gray fungus (*Agaricus fimiputris*), exceedingly
delicate and frosted exquisitely. Some ligneous fungi (*Clavaria*),
will be found attached to the decaying stems of *Salix reticulata*
(creeping willows). The irregularity of the annual growing of the
agarics, and their rapid growth when they do appear, makes their

of adding greatly to the comfort of the inhabitants ; for the Oonalashka huckle-
berry is an exceeding pleasant, juicy fruit, large and well adapted for canning
and preserving. Having less sunshine here than at Illoolook, it may not ripen
up as well flavored, but will, I think, succeed. The roots of the bushes when
brought up from Oonalashka in April or early May should be kept moist by
wet-moss wrappings from the moment they are first taken up until they are
reset, with the tops well pruned back, on the Pribylov Islands. The experiment is surely worth all the trouble of making, and I hope it will be done.

determination excessively difficult; they are as unstable in their visits as are several of the *Lepidoptera*. The cool humidity of climate during the summer season on the Pribylov Islands is especially adapted to that mysterious, but beautiful growth of these plants—the apotheosis of decay. The coloring of several varieties is very bright and attractive, shading from a purplish-scarlet to a pallid white.

A great many attempts have been made, both here and at St. George, to raise a few of the hardy vegetables. With the exception of growing lettuce, turnips, and radishes on the Island of St. Paul, nothing has been or can be done. On the south shore of St. George, and at the foot of a mural bluff, is a little patch of ground less in area than one-sixteenth of an acre, which appears to be so drained and so warmed by the rarely-reflected sunlight from this cliff, every ray of which seems to be gathered and radiated from the rocks, as to allow the production of fair turnips; and at one season there were actually raised potatoes as large as walnuts. Gardening, however, on either island involves so much labor and so much care, with so poor a return, that it has been discontinued. It is a great deal better, and a great deal easier, to have the "truck" come up once a·year from San Francisco on the steamer.

There is one comfort which nature has vouchsafed to civilized man on these islands. There are very few indigenous insects. A large flesh-fly, *Bombylius major*, appears during the summer and settles in a striking manner on the backs of quiet, loafing natives, or strings itself in rows of millions upon the long grass-blades which flourish about the killing-grounds, especially on the leaf-stalks of an elymus, causing this vegetation, over the whole slaughtering-field and vicinity, to fairly droop to earth as if beaten down by a tornado of wind and rain. It makes the landscape look as though it had moulded over night, and the fungoid spores were blue and gray. Our common house-fly is not present; I never saw one while I was up there. The flesh-flies which I have just mentioned never came into the dwellings unless by accident: the natives say they do not annoy them, and I did not notice any disturbance among the few animals which the resident company had imported for beef and for service.

Then, again, this is perhaps the only place in all Alaska where man, primitive and civilized, is not cursed by mosquitoes. There are none here. A gnat, that is disagreeably suggestive of the real enemy just referred to, flits about in large swarms, but it is inoffensive, and seeks shelter in the grass. Several species of beetles are also numerous here. One of them, the famous green and gold "carabus," is exceedingly common, crawling everywhere, and is just as bright in the rich bronzing of its wing-shields as are its famous prototypes of Brazil. One or two species of *Itemosa*, a *Cymindis*, several representatives of the *Aphidiphaga*, one or two of *Dytiscidæ*, three or four *Cicindelidæ*—these are nearly all that I found. A single dragon-fly, *Perla bicaudata*, flitted over the lakes and ponds of St. Paul. The familiar form to our eyes, of the bumble-bee, *Bombus borealis*, passing from flower to flower, was rarely seen ; but a few are here resident. The *Hydrocorisæ* occur in great abundance, skipping over the water in the lakes and pools everywhere, and a very few species of butterflies, principally the yellow *Nymphalidæ*, are represented by numerous individuals.

Aside from the seal-life on the Pribylov Islands, there is no indigenous mammalian creature, with the exception of the blue and white foxes, *Vulpes lagopus*,* and a lemming, *Myodes obensis*. The latter is restricted, for some reason or other, to the Island of St. George, where it is, or at least was, in 1874, very abundant. Its

---

* Blue foxes were also, and are, natives of the Commander Islands. Steller describes their fearlessness when the shipwrecked crew of the *St. Peter* landed there, November 6, 1741.

In regard to these foxes the Pribylov natives declare that when the islands were first occupied by their ancestors, 1786–87, the fur was invariably *blue ;* that the present smoky blue, or ashy indigo color, is due to the coming of white foxes across on the ice from the mainland to the eastward. The white-furred *vulpes* is quite numerous on the islands to-day. I should judge that perhaps one-fifth of the whole number were of this color ; they do not live apart from the blue ones, but evidently breed "in and in." I notice that Veniaminov, also, makes substantially the same statement ; only differing by charging this deterioration of the blue foxes' fur to a deportation from outside of red foxes, on ice-floes, and adds that the natives always hunted down these "krassnie peeschee" as soon as their presence was known ; hence my inability, perhaps, to see any sign of their posterity in 1872–76.

The presence of these animals on the Pribylov Islands is a real source of happiness to the natives, especially so to the younger ones. The little pup-foxes make pets and playfellows for the children, while hunting the adults during the winter gives wholesome employment to the mind and body of the native who does so. They are trapped in common dead-falls, steel spring-clips, or beaver traps, and shot. A very large portion of the gossip on the islands is in relation to this business.

burrows and paths, under and among the grassy hummocks and mossy flats, checkered every square rod of land there covered with this vegetation. Although the Island of St. Paul is but twenty-nine or thirty miles to the northwest, not a single one of these active, curious little animals is found there, nor could I learn from the natives that it had ever been seen there. The foxes are also restricted to these islands; that is, their kind, which are not found elsewhere, except the stray examples on St. Matthew seen by myself, and those which are carefully domesticated and preserved at Attoo, the extreme westernmost land of the Aleutian chain. These animals find comfortable holes for their accommodation and retreat on the Seal Islands, among the countless chinks and crevices of the basaltic formation. They feed and grow fat upon sick and weakly seals, also devouring many of the pups, and they vary this diet by water-fowl and eggs* during the summer, returning for their subsistence during the long winter to the bodies of seals upon the breeding-grounds and the skinned carcasses left upon the killing-fields. Were they not regularly hunted from December until April, when their fur is in its prime beauty and condition, they would swarm like the lemming on St. George, and perhaps would soon be obliged to eat one another. The natives, however, thin them out by incessant trapping and shooting during the period when the seals are away from the islands.

The Pribylov group is as yet free from rats; at least none have

---

* The temerity of the fox is wonderful to contemplate, as it goes on a full run or stealthy tread up and down and along the faces of almost inaccessible bluffs, in search of old and young birds and their nests and eggs, for which the "peeschee" have a keen relish. The fox always brings an egg up in its mouth, and, carrying it back a few feet from the brink of the precipice, leisurely and with gusto breaks the larger end and sucks the contents from the shell. One of the curious sights of my notice, in this connection, was the sly, artful, and insidious advances of Reynard at Tolstoi Mees, St. George, where, conspicuous and elegant in its fluffy white dress, it cunningly stretched on its back as though dead, making no sign of life whatever, save to gently hoist its thick brush now and then; whereupon many dull, curious sea-birds, *Graculus bicristatus*, in their intense desire to know all about it, flew in narrowing circles overhead, lower and lower, closer and closer, until one of them came within the sure reach of a sudden spring and a pair of quick snapping jaws, while the gulls and others, rising safe and high above, screamed out in seeming contempt for the struggles of the unhappy "shag," and rendered hideous approbation.

got off from the ships. There is no harbor on either of these islands, and vessels lie out in the roadstead, so far from land that those pests do not venture to swim to the shore. Mice were long ago brought to shore in ships' cargoes, and they are a great nuisance to the white people as well as the natives throughout the islands. Hence cats also are abundant. Nowhere, perhaps, in the wide world are such cats to be seen as these. The tabby of our acquaintance, when she goes up there and lives upon the seal-meat spread everywhere under her nose, is metamorphosed, by time of the second generation, into a stubby feline ball. In other words, she becomes thickened, short, and loses part of the normal length of her tail; also her voice is prolonged and resonant far beyond the misery which she inflicts upon our ears here. These cats actually swarm about the natives' houses, never in them much, for only a tithe of their whole number can be made pets of; but they do make night hideous beyond all description. They repair for shelter often to the chinks of precipices and bluffs; but although not exactly wild, yet they cannot be approached or cajoled. The natives, when their sluggish wits are periodically aroused and thoroughly disturbed by the volume of cat-calls in their village, sally out and by a vigorous effort abate the nuisance for the time being. Only the most fiendish caterwauling will or can arouse this Aleutian ire.

On account of the severe climatic conditions it is, of course, impracticable to keep stock here with any profit or pleasure. The experiment has been tried faithfully. It is found best to bring beef-cattle up in the spring on the steamer, turn them out to pasture until the close of the season in October and November, and then, if the snow comes, to kill them and keep the carcasses refrigerated until consumed. Stock cannot be profitably raised here; the proportion of severe weather annually is too great. From three to perhaps six months of every year they require feeding and watering, with good shelter. To furnish an animal with hay and grain up there is a costly matter, and the dampness of the growing summer season on both islands renders hay-making impracticable. Perhaps a few head of hardy Siberian cattle might pick up a living on the north shore of St. Paul, among the grasses and sand-dunes there, with nothing more than shelter and water given them; but they would need both of these attentions. Then the care of them would hardly return expenses, as the entire grazing-ground could not support any number of animals. It is less than two square miles in extent, and half of this area is unproductive. Then, too, a struggle for existence would reduce the flesh and vitality of these cattle to so low an ebb that it is doubtful whether they could be put through another winter alive, especially if severe. I was then and am now strongly inclined to think that if a few of those Siberian reindeer could be brought over to St. Paul and to St. George they would make a very successful struggle for existence, and be a source of a good supply, summer and winter, of fresh meat for the agents of the Government and the company who may be living upon the islands. I do not think that they would be inclined to molest or visit the seal-grounds; at least, I noticed that the cattle and mules of the company running loose on St. Paul were careful never to poke around on the outskirts of a rookery, and deer would be more timid and less obtrusive than our domesticated animals. But I did notice on St. George that a little squad of sheep, brought up and turned out there for a summer's feeding, seemed to be so attracted by the quiet calls of the pups on the rookeries that they were drawn to and remained by the seals without disturbing them at all, to their own physical detriment, for they lost better pasturage by so doing. The natives of St. Paul have a strange passion for seal-fed pork, and there are quite a large number of pigs on the Island of St. Paul and a few on St. George. Such hogs soon become entirely carnivorous, living, to the practical exclusion of all other diet, on the carcasses of seals.

Chickens are kept with great difficulty. In fact, it is only possible to save their lives when the natives take them into their own rooms or keep them over their heads, in their dwellings, during winter.

While the great exhibition of pinnipedia preponderates over every other feature of animal life at the Seal Islands, still there is a wonderful aggregate of ornithological representation thereon. The spectacle of birds nesting and breeding, as they do on St. George Island, to the number of millions, flecking those high basaltic bluffs of its shore-line, twenty-nine miles in length, with color-patches of black, brown, and white, as they perch or cling to the mural cliffs in the labor of incubation, is a sight of exceeding attraction and constant novelty. It affords a naturalist an opportunity of a lifetime for minute investigation into all the details of the reproduction of these vast flocks of circumboreal water-fowl. The Island of St. Paul, owing to the low character of its shore-line, a large pro-

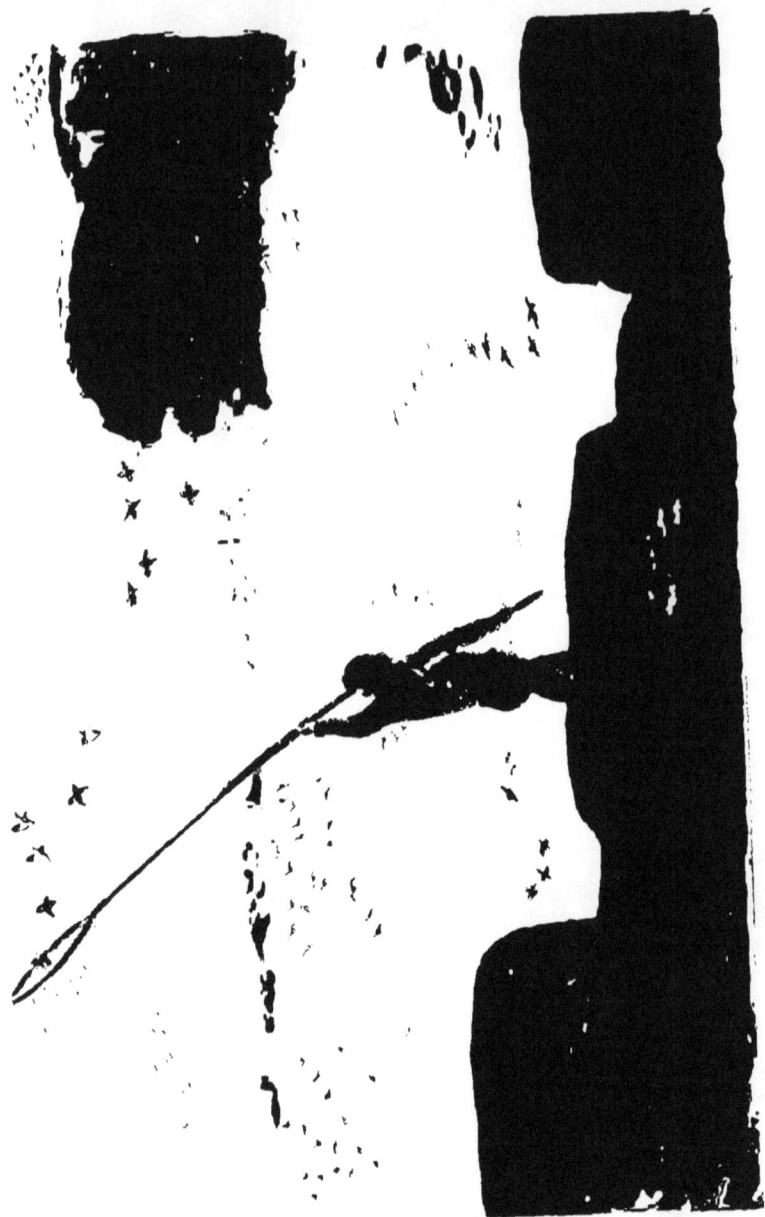

portion of which is but slightly elevated above the sea and is sandy, is not visited and cannot be visited by such myriads of birds as are seen at St. George ; but the small rocky Walrus Islet is fairly covered with sea-fowls, and the Otter Island bluffs are crowded by them to their utmost capacity of reception. The birds string themselves anew around the bluffs with every succeeding season, like endless ribbons stretched across their rugged faces, while their numbers are simply countless. The variety is not great, however, in these millions of breeding-birds. It consists of only ten or twelve names, and the whole list of birds belonging to the Pribylov Islands, stragglers and migratory, contains but forty species. Conspicuous among the last-named class is the robin, a straggler which was brought from the mainland, evidently against its own effort, by a storm or a gale of wind, which also brings against their will the solitary hawks, owls, and waders occasionally noticed here.

After the dead silence of a long ice-bound winter, the arrival of large flocks of those sparrows of the north, the "choochkies," *Phaleris microceros*, is most cheerful and interesting. These plump little auks are bright, fearless, vivacious birds, with bodies round and fat. They come usually in chattering flocks on or immediately after May 1st, and are caught by the people with hand-scoops or dip-nets to any number that may be required for the day's consumption, their tiny, rotund forms making pies of rare savory virtue, and being also baked and roasted and stewed in every conceivable shape by skilful cooks. Indeed, they are equal to the reed-birds of the South. These welcome visitors are succeeded along about July 20th by large flocks of fat, red-legged turn-stones, *Strepsilas interpres*, which come in suddenly from the west or north, where they have been breeding, and stop on the islands for a month or six weeks, as the case may be, to feed luxuriantly upon the flesh-flies, which we have just noticed, and their eggs. These handsome birds go in among the seals, familiarly chasing the flies, gnats, etc. They are followed as they leave in September by several species of jack-snipe and a plover, *Tringa* and *Charadrius*. These, however, soon depart, as early as the end of October and the beginning of November, and then winter fairly closes in upon the islands. The loud, roaring, incessant seal-din, together with the screams and darkening flight of innumerable water-fowl, are replaced in turn again by absolute silence, marking out, as it were, in lines of sharp and vivid contrast, summer's life and winter's death.

14

The author of that quaint old saying, "Birds of a feather flock together," might well have gained his inspiration had he stood under the high bluffs of St. George at any season, prehistoric or present, during the breeding of the water-birds there, where myriads of croaking murres and flocks of screaming gulls darken the light of day with their fluttering forms, and deafen the ear with their shrill, harsh cries as they do now, for music is denied to all those birds of the sea. Still, in spite of the apparent confusion, he would have taken cognizance of the fact that each species had its particular location and kept to its own boundary, according to the precision of natural law. The dreary expanse and lonely solitudes of the North owe their chief enlivenment, and their principal attractiveness for man, to the presence of those vast flocks of circumboreal water-fowl, which repair thither annually.

Over fifteen miles of the bold, basaltic, bluff line of St. George Island is fairly covered with nesting gulls (*Rissa*) and "arries" (*Uria*), while down in the countless chinks and holes over the entire surface of the north side of this island millions of "choochkies" (*Simorhyncus microceros*) breed, filling the air and darkening the light of day with their cries and fluttering forms. On Walrus Islet the nests of the great white gull of the north (*Larus glaucus*) can be visited and inspected, as well as those of the sea-parrot or puffin (*Fratercula*), shags or cormorants (*Graculus*), and the red-legged kittiwake (*Larus brevirostris*). These birds are accessible on every side, can be reached, and afford the observer an unequalled opportunity of taking due notice of them through the breeding-season of their own, as it begins in May and continues until the end of September.

Not one of the water-birds found on and around the islands is exempted from a place in the native's larder; even the delectable "oreelie" are unhesitatingly eaten by the people, and indeed these birds furnish, during the winter season in especial, an almost certain source of supply for fresh meat. But the heart of the Aleut swells to its greatest gastronomic happiness when he can repair, in the months of June and July, to the basaltic cliffs of St. George, or the lava table-bed of Walrus Islet, and lay his grimy hands on the gayly-colored eggs of the "arrie" (*Lomvia arra*); and if he were not the most improvident of men, instead of taking only enough for the day, he would lay up a great store for the morrow, but he never does. On the occasion of one visit, and my first one there, July 5, 1872, six men loaded a bidarrah at Walrus Islet, capable of carrying four tons, exclusive of our crew, down to the water's edge with eggs, in less than three working hours.*

During winter months these birds are almost wholly absent, especially so if ice-floes shall have closed in around the islands ; then there is nothing of the feathered kind save a stupid shag (*P. bicristatus*) as it clings to the leeward cliffs, or the great burgomaster gull, which sweeps in circling flight high overhead ; but, early in May they begin to make their first appearance, and they come up from the sea overnight, as it were, their chattering and their harsh carolling waking the natives from slothful sleeping, which, however, they gladly break, to seize their nets and live life anew, as far as eating is concerned. The stress of severe weather in the winter months, the driving of the snow "boorgas," and the floating ice-fields closing in to shut out the open water, are cause enough for a disappearance of all water-fowl, *pro tem.*

Again, the timid traveller here is delighted ; he has been re-

---

* Using the egg of our domestic fowl, the hen, as a standard, the following note made in regard to the size and quality of the eggs of the sea-birds of these waters may not be uninteresting to many. When daily served on St. George, during June and July, with eggs of indigenous sea-fowl, I recorded my gastronomic comparisons which occurred then as I ate them. Here follows a re-capitulation :

Fresh-laid eggs of "lupus," or *F. glacialis :* Best eggs known to the islands ; can be soft-boiled or fried, etc., and are as good as our own hens' eggs ; the yolk is light and clear ; the size thereof is in shape and bulk like a duck's egg ; it has a white shell. Season : June 1st to 15th, inclusive ; scarce on St. Paul, and not abundant on St. George.

Fresh-laid eggs of " arrie," or *L. arra :* Very good ; can be soft-boiled or fried ; are best scrambled ; yolks are dark ; no strange taste whatever to them ; pyriform in shape ; large as a goose-egg ; shell gayly-colored ; they are exceedingly abundant on Walrus Island and St. George ; tons of them. Season : June 25th to July 10th, inclusive.

Fresh-laid eggs of gulls, *Laridæ :* Perceptibly strong ; cannot be relished unless in omelettes ; yolks very dark ; size and shape of our hen's egg ; shell dark, clay-colored ground, mottled. Season : June 5th to July 20th, inclusive ; they are in moderate supply only. The other eggs in the list, such as those of the "choochkie," the "shag," and the several varieties of water-fowl which breed here, are never secured in sufficient quantity to be of any consideration as articles of diet. It is, perhaps, better that a scarcity of their kind continue, judging from the strong smack of the choochkie's, the repulsive taint of the shag's, and the "twang" of the sea-parrot's, all of which I tasted as a matter of investigation.

lieved of the great Alaskan curse of mosquitoes: he also walks the
moors and hillsides secure in never finding a reptile of any sort
whatever—no snakes, no lizards, no toads or frogs—nothing of the
sort to be found on the Seal Islands.

Fish are scarce in the vicinity of these islands. Only a few representatives of those families which can secrete themselves with rare
cunning are safe in visiting the Pribylovs in summer. Naturally
enough, the finny tribes avoid the seal-churned waters for at least
one hundred miles around. Among a few specimens, however, which

Aleutes catching Halibut, Akootan Pass, Bering Sea.

I collected, three or four species new to natural science were found,
and have since been named by experts in the Smithsonian Institution.

The presence of such great numbers of amphibian mammalia
about the waters during five or six months of every year renders
all fishing abortive, and unless expeditions are made seven or eight
miles at least from the land, unless you desire to catch large halibut, it is a waste of time to cast your line over the gunwale of the
boat. The natives capture "poltoos" or halibut, *Hippoglossus vul-
garis*, within two or three miles of the reef-point on St. Paul and
the south shore during July and August. After this season the
weather is usually so stormy and cold that fishermen venture no
more until the ensuing summer. *

---

* The St. George natives have caught codfish just off the Tolstoi Head
early in June; but it is a rare occurrence. By going out two or three miles

With regard to the *Mollusca* of the Pribylov waters, the characteristic forms of *Toxoglossata* and *Rhachiglossata* peculiar to this north latitude are most abundant ; of the *Cephalopoda* I have seen only a species of squid, *sepia*, or *loligo*. The clustering whelks (*Buccinum*) literally conceal large areas of the boulders on the beaches here and there. They are in immense numbers, and are crushed under your foot at almost every step when you pass over long reaches of rocky shingle at low tide. A few of the *Neptunea* are found, and the live and dead shells of *Limacina* are in great abundance wherever the floating kelp-beds afford them shelter.

On land a very large number of shells of the genera *Succinea* and *Pupa* abound all over the islands. On the bluffs of St. George, just over Garden Cove, I gathered a beautiful *Helix*.

The little fresh-water lakes and ponds contain a great quantity of representatives of the characteristic genera *Planorbis*, *Melania*, *Limnea*, and that pretty little bivalve, the *Cyclas*.

Of the *Crustacea*, the *Annelidæ*, and *Echinodermata*, there is abundant representation here. The sea-urchins, "repkie" of the natives, are eagerly sought for at low tide and eaten raw by them. The arctic sea-clam, *Mya truncata*, is once in a long time found here (it is the chief food of the walrus of Alaska), and the species of *Mytilus*, the mussels, so abundant in the Aleutian archipelago, are almost absent here at St. Paul and only sparingly found at St. George. Frequently the natives have brought a dish of sea-urchins' viscera for our table, offering it as a great delicacy. I do not think any of us did more than to taste it. The native women are the chief hunters for echinoidæ, and during the whole spring and summer seasons they will be seen at both islands, wading in the pools at low water, with their scanty skirts high up, eagerly laying possessive hands upon every "bristling egg" that shows itself. They

---

from the village at either island during July and August the native fisherman usually captures large halibut—not in abundance, however. The St. Paul people, as well as their relatives on St. George, fish in small "two-hole" bidarkies. They go out together in squads of four to six. One man alone in the kyack is not able to secure a "bolshoi poltoos." The method, when the halibut is hooked, is to call for your nearest neighbor in his bidarka, who paddles swiftly up. You extend your paddle to him, retaining your own hold, and he grasps it, while you seize his in turn, thus making it impossible to capsize, while the large and powerfully struggling fish is brought to the surface between the canoes and knocked on the head. It is then towed ashore and carried in triumph to its lucky captor's house.

vary this search by poking, with a short-handled hook, into holes
and rocky crevices for a small cottoid fish, which is also found here
at low water in this manner. Specimens of this cottoid which I
brought down declared themselves as representatives of a new departure from all other recognized forms in which the sculpin is
known to sport; hence the name, generic and specific. The "sand-
cake," *echinarachinus*, is also very common here.

By May 28th to the middle of June a fine table-crab, large, fat,
and sweet, with a light, brittle shell, is taken while it is skurrying
in and out of the lagoon as the tide ebbs and flows. It is the best-
flavored crustacean known to Alaskan waters. They are taken nowhere else at St. Paul, and when on St. George I failed to see one.
I am not certain as to the accuracy of the season of running, viz. :
May 28th to June 15th, inasmuch as one of my little note-books
on which this date is recorded turns out to be missing at the present writing, and I am obliged to give it from memory. The only
economic shell-fish which the islands afford is embodied in this
Chionoecetes opilio (?). The natives affirm an existence of mussels
here in abundance when the Pribylov group was first discovered;
but now only a small supply of inferior size and quality is to be
found.

With reference to the jelly-fishes, *Medusæ*, which are so abundant in the waters around these islands, their exceeding number and
variety and beauty startled and enchanted me. An enormous aggregate of these creatures, some of them exquisitely delicate and
translucent, ride in and out of the lagoon at St. Paul when the
spring-tides flow and ebb. Myriads of them are annually stranded,
to decay on the sandy flats of this estuary.

As to sea-weeds, or mosses,—the extent, luxuriance, variety, and
beauty of the algæ forests of those waters of Bering Sea which
lave the coasts of the Pribylov Islands, call for more detail of description than space in this volume will allow, since anything like a
fair presentation of the subject would require the reproduction of
my water-colored drawings. After all heavier gales, especially the
southeasters in October, if a naturalist will take the trouble to
walk the sand-beach between Lukannon and northeast point of St.
Paul Island, he will be rewarded by the memorable sight. He
will find thrown up by the surf a vast windrow of kelp along the
whole eight or ten miles of this walk—heaped, at some spots, nearly
as high as his head ; the large trunks of *Melanospermæ*, the small,

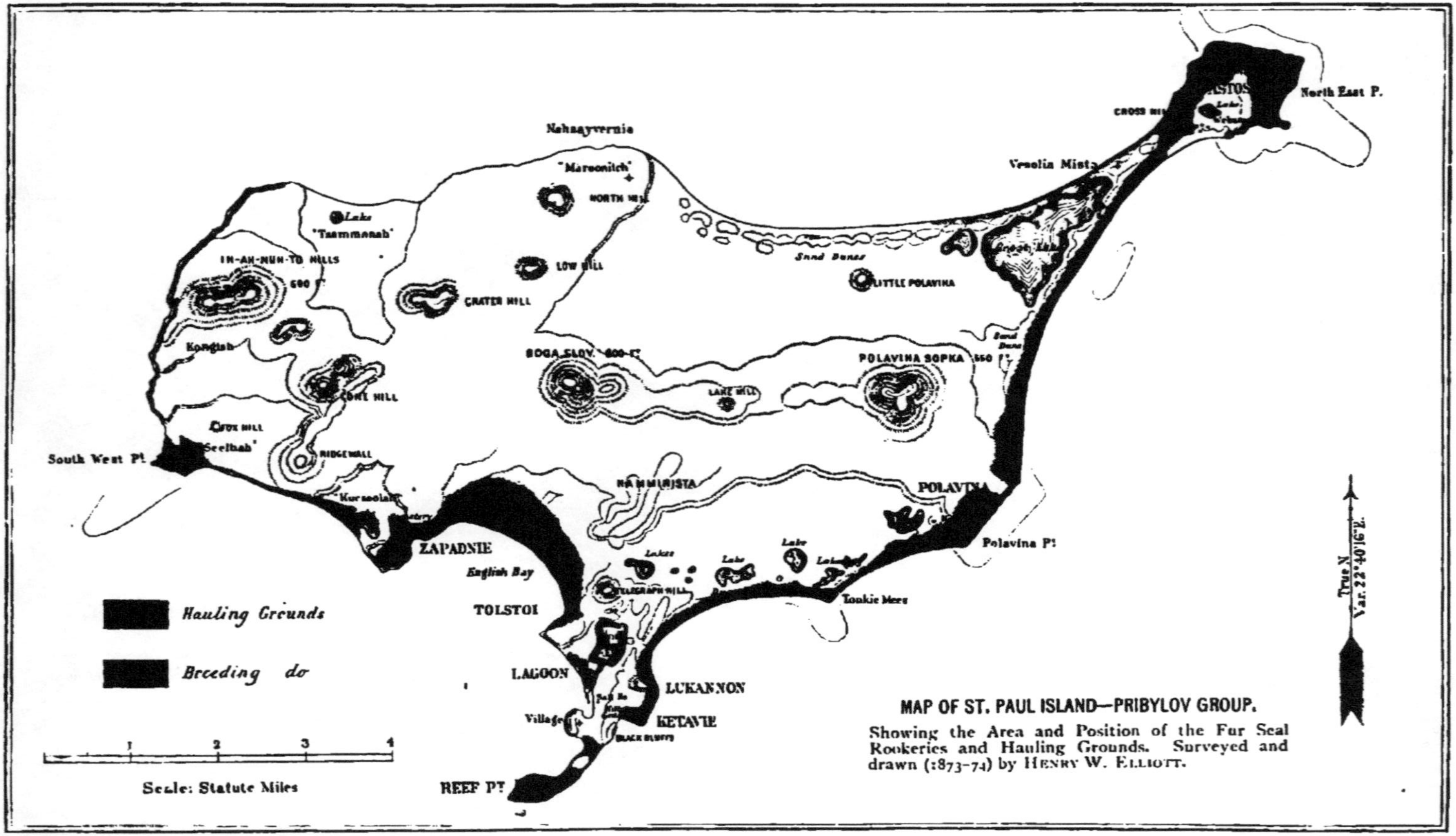

MAP OF ST. PAUL ISLAND—PRIBYLOV GROUP.

Showing the Area and Position of the Fur Seal Rookeries and Hauling Grounds. Surveyed and drawn (1873-74) by HENRY W. ELLIOTT.

True N.
Var. 22° 40.16′ E.

Hauling Grounds

Breeding do

Scale: Statute Miles

North East P.
CROSS HILL
CASTOR
Vrooia Mints
LITTLE POLAVINA
Sand Dunes
POLAVINA SOPKA 560
POLAVINA
Polavina P.
Tonkie Mees
Lake
Lake
Nahaevernie
"Marovnia"
NORTH HILL
LOW HILL
CRATER HILL
BOGASLAV 600 FT.
LAKE HILL
Lake
Lakes
HAMILISTIA
LUKANNON
KETAVIE
Lake "Tammanah"
IN-AH-NUH-TO HILLS 660
Kongoss
CONE HILL
RIDGE HILL
BIRCHWILL
DUN HILL Seethah
Kar seahia
ZAPADNIE
English Bay
TOLSTOI
LAGOON
Village
REEF P.
South West Pt.

but brilliant red and crimson fronds of *Rhodospermœ* interwoven with the emerald-green leaves of the *Chlorospermœ*. The first-named group is by far the most abundant, and upon its decaying, fermenting brown and ochre heaps, he will see countless numbers of a buccinoid whelk, and a limnaea, feeding as they bore or suck out myriads of tiny holes in the leaf-fronds of the strong growing species. *Actinia* or sea-anemones, together with asteroids or star-fishes, *Discophorœ* or jelly-fishes, are also interwoven and heaped up with the "kapoosta" or sea-cabbages just referred to ; also, many rosy "sea-squirts," yellow "cucumbers," and other forms of *Holo-thuridœ*.

On the old killing-fields, on those spots where the sloughing carcasses of repeated seasons have so enriched the soil as to render it like fire to most vegetation, a silken green *Confervœ* grows luxu-riantly. This terrestrial algoid covering appears here and there, on these grounds, like so many door-mats of pea-green wool. That confervoid flourishes only on those spots where nothing but pure decaying animal matter is found. · An admixture of sand or earth will always supplant it by raising up instead those strong growing grasses which I have alluded to elsewhere, and which constitute the chief botanical life of the killing-grounds.

In order that the reader can follow easily the narrative of that remarkable life-system which is conducted by the fur-seal as it an-nually rests and breeds upon the Pribylov group, I present a care-ful chart of each island and the contiguous islets, which are the only surveys ever made upon the ground. The reader will observe, as he turns to these maps, the striking dissimilarity which exists be-tween them, not only in contour but in physical structure, the Island of St. Paul being the largest in superficial area, and receiv-ing a vast majority of the *Pinnipedia* that belong to both. As it lies in Bering Sea to-day, this island is, in its greatest length, be-tween northeast and southwest points, thirteen miles, air-line ; and a little less than six at points of greatest width. It has a super-ficial area of about thirty-three square miles, or twenty-one thou-sand one hundred and twenty acres, of diversified, rough and rocky uplands, rugged hills, and smooth, volcanic cones, which either set down boldly to the sea or fade out into extensive wet and mossy flats, passing at the sea-margins into dry, drifting, sand-dune tracts. It has forty-two miles of shore-line, and of this coast sixteen and a half miles are hauled over by fur-seals *en masse*. At the time of its first upheaval above the sea, it doubtless presented the appearance of ten or twelve small rocky, bluffy islets and points, upon some of which were craters that vomited breccia and cinders, with little or no lava overflowing. Active plutonic agency must have soon ceased after this elevation, and then the sea round about commenced a work which it is now engaged in—of building on to the skeleton thus created; and it has progressed to-day so thoroughly and successfully in its labor of sand-shifting, together with the aid of ice-floes, in their action of grinding, lifting, and shoving, that nearly all of these scattered islets within the present area of the island, and marked by its bluffs and higher uplands, are completely bound together by ropes of sand, changed into enduring bars and ridges of water-worn boulders. These are raised above the highest tides by winds that whirl the sand up, over and on them, as it dries out from the wash of the surf and from the interstices of rocks, which are lifted up and pushed by ice-fields.

The sand that plays so important a part in the formation of the Island of St. Paul, and which is almost entirely wanting in and around the others of this Pribylov group, is principally composed of *Foraminifera*, together with *Diatomacea*, mixed in with a volcanic base of fine comminuted black and reddish lavas and old friable gray slates. It constitutes the chief beauty of the sea-shore here, for it changes color like a chameleon, as it passes from wet to dry, being a rich steely-black at the surf-margin and then drying out to a soft purplish-brown and gray, succeeding to tints most delicate of reddish and pale neutral, when warmed by the sun and drifting up on to the higher ground with the wind. The sand-dune tracts on this island are really attractive in the summer, especially so during those rare days when the sun comes out, and the unwonted light shimmers over them and the most luxuriant grass and variety of beautiful flowers which exist in profusion thereon. In past time, as these sand and boulder bars were forming on St. Paul Island, they, in making across from islet to islet, enclosed small bodies of sea-water. These have, by evaporation and time, by the flooding of rains and annual melting of snow, become, nearly every one of them, fresh; they are all, great and small, well shown on my map, which locates quite a large area of pure water. In them, as I have hinted, are no reptiles; but an exquisite species of a tiny fish* ex-

---

* *Gasterosteus cataphractus;* and *pungitius; beautiful sticklebacks.*

ists in the lagoon-estuary near the village, and the small pure-water lakes of the natives just under the flanks of Telegraph Hill. The Aleutes assured me that they had caught fish in the big lake toward Northeast Point, when they lived in their old village out there; but, I never succeeded in getting a single specimen. The waters of these pools and ponds are fairly alive with vast numbers of minute *Rotifera*, which sport about in all of them wherever they are examined. Many species of water-plants, pond-lilies, algæ, etc., are found in those inland waters, especially in that large lake "Meesulk-mah-nee," which is very shallow.

The backbone of St. Paul, running directly east and west, from shore to shore, between Polavina Point and Einahnuhto Hills, constitutes the high land of that island: Polavina Sopka, an old extinct cinder-crater, five hundred and fifty feet; Bogaslov, an upheaved mass of splinted lava, six hundred feet; and the hills frowning over the bluffs there, on the west shore, are also six hundred feet in elevation above the sea. But the average height of the upland between is not much over one hundred to one hundred and fifty feet above water-level, rising here and there into little hills and broad, rocky ridges, which are minutely sketched upon the map. From the northern base of Polavina Sopka a long stretch of low sand-flats extends, enclosing the great lake, and ending in a narrow neck where it unites with Novastoshnah, or Northeast Point. Here that volcanic nodule known as Hutchinson's Hill, with its low, gradual slopes, trending to the east and southward, makes a rocky foundation secure and broad, upon which the great single rookery of the island, the greatest in the world, undoubtedly, is located. The natives say that when they first came to these islands Novastoshna was an island by itself, to which they went in boats from Vesolia Mista; and the lagoon now so tightly enclosed was then an open harbor in which the ships of the old Russian Company rode safely at anchor. To-day, no vessel drawing ten feet of water can safely get nearer than half a mile of the village, or a mile from this lagoon at low tide.

The total absence of a harbor at the Pribylov Islands is much to be regretted. The village of St. Paul, as will be seen by reference to the map, is so located as to command the best landings for vessels that can be made during the prevalence of any and all winds, except those from the south. From these there is no shelter for ships, unless they run around to the north side, where they are

unable to hold practicable communication with the people or to
discharge.   At St. George matters are still worse, for the prevailing
northerly, westerly, and easterly winds drive the boats away from
the village roadstead, and weeks often pass at either island, but
more frequently at the latter, ere a cargo is landed at its destination.   Under the very best circumstances, it is both hazardous and
trying to unload a ship at any of these places.   The approach to St.
Paul by water during thick weather is doubtful and dangerous, for
the land is mostly low at the coast, and the fogs hang so dense and
heavy over and around the hills as to completely obliterate their
presence from vision.   The captain fairly feels his way in by throwing his lead-line and straining his ear to catch the muffled roar of
the seal-rookeries, which are easily detected when once understood,
high above the booming of the surf.   At St. George, however, the
bold, abrupt, bluffy coast everywhere all around, with its circling
girdle of flying water-birds far out to sea, looms up quite prominently, even in the fog ; or, in other words, the navigator can
notice it before he is hard aground or struggling to haul to windward from the breakers under his lee.   There are no reefs making
out from St. George worthy of notice, but there are several very
dangerous and extended ones peculiar to St. Paul, which Captain
John G. Baker, in command of the vessel * under my direction,
carefully sounded out, and which I have placed upon my chart for
the guidance of those who may sail in my wake hereafter.

When the wind blows from the north, northwest, and west to
southwest, the company's steamer drops her anchor in eight fathoms
of water abreast of the black bluffs opposite the village, from
which anchorage her stores are lightered ashore ; but in the north-
easterly, easterly, and southeasterly winds, she hauls around to the
lagoon bay west of the village, and there, little less than half a
mile from the landing, she drops her anchor in nine fathoms of
water, and makes considerable headway at discharging her cargo.
Sailing-craft come to both anchorages, but, however, keep still
farther out, though they choose relatively the same positions, yet
seek deeper water to swing to their cables in : the holding-ground
is excellent.   At St. George the steamer comes, wind permitting,
directly to the village on the north shore, close up, and finds her
anchorage in ten fathoms of water, over poor holding-ground ; still

---

* United States Revenue-marine cutter *Reliance*, June to October, 1874.

it is only when three or four days have passed, free from northerly, westerly, or easterly winds that she can make the first attempt to safely unload. The landing here is a very bad one, surf breaking most violently upon the rocks from one end of the year to the other.

The observer will notice that six miles southward and westward of the reef of St. Paul Island is a bluffy islet, called by the Russians Bobrovia, because in olden time the promishlyniks are said to have captured many thousands of sea-otters on its rocky coast. It rises from the ocean, sheer and bold, an unbroken mural precipice extending nearly all around, of sea-front, but dropping on its northern margin, at the water, low, and slightly elevated above the surf-wash, with a broken, rocky beach and no sand. The height of

"Bobrovia," or Otter Island six miles south of St. Paul Island.
[*The North Shore and landing, viewed from St. Paul.*]

the bluffs at their greatest elevation over the west end is three hundred feet, while the eastern extremity is quite low, and terminated by a queer, funnel-shaped crater-hill, which is as distinctly defined, and as plainly scorched and devoid of the slightest sign of vegetation within as though it had burned up and out yesterday. That crater-point on Otter Island is the only unique feature of the place, for with the exception of this low north shore, before mentioned, where a few thousand of "bachelor seals" haul out during the season every year, there is nothing else worthy of notice concerning it. A bad reef makes out to the westward, which I have indicated from my observation of the rocks awash, looking down upon them from the bluffs. Great numbers of water-fowl roost upon the cliffs, and there are here about as many blue foxes to the acre as the law of life allows. A small, shallow pool of impure water

lies close down to the north shore, right under a low hill upon which
the Russians in olden times posted a huge Greek cross, that is still
standing ; indeed it was the habit of those early days of occupation
in Alaska. to erect such monuments everywhere on conspicuous
elevations adjacent to the posts or settlements. One of these is
still standing at Northeast Point, on the large sand-dune there which
overlooks the killing-grounds, and another sound stalwart cross yet
faces the gales and driving " boorgas " on the summit of Bogoslov
Mountain, as it has withstood them during the last sixty years.

To the eastward, six miles from Northeast Point, will be noticed
a small rock named Walrus Island. It is a mere ledge of lava, flat-
capped, lifted just above the wash of angry waves ; indeed, in
storms of great power, the observer, standing on either Cross or
Hutchinson's Hills, with a field-glass, can see the water breaking
clear over it : these storms, however, occur late in the season,
usually in October or November. This island has little or no commercial importance, being scarcely more than a quarter of a mile in
length and one hundred yards in point of greatest width, with bold
water all around, entirely free from reefs or sunken rocks. As
might be expected, there is no fresh water on it. In a fog it makes
an ugly neighbor for the sea-captains when they are searching for
St. Paul ; they all know it, and they all dread it. It is not resorted
to by the fur-seals or by sea-lions in particular ; but, singularly
enough, it is frequented by several hundred male walrus, to the
exclusion of females, every summer. A few sea-lions, but only a very
few, however, breed here. On account of the rough weather, fogs,
etc., this little islet is seldom visited by the natives of St. Paul, and
then only in the egging season of late June and early July when
that surf-beaten breakwater literally swarms with breeding sea-fowl.

This low, tiny, island is, perhaps, the most interesting single
spot now known to the naturalist who may land in northern seas,
to study the habits of bird-life ; for here, without exertion or
risk, he can observe and walk among tens upon tens of thousands
of screaming water-fowl ; and, as he sits down upon the polished
lava rock, he becomes literally ignored and environed by these
feathered friends, as they reassume their varied positions of incubation, from which he disturbs them by his arrival. Generation
after generation of their kind have resorted to this rock unmolested,
and to-day, when you get among them, all doubt and distrust seem
to have been eliminated from their natures. The island itself is

rather unusual in those formations which we find peculiar to Alaskan waters. It is almost flat, with slight, irregular undulations on top, spreading over an area of five acres perhaps. It rises abruptly, though low, from the sea, and it has no safe beach upon which a person can land from a boat; not a stick of timber or twig of shrubbery ever grew upon it, though the scant presence of low, crawling grasses in the central portions prevents the statement that *all* vegetation is absent. Were it not for the frequent rains and dissolving fog characteristic of summer weather here, the accumulation of guano would be something wonderful to contemplate—Peru would have a rival. As it is, however, the birds, when they return, year after year, find their nesting-floor swept as clean as though they had never sojourned there before. The scene of confusion and uproar that presented itself to my astonished senses when I approached this place in search of eggs, one threatening, foggy July morning, may be better imagined than described, for, as the clumsy bidarrah came under the lee of the low cliffs, swarm upon swarm of thousands of murres or "arries," dropped in fright from their nesting-shelves, and, before they had control of their flight, they struck to the right and left of me, like so many cannon-balls. I was forced, in self-protection, to instantly crouch for a few moments under the gunwale of the boat until the struggling, startled flock passed, like an irresistible, surging wave, over my head. Words cannot depict the amazement and curiosity with which I gazed around after climbing up to the rocky plateau, and stood among myriads of breeding-birds; they fairly covered the entire surface of the island with their shrinking forms, while others whirled in rapid flight over my head, as wheels within wheels, so thickly inter-running that the blue and gray of the sky was hidden from my view. Add to this impression the stunning whir of hundreds of thousands of strong, beating wings, the shrill screams of the gulls, and a muffled croaking of the "arries," coupled with an indescribable, disagreeable smell which arose from broken eggs and other decaying substances—then a faint idea may be evoked of the strange reality spread before me. Were it not for this island and the ease with which the natives can gather, in a few hours, tons upon tons of sea-fowl eggs, the people of St. Paul would be obliged to go the westward, and suspend themselves from the lofty cliffs of Einahnuhto, dangling over the sea by ropes, as their less favored neighbors are only too glad and willing to do at St. George.

I am much divided in my admiration of the two great bird-rook-
eries of this Pribylov group, the one on the face of the high bluffs
at St. George, and the other on the table-top of Walrus Islet ; but
perhaps the latter place gives, within the smallest area, the greatest
variety of nesting and breeding birds, for here the "arrie" and
many gulls, cormorants, sea-parrots, and auks come to lay their
eggs in countless numbers. The face and brow of the low, cliff-
like sea-front to this island are occupied almost exclusively by the
"arries," *Lomvia arra,* which lay a single egg each on the surface
of the bare rock, and stand, just like so many champagne bottles,
straddling over them while hatching, only leaving at irregular intervals to feed, and then not until their mates relieve them. Hundreds of thousands of these birds alone are thus engaged about
the 29th of every June on this little rocky island, roosting stacked
up together as tight as so many sardines in a box, as compactly as
they can be stowed, each and all of them uttering an incessant,
muffled, hoarse, grunting noise. How fiercely they quarrel among
themselves—everlastingly ! and in this way thousands of eggs are
rolled off into the sea, or into crevices, or into fissures, where they
are lost and broken.

The "arrie" lays but one egg. If it is removed or broken, she
will soon lay another ; but if undisturbed after depositing the first,
she undertakes its hatching at once. The size, shape, and coloration of this egg, among the thousands which came under my observation, are exceedingly variable. A large proportion of the eggs
become so dirty by rolling here and there in the guano while the
birds tread and fight over them as to be almost unrecognizable. I
was struck by a happy adaptation of nature to their rough nesting. It is found in the toughness of this shell of the egg, so tough
that the natives, when gathering them, throw them as farmers do
apples into their tubs, baskets, etc., on the cliff, and then carry
them down to a general heap of collection near the boats' landing, where they are poured out upon the rocks with a single flip of
the hand, just as a sack of potatoes would be emptied ; and then
again, after this, they are quite as carelessly handled when loaded
into the "bidarrah," sustaining through it all a very trifling loss
from crushed or broken specimens. *

* To visit Walrus Island in a boat, pleasantly and successfully, it is best to
submit to the advice and direction of the natives. They leave the village in
the evening, and, taking advantage of the tide, proceed along the coast as far

These "arries" seem to occupy a ribbon strip in width : it is drawn around the outward edges of the flat table-top to Walrus Island as a regular belt, reserved all to themselves : while the small grassy interior from which they are thus self-excluded is the only place, I believe, in Bering Sea where the big white gull, *Larus glaucus*, breeds. Here I found among grassy tussocks the white burgomaster building a nest of dry grass, sea-ferns, *Sertularidæ*, etc., very nicely laid up and rounded, and in which it laid usually three eggs, sometimes only a couple ; occasionally I would look into a nest with four. These heavy gulls could not breed on either of the other islands in this manner, for the glaucous gull is too large to settle on the narrow shelf-ledges of the cliffs, as the smaller gulls do, and lesser water-fowls, and those places which could receive it would also be a happy hunting-ground and footing to the foxes.

The red-legged kittiwake, *Rissa brevirostris*, and its cousin, *Rissa tridactyla*, build in the most amicable manner together on the faces of those cliffs, for they are little gulls, and they associate with cormorants, sea-parrots, and tiny auks, all together, and, with the exception of the last, their nests are very easy of access. All birds, especially the "arries," have an exceedingly happy time of it on this Walrus Islet—nothing to disturb them, in my opinion, free from the ravenous maw of blue foxes over on St. Paul, and from the piratical and death-dealing sweep of owls and hawks, which infest the Aleutian chain and the mainland.

The position of the islands is such as to be somewhat outside of that migratory path pursued by the birds on the mainland, and owing to this reason they are only visited by a few stragglers from

as the bluffs of Polavina, where they rest on their oars, doze, and smoke until the dawning of daylight, or later, perhaps, until the fog lifts enough for them to get a glimpse of the islet which they seek. They row over then in about two hours with their bidarrah. They leave, however, with perfect indifference as to daylight or fog. Nothing but a southeaster can disturb their tranquillity when they succeed in landing on Walrus Island. They would find it as difficult to miss striking the extended reach of St. Paul on their return, as they found it well-nigh impossible to push off from Polavina and find "Morzovia" in a thick, windy fog and running sea.

Otter Island, or "Bobrovia," is easily reached in almost any weather that is not very stormy, for it looms up high above the water. It takes the bidarrah about two hours to row over from the village, while I have gone across once in a whale-boat with less than one hour's expenditure of time, sail, and oars *en route*.

that quarter, a few from the Asiatic side, and by the millions of
their own home-bred and indigenous stock.   One of these migratory species, a turnstone, however, comes here every summer, for
three or four weeks' stay, in great numbers, and actually gets so
fat in feeding upon the larvæ which abound in the decaying carcasses over the killing-grounds that it usually bursts open when it
falls, shot on the wing.   A heavy easterly gale once brought a
strange bird to the islands from the mainland—a grebe, *P. grisei-
gena*.   It was stranded on St. George in 1873, whereupon the natives
declared the like of which they had never seen before; again, I
found a robin one cool morning in October, the 15th : the natives
told me that it was an accident—brought over by some storm or
gale of wind that took it up and off from its path across the tundra
of Bristol Bay.   The next fair wind sweeping from the north or
the west could be so improved by this robin, *M. migratoria*, that
· it would spread its wings and as abruptly return.   Thus hawks,
owls, and a number of strange water-fowls visit the islands, but
never remain there long.

   The Russians tried the experiment of bringing up from Sitka
and Oonalashka a flock of ravens, as scavengers, a number of years
ago, and when they were very uncleanly in the village, in contrast with the practice of the present hour.   They reasoned that
they would—these ill-omened birds—be invaluable as health officers ; but the *Corvidæ* invariably, sooner or later, and within a very
short time, took the first wind-train or lightning-express back to
the mainland or the Aleutian islands.   Yet the natives say that if
the birds had been young ones instead of old fellows they would
have remained.   I saw a great many, however, at St. Matthew
Island in August, 1874.

   A glance at the map of St. Paul shows that nearly half of its
superficial area is low and quite flat, not much elevated above the
sea.   Wherever the sand-dune tracts are located, and that is right
along the coast, will be found an irregular succession of hummocks
and hillocks, drifted by the wind, which are very characteristic.   On
the summits of these hillocks an *Elymus* has taken root in times
past, and, as the sand drifts up, it keeps growing on and up too,
so that a quaint spectacle is presented of large stretches to the view
wherein sand-dunes, entirely bare of all vegetation at their base
and on their sides, are crowned with a living cap of the brightest
green—a tuft of long, waving grass blades which will not down.

None of this peculiar landscaping, however, is seen on St. George, not even in the faintest degree. Travel about St. Paul, with the exception of that trail to Northeast Point, where the natives take advantage of low water to run on the hard, wet sand, is exceedingly difficult, and there are examples of only a few white men who have ever taken the trouble and expended the physical energy necessary to accomplish a comparatively short walk from the village to Nahsayvernia, or the north shore. Walking upon the moss-hidden and slippery rocks, or tumbling over slightly uncertain tussocks, is a task and not a pleasure. On St. George, with the exception of a half-mile path to the village cemetery and back, nobody pretends to walk, except the natives who go to and from the rookeries in their regular seal-drives. Indeed, I am told that I am ·the only white man who has ever traversed the entire coast-line of both islands.*

---

* That profile of the south shore, between the village hill and Southwest Point, taken from the steamer's anchorage off the village cove, shows its characteristic and remarkable alternation of rookery slope and low sea-level flats. This point of viewing is slightly more than half a mile true west of the village hill, to a sight which brings Bogaslov summits and Tolstoi Head nearly in line. At Zapadnie is the place where the Russian discoverers first landed in 1787, July 10th. With the exception of that bluffy west-end Ein-ahnuh-to cliffs, the whole coast of St. Paul is accessible, and affords an easy landing, except at the short reach of "Seethah" and the rookery points, as indicated. The great sand beach of this island extends from Lukannon to Polavina, thence to Webster's house, Novastoshnah; from there over, and sweeping back and along the north shore to Nahsayvernia headland, then between Zapadnie and Tolstoi, together with the beautiful though short sand of Zoltoi. This extensive and slightly broken sandy coast is not described as peculiar to any other island in Alaska, or of Siberian waters.

There are no running streams at any season of the year on St. Paul; but an abundance of fresh water is plainly afforded by the numerous lakes, all of which are "svayjoi," save the lagoon estuary. The four big reefs which I have located are each awash in every storm that blows from seaward over them; they are all rough, rocky ledges. That little one indicated in English Bay caused the wrecking of a large British vessel in 1847, which was coming in to anchor just without Zapadnie; a number of the crew were "massluck-en," so my native informant averred. Most of the small amount of drift-wood that is found on this island is procured at Northeast Point and Polavina; the north shore from Maroonitch to Tsammanah has also been favored with sea-waif logs in exceptional seasons, to the exclusion of all other sections of the coast. The natives say that the St. George people get much more drift-wood every year, as a rule, than they do on St. Paul. From what I could see

Turning to St. George and its profile, presented by the accompanying map, the observer will be struck at once by the solidity of that little island and its great boldness, rising, as it does, sheer and precipitous from the sea all around, except at the three short reaches of the coast indicated on my chart, and where the only chance to come ashore exists.

The seals naturally have no such opportunity to gain a footing here as they have on St. Paul, hence their comparative insignificance as to number. The island itself is a trifle over ten miles in extreme length, east and west, and about four and a quarter miles in greatest width, north and south. It looks, when plotted, somewhat like an old stone axe ; and, indeed, when I had finished my initial contours from my field-notes, the ancient stone-axe outline so disturbed me that I felt obliged to resurvey the southern shore, in order that I might satisfy my own mind as to the accuracy of my first work. It consists of two great plateaus, with a high upland valley between, the western table-land dropping abruptly to the sea at Dalnoi Mees, while the eastern falls as precipitately at Waterfall Head and Tolstoi Mees. There are several little reservoirs of fresh water—I can scarcely call them lakes—on this island ; pools, rather, that the wet sphagnum seems to always keep full, and from which drinking-water in abundance is everywhere found. At Garden Cove is a small, living stream : it is the only one on the Pribylov group.

St. George has an area of about twenty-seven square miles ; it has twenty-nine miles of coast-line, of which only two and a quarter are visited by the fur-seals, and which is in fact all the eligible landing-ground afforded them by the structure of the island. Nearly half of the shore of St. Paul is a sandy beach, while on St. George there is less than a mile of it all put together, namely : a few hundred yards in front of the village, the same extent on the

---

during my four seasons of inspection, they never have got much, under the best of circumstances, on either island. They pay little attention to it now, and gather what they do during the winter season, going to Polavina and the north shore with sleds, on which they hoist sails after loading there, and scud home before strong northerly blasts.

Captain Erskine informs me that the water is free and bold all around the north shore, from Cross Hill to Southwest Point ; no reefs or shoals up to within in half a mile of land anywhere. English Bay is very shallow, and no sea-going vessel should attempt to enter it that draws over six feet.

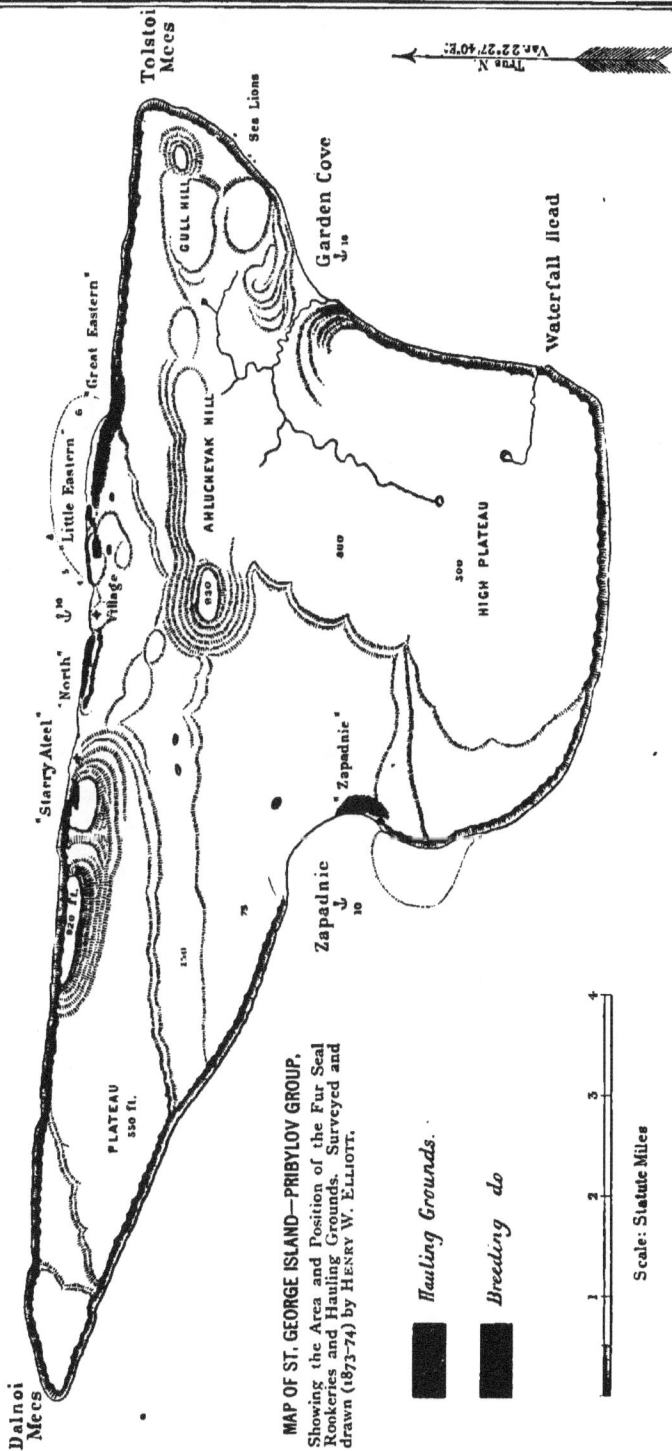

MAP OF ST. GEORGE ISLAND—PRIBYLOV GROUP.

Showing the Area and Position of the Fur Seal Rookeries and Hauling Grounds. Surveyed and drawn (1873-74) by HENRY W. ELLIOTT.

Hauling Grounds.

Breeding do

Scale: Statute Miles

Dalnoi Mees

PLATEAU 350 ft.

"Starry Aleet"

"North"

"Little Eastern"

"Great Eastern"

Village

Zapadnie

"Zapadnie"

AHLUCHEYAK HILL

HIGH PLATEAU

GULL HILL

Sea Lions

Garden Cove

Waterfall Head

Tolstoi Mees

True N.

Var. 22° 27' 40" E.

ST. GEORGE'S ISLAND, PRIBYLOV GROUP

Garden Cove beach, southeast side, and less than half a mile at Zapadnie on the south side.

Just above the Garden Cove, under the overhanging bluffs, several thousand sea-lions hold exclusive, though shy, possession. Here there is a half-mile of good landing. On the north shore of the island, three miles west from the village, a grand bluff wall of basalt and tufa intercalated rises abruptly from the sea to a sheer height of nine hundred and twenty feet at its reach of greatest elevation : thence, dropping a little, runs clear around the island to Zapadnie, a distance of nearly ten miles, without affording a single passage-way up or down to the sea that thunders at its base. Upon its innumerable narrow shelf-margins, and in its countless chinks and crannies, and back therefrom over an extended area of lava-shingled inland ridges and terraces, millions upon millions of water-fowl breed during the summer months.

The general altitude of St. George, though in itself not great, has, however, an average three times higher than that of St. Paul, the elevation of which is quite low, and slopes gently down to the sea east and north ; St. George rises abruptly, with exceptional spots for landing. The loftiest summit on St. George, the top of the hill right back to the southward of the village, is nine hundred and thirty feet, and is called by the natives Ahluckeyak. That on St. Paul, as I have before said, is Bogaslov Hill, six hundred feet. All elevations on either island, fifteen or twenty feet above sea-level, are rough and hummocky, with the exception of those sand-dune tracts at St. Paul and the summits of the cinder hills, on both islands. Weathered out, or washed from the basalt and pockets of olivine on either island, are aggregates of augite, seen most abundant on the summit slopes of Ahluckeyak Hill, St. George. Specimens from stratified bands of old, friable, gray lavas, so conspicuous on the shore of this latter island, show an existence of hornblende and vitreous felspar in considerable quantity, while on the south shore, near Garden Cove, is a large dike of a bluish and greenish gray phonolite, in which numerous small crystals of spinel are found. A dike, with well-defined walls, of old close-grained, clay-colored lava, is near the village of St. George, about a quarter of a mile east from the landing, in the face of those reddish breccia bluffs that rise from the sea. It is the only example of the kind on the islands. The bases or foundations of the Pribylov Islands are, all of them, basaltic ; some are compact and grayish-white, but most of them exceedingly porous and ferruginous.* Upon this solid floor are many hills of brown and red tufa, cinder-heaps, etc. Polavina Sopka, the second point in elevation on St. Paul Island, is almost entirely built up of red scoria and breccia ; so is Ahluckeyak Hill, on St. George, and the cap to the high bluffs opposite. The village hill at St. Paul, Cone hill, the Einahnuhto peaks, Crater Hill, North Hill, and Little Polavina are all ash-heaps of this character. The bluffs at the shore of Polavina Point, St. Paul, show in a striking manner a section of the geological struct-ure of the island. The tufas on both islands, at the surface, de-compose and weather into the base of good soil, which the severe climate, however, renders useless to good husbandmen. There is not a trace of granitic or of gneissoid rocks found *in situ*. Meta-morphic boulders have been collected along the beaches and pushed up by heavy ice-floes which have brought them down from

---

* The profile of the coast of St. George's Island, which I give on the map, presents clearly an idea of its characteristic, bold, abrupt elevation from the sea. From the Garden Cove around to Zapadnie beach there is no natural opportunity for a man to land ; then, again, from Zapadnie beach round to Starry Arteel there is not a sign of a chance for an agile man to come ashore and reach the plateau above. From Starry Arteel to the Great Eastern rookery there is an alternation, between the several breeding-grounds, of three low and gradual slopes of the land to sea-level ; these, with the landing at Garden Cove and at Zapadnie, are the only spots of the St. George coast where we can come ashore. An active person can scramble up at several steep places be-tween the Sea Lion rookery and Tolstoi Mees, but the rest of that extended bluffy sea-wall, which I have just defined, is wholly inaccessible from the water. A narrow strip of rough, rocky shingle, washed over by every storm-beaten sea, is all that lies beneath the mural precipices.

In the spring, when snow melts on the high plateau, a beautiful cascade is seen at Waterfall Head ; its feathery, filmy, silver ribbon of plunging water is thrown out into exquisite relief by the rich background of that brownish basalt and tufa over which it drops. Another pretty little waterfall is to be seen just west of the village, at this season only, where it leaps from a low range of bluffs to the sea. The first-named cascade is more than four hun-dred feet in sheer unbroken precipitation.

One or two small, naked, pinnacle rocks, standing close in, and almost joined to the beach at the Sea Lion rookery, constitute the only outlying islets or rocks ; a stony kelp-bed at Zapadnie, and one off the Little Eastern rook-ery, both of limited reach seaward, are the only hindrances to a ship's sailing boldly round the island, even to scraping the bluffs, at places, safely with her yard-arms. I have located the Zapadnie shoal by observation from the bluffs above ; while Captain Baker, of the *Reliance*, sounded out the other.

Siberian coasts far away to the northwest. The dark-brown tufa bluffs and the breccia walls at the east landing of St. Paul Island, known as "Black Bluffs," rise suddenly from the sea sixty to eighty feet, with stratified horizontal lines of light-gray calcareous conglomerate, or cement, in which are embedded sundry fossils characteristic of and belonging to the Tertiary Age, such as *Cardium grœnlandicum, C. decoratum,* and *Astarte pectunculata,* etc. This is the only locality within the purview of the Pribylov Islands where any palæntological evidence of their age can be found. These specimens, as indicated, are exceedingly abundant ; I brought down a whole series, gathered there at the east landing or "Navastock," in a short half-hour's search and labor.

Although small quantities of drift-wood lodge at all points of the coast, yet the greatest amount is found on the south shore, and thence around to Garden Cove ; this drift-timber is usually wholly stripped of its bark, principally pine and fir sticks, some of them quite large, eighteen inches or two feet in diameter. Several years occur when a large driftage will be thrown or stranded here ; then long intervals of many seasons will elapse with scarcely a log or stick coming ashore. I found at Garden Cove, in June, 1873, the well-preserved husk of a cocoanut, cast up by the surf on the beach : did I not know that it was most undoubtedly thrown over by some whaler in these waters, not many hundred miles away at the farthest, I should have indulged in a pretty reverie as to its path in drifting from the South Seas to this lonely islet. I presume, however, that the timber which the sea brings for the Pribylov Islands is that borne down upon the annual floods of the Kuskokvim and Nooshagak Rivers on the mainland, and to the east-northeastward, a trifle more than two hundred and twenty-five miles ; it comes, however, in very scant supply. I saw very little drift-wood on St. Matthew Island ; but on the eastern shore of St. Lawrence there was an immense aggregate, which unquestionably came from the Yukon mouth.

The fact that fur-seals frequent these islands and those of Bering and Copper, on the Russian side, to the exclusion of other land, seems at first odd or singular, to say the least ; but when we come to examine the subject we find that those animals, when they repair hither to rest for two or three months on the land, as they must do by their habit during the breeding-season, require a cool, moist atmosphere, imperatively coupled with firm, well-drained land, or dry broken rocks or shingle, rather, upon which to take their positions and remain undisturbed by the weather and the sea for a lengthy period of reproduction. If the rookery-ground is hard and flat, with an admixture of loam or soil, puddles are speedily formed in this climate, where it rains almost every day, and when not raining, rain-fogs take rapid succession and continue the saturation, making thus a muddy slime, which very quickly takes the hair off the animals whenever it plasters or wherever it fastens on them ; hence they carefully avoid any such landing. If they occupy a sandy shore the rain beats that material into their large, sensitive eyes, and into their fur, so they are obliged, from simple irritation, to leave and return to the sea for relief.

This inspection of some natural characteristics of the Pribylov group renders it quite plain that the Seal Islands, now under discussion, offer to the *Pinnipedia* very remarkable advantages for landing, especially so at St. Paul, where the ground of basaltic rock and of volcanic tufa or cement slopes up from many points gradually above the sea, making thereby a perfectly adapted resting-place for any number, from a thousand to millions, of those intelligent animals, which can lie out here from May until October every year in perfect physical peace and security. There is not a rod of ground of this character offered to these animals elsewhere in all Alaska, not on the Aleutian chain, not on the mainland, not on St. Matthew or St. Lawrence. Both of the latter islands were surveyed by myself, with special reference to this query, in 1874 ; every foot of St. Matthew shore-line was examined, and I know that the fur-seal could not rest on the low clayey flats there in contentment a single day ; hence he never has rested there, nor will he in the future. As to St. Lawrence, it is so ice-bound and snow-covered in spring and early summer, to say nothing of numerous other physical disadvantages, that it never becomes of the slightest interest to fur-seals.

When Pribylov, in taking possession, landed on St. George a part of his little ship's crew, July, 1786, he knew that, as it was uninhabited, it would be necessary to establish a colony there from which to draft laborers to do all killing, skinning, and curing of the peltries ; therefore he and his associates, and his rivals after him, imported natives of Oonalashka and Atkha—passive, docile Aleutes. They founded their first village a quarter of a mile to the eastward of one of the principal rookeries on St. George, now called "Starry Arteel," or "Old Settlement"; a village was also located at Zapadnie, and a succession of barraboras planted at Garden Cove. Then, during the following season, more men were brought up from Atkha and taken over to St. Paul, where five or six rival traders posted themselves on the north shore, near and at "Maroonitch," and at the head of the Big Lake, among the sand-dunes there. They were then, as they are now, somewhat given to riotous living if they only had the chance, and the ruins of the Big Lake settlement are pleasantly remembered by the descendants of those pioneers to-day, on St. Paul, who take off their hats as they pass by to affectionately salute, and call the place "Vesolia Mista," or "Jolly Spot"—the aged men telling me, in a low whisper, that "in those good old days they had plenty of rum." But, when the pressure of competition became great, another village was located at Polavina, and still another at Zapadnie, until the activity and unscrupulous energy of all these rival settlements well-nigh drove out and eliminated the seals in 1796. Three years later the whole territory of Alaska passed into the hands of the absolute power vested in the Russian American Company. These islands were in the bill of sale, and early in 1799 the competing traders were turned off neck and heels from them, and the Pribylov group passed under the control of a single man, the iron-willed Baranov. The people on St. Paul were then all drawn together, for economy and warmth, into a single settlement at Polavina. Their life in those days must have been miserable. They were mere slaves, without the slightest redress from any insolence or injury which their masters might see fit, in petulance or brutal orgies, to inflict upon them. Here they lived and died, unnoticed and uncared for, in large barracoons half under ground and dirt-roofed, cold and filthy. Along toward the beginning or end of 1825, in order that they might reap the advantage of being located best to load and unload ships, the Polavina settlement was removed to the present village site, as indicated on the map, and the natives have lived there ever since.

On St. George the several scattered villages were abandoned, and consolidated at the existing location some years later, but for a different reason. The labor of bringing the seal-skins over to Garden Cove, which is the best and surest landing, was so great, and that of carrying them from the north shore to Zapadnie still greater, that it was decided to place the consolidated settlement at such a point between them, on the north shore, that the least trouble and exertion of conveyance would be necessary. A better place, geographically, for the business of gathering the skins and salting them down at St. George cannot be found on the island, but a poorer place for a landing it is difficult to pick out, though in this respect there is not much choice outside of Garden Cove.

Up to the time of the transfer of the territory and leasing of the islands to the Alaska Commercial Company, in August, 1870, these native inhabitants all lived in huts or sod-walled and dirt-roofed houses, called "barrabkies," partly under ground. Most of these huts were damp, dark, and exceedingly filthy : it seemed to be the policy of a short-sighted Russian management to keep them so, and to treat the natives not near so well as they treated the few hogs and dogs which they brought up there for food and for company. The use of seal-fat for fuel, caused the deposit upon every-thing within doors of a thick coat of greasy, black soot, strongly im-pregnated with a damp, moldy, and indescribably offensive odor. They found along the north shore of St. Paul and at Northeast Point, occasionally scattered pieces of drift-wood, which was used, carefully soaked anew in water if it had dried out, split into little fragments, and, trussing the blubber with it when making their fires, the combination gave rise to a roaring, spluttering blaze. If this drift-wood failed them at any time when winter came round, they were obliged to huddle together beneath skins in their cold huts, and live or die, as the case might be. But the situation to-day has changed marvellously. We see here now at St. Paul, and on St. George, in the place of the squalid, filthy habitations of the imme-diate past, two villages, neat, warm, and contented. Each family lives in a snug frame-dwelling ; every house is lined with tarred paper, painted, furnished with a stove, with out-houses, etc., com-plete ; streets laid out, and the foundations of these habitations reg-ularly plotted thereon. There is a large church at St. Paul, and a less pretentious but very creditable structure of the same character on St. George ; a hospital on St. Paul, with a full and complete stock of drugs, and skilled physicians on both islands to take care of the people, free of cost. There is a school-house on each island, in which teachers are also paid by the company eight months in the year, to instruct the youth, while the Russian Church is sustained entirely by the pious contributions of the natives themselves on these two islands, and sustained well by each other. There are eighty families, or eighty houses, on St. Paul, in the village, with twenty or twenty-four such houses to as many families at St. George, and eight other structures. The large ware-houses and salt-sheds of the Alaska Commercial Company, built by skilful mechanics, as have been the dwellings just referred to, are also neatly painted ; and, taken in combination with the other features, constitute a picture fully equal to the average presentation of any one of our small eastern towns. There is no misery, no downcast, dejected, suffering humanity here to-day. These Aleutes, who enjoy as a price of their good behavior, the sole right to take and skin seals for the company, to the exclusion of all other people, are known to and by their less fortunate neighbors elsewhere in Alaska as the "Bogatskie Aloutov," or the " rich Aleutes." The example of the agents of the Alaska Commercial Company, on both islands, from the beginning of its lease, and the course of the Treasury agents during the last eight or nine years, have been silent but powerful promoters of the welfare of these people. They have maintained perfect order ; they have directed neatness, and cleanliness, and stimulated industry, such as those natives had never before dreamed of.* The chief source of sickness used to arise from the wretched character of the barrabkies in which they lived ; but it was, at first, a very difficult matter to get frame-houses to supplant successfully the sod-walled and dirt-roofed huts of the islands.

Many experiments, however, were made, and a dozen houses built, ere the result was as good as the style of primitive housing, when it had been well done and kept in best possible repair. In such a damp climate, naturally, a strong moldy smell pervades all inclosed rooms which are not thoroughly heated and daily dried by

---

* Surprise has often been genuine among those who inquire, over the fact that there is no law officer here at either village, and wonder is expressed why such provision is not made by the Government. But when the following facts relative to this subject are understood, it is at once clear that a justice of the peace and his constabulary would be entirely useless if established on the seal-islands. As these natives live here, they live as a single family in each settlement, having one common purpose in life and only one ; what one native does, eats, wears, or says, is known at once to all the others, just as whatsoever any member of our household may do will soon be known to us all who belong to its organization ; hence if they steal or quarrel among themselves, they keep the matter wholly to themselves, and settle it to their own satisfaction. Were there rival villages on the islands and diverse people and employment, then the case would be reversed, and the need of legal machinery apparent.

fires ; and, in the spring and fall, frost works through and drips and trickles like rain adown the walls. The present frame-houses occupied by the natives owe their dryness, their warmth, and protection from the piercing "boorgas" to the liberal use of stout tarred paper in the lining. An overpowering mustiness of the hallways, out-houses, and, in fact, every roofed-in spot, where a stove is not regularly used, even in the best-built residences, is one of the first disagreeable sensations which the new arrivals always experience when they take up their quarters here. Perhaps, if it were not for the nasal misery that floats in from the killing-grounds to the novice, this musty, moldy state of things up here would be far more acute, as an annoyance, than it is now. The greater grief seems to soon fully absorb the lesser one ; at least, in my own case, I can affirm the result.

As they lived in early time, it was a physical impossibility for them to increase and multiply ; * but, since their elevation and their sanitary advancement are so marked, it may be reasonably expected that those people for all time to come will at least hold their own, even though they do not increase to any remarkable degree. Perhaps it is better that they should not. But it is exceedingly fortunate that they do sustain themselves so as to be, as it were, a prosperous corporate factor, entitled to the exclusive privilege of labor on these islands. As an encouragement for their good behavior the Alaska Commercial Company, in pursuance of its enlightened treatment of the whole subject, so handsomely exhibited by its housing of these people, has assured them that so long as they are capable and willing to perform the labor of skinning the seal-catch every year, so long will they enjoy the sole privilege of participating in that toil and its reward. This is wise on the part of the company, and it is exceedingly happy for the people. They are, of all men, especially fitted for the work connected with the seal-business—no comment is needed—nothing better in the way

* The population of St. Paul in 1880 was 298. Of these, 14 were whites (13 males and 1 female), 128 male Aleutians, and 156 females. On St. George we have 92 souls: 4 white males, 35 male Aleutians, and 53 females, a total population on these islands of 390. This is an increase of between thirty and forty people since 1873. Prior to 1873 they had neither much increased nor diminished for fifty years, but would have fallen off rapidly (since the births were never equal to the deaths) had not recruits been regularly drawn from the mainland and other islands every season when the ships came up.

of manual labor, skilled and rapid, could be rendered by any body of men, equal in numbers, living under the same circumstances, all the year round. They appear to shake off the periodic lethargy of winter and its forced inanition, to rush with the coming of summer into the severe exercise and duty of capturing, killing, and skinning the seals, with vigor and with persistent and commendable energy.

To-day only a very small proportion of the population are descendants of the pioneers who were brought here by the several Russian companies in 1787 and 1788 ; a colony of one hundred and thirty-seven souls, it is claimed, principally recruited at Oonalashka and Atkha.

The Aleutes on the islands as they appear to-day have been so mixed in with Russian, Koloshian, and Kamschadale blood that they present characteristics, in one way or another, of all the various races of men from the negro up to the Caucasian. The predominant features among them are small, wide-set eyes, broad and high cheek-bones, causing the jaw, which is full and square, to often appear peaked ; coarse, straight, black hair, small, neatly-shaped feet and hands, together with brownish-yellow complexion. The men will average in stature five feet four or five inches ; the women less in proportion, although there are exceptions to this rule among them, some being over six feet in height, and others are decided dwarfs. The manners and customs of these people to-day possess nothing in themselves of a barbarous or remarkable character aside from that which belongs to an advanced state of semi-civilization. They are exceedingly polite and civil, not only in their business with the agents of the company on the seal-islands, but among themselves, and they visit, the one with the other, freely and pleasantly, the women being great gossips ; but, on the whole, their intercourse is subdued, for the simple reason that the topics of conversation are few : and, judging from their silent but unconstrained meetings, they seem to have a mutual knowledge, as if by sympathy, as to what may be occupying each other's minds, rendering speech superfluous. It is only when under the influence of beer or strong liquor that they lose their naturally quiet and amiable disposition. They then relapse into low, drunken orgies and loud, brawling noises. * Having been so long

---

* This evil of habitual and gross intoxication under Russian rule was not characteristic of these islands alone. It was universal throughout Alaska. Sir

under the control and influence of the Russians, they have adopted many Slavic customs, such as giving birthday-dinners, naming their children, etc. They are remarkably attached to their church, and no other form of religion could be better adapted or have a firmer hold upon the sensibilities of the people. Their inherent chastity and sobriety cannot be commended. They have long since thrown away the uncouth garments of Russian rule—those shaggy dog-skin caps, with coats half seal and half sea-lion—for a complete outfit, *cap-à-pie*, such as our own people buy in any furnishing house, the same boots, socks, underclothing, and clothing, with ulsters and ulsterettes; but the violence of the wind prevents their selecting the hats of our fashion and sporting fraternity. As for the women, they, too, have kept pace and even advanced to the level of the men, for in these lower races there is usually more vanity displayed by the masculine element than the feminine, according to my observation. In other words, I have noticed a greater desire among the young men than among the young women of savage and semi-civilized people to be gaily dressed, and to look fine; but the visits of the wives of our treasury officials and the company's agents to these islands during the last ten years, bringing with them a full outfit, as ladies always do, of everything under the sun that women want to wear, has given the native female mind an undue expansion up there and stimulated it to unwonted activity. They watch the cut of the garments and borrow the patterns, and some of them are very expert dressmakers to-day. When the Russians controlled affairs, the women were the hewers of drift-wood and the drawers of water. At St. Paul there was no well of drinking-fluid about the village, nor within half a mile of the village. There was no drinking-water unless it was caught in reservoirs, and the cistern-water, owing to those particles of seal-fat soot which fall upon the roofs of the houses, is rendered undrinkable, so that the supply for the town until quite recently used to be carried by women from two little lakes at the head of the lagoon, a mile and a half as the

George Simpson, speaking of the subject when in Sitka, April, 1842, says: "Some reformation certainly was wanted in this respect, for of all the drunken as well as of all the dirty places that I had visited, New Archangel (Sitka) was the worst. On the holidays in particular, of which, Sundays included, there are one hundred and sixty-five in the year, men, women, and even children were to be seen staggering about in all directions."—Simpson: Journey Around the World, 1841–42, p. 88.

crow flies from the village, and right under Telegraph Hill. This is quite a journey, and when it is remembered that they drink so much tea, and that water has to go with it, some idea of the labor of the old and young females can be derived from an inspection of the map. Latterly, within the last four or five years, the company have opened a spring less than half a mile from the "gorode," which they have plumbed and regulated, so that it supplies them with water now and renders the labor next to nothing, compared with all former difficulty. But to-day, when water is wanted in the Aleutian houses at St. Paul, the man has to get it—the woman does not; he trudges out with a little wooden firkin or tub on his back and brings it to the house.

Some of the natives save their money; yet there are very few among them, perhaps not more than a dozen, who have the slightest economical tendency. What they cannot spend for luxuries, groceries, and tobacco they manage to get away with at the gaming-table. They have their misers and their spendthrifts, and they have the usual small proportion who know how to make money, and then how to spend it. A few among them who are in the habit of saving have opened a regular bank-account with the company. Some of them have to-day two or three thousand dollars saved, drawing an interest of nine per cent.

When the ships arrive and go, the severe and necessary labor of lightering their cargoes off and on from the roadsteads where they anchor is principally performed by these people, and they are paid so much a day for their labor: from fifty cents to one dollar, according to the character of the service they render. This operation, however, is much dreaded by the ship-captains and sea-going men, whose habits of discipline and automatic regularity and effect of working render them severe critics and impatient coadjutors of the natives, who, to tell the truth, hate to do anything after they have pocketed their reward for sealing; and when they do labor after this, they regard it as an act of very great condescension on their part.

As they are living to-day up there, there is no restraint, such as the presence of policemen, courts of justice, fines, etc., which we employ for the suppression of disorder and maintenance of the law in our own land. They understand that if it is necessary to make them law-abiding, and to punish crime, such officers will be among them, and hence, perhaps, is due the fact that from the time that the Alaska Commercial Company has taken charge, in 1870, there has not been one single occasion where the simplest functions of a justice of the peace would or could have been called in to settle any difficulty. This speaks eloquently for their docile nature and their amiable disposition.

These people are singularly affectionate and indulgent toward their children.   There are no "bald-headed" tyrants in our homes as arbitrary and ruthless in their rule as are those snuffly babies and young children on the Seal Islands. While it is very young, the Aleut gives up everything to the caprice of his child, and never crosses its path or thwarts its desire; the "deetiah" literally take charge of the house; but as soon as these callow members of the family become strong enough to bear burdens and to labor, generally between twelve and fifteen years of age, they are then pressed into hard service relentlessly by their hitherto indulgent parents.   The extremes literally meet in this application.

They have another peculiarity: when they are ill, slightly or seriously, no matter which, they maintain or affect a stolid resignation, and are patient to positive apathy.   This is not due to deficiency of nervous organization, because those among them who exhibit examples of intense liveliness and nervous activity behave just as stolidly when ill as their more lymphatic townsmen do. Boys and girls, men and women, all alike, are patient and resigned when ailing and under treatment; but it is a bad feature after all, inasmuch as it is well-nigh impossible to rally a very sick man who himself has no hope, and who seems to mutely deprecate every effort to save his life.   The principal cause of death among the people, by natural infirmity, on the Seal Islands is the varying forms of consumption and bronchitis, always greatly aggravated by that inherited scrofulous taint or stain of blood which was, in one way or another, flowing through the veins of their recent progenitors, both here and throughout the Aleutian Islands.   There is nothing worth noticing in the line of nervous diseases, unless it be now and then the record of a case of alcoholism superinduced by excessive quass drinking.   The "makoolah" intemperance among these people, which was not suppressed until 1876, was a chief factor to an immediate death of infants; for, when they were at the breast, their mothers would drink quass to intoxication, and the stomachs of newly-born Aleutes or Creoles could not stand the infliction which they received, even second-hand.   Had it not been for this wretched spectacle, so often presented to my eyes in 1872-73, I should hardly have taken the active steps which I did to put the nuisance down ; for it involved me, at first, in a bitter personal controversy, which, although I knew at the outset was inevitable, still it weighed nothing in the scales against the evil itself. A few febrile disorders are occurring, yet they yield readily to good treatment.

The inherent propensity of man to gamble is developed here to a very appreciable degree, but it in no way whatever suggests the strange gaming love and infatuation with which all Indians and Eskimo elsewhere of Alaska are possessed. The chief delight of men and boys in the two villages is to stand on the street corners "pitching" half-dollars. So devoted, indeed, have I found the native mind to this hap-hazard sport, that frequently I would detect groups of them standing out in pelting gales of wind and of rain, "shying" silver coins at the little dirt-driven pegs. A few of them, men and women, play cards with much skill and intelligence.

One of the peculiarities * of these people is that they seldom undress when they go to bed—neither the men, women, nor children ; and also that at any and all hours of the night during the summer season, when I have passed in and out of the village to and from the rookeries, I always found several of the natives squatting before their house-doors or leaning against the walls, stupidly staring out into the misty darkness of the fog, or chatting one with the other over their pipes. A number of the inhabitants, by this disposition, are always up and around throughout the settlement during the entire night and day. In olden times, and even recently,

---

* I was told by a very bright Russian, who spent a season here, 1871-72, as special agent of the Treasury Department, that the Aleutian ancestors of these people when they were converted and baptized into the Greek Catholic Church received their names, brand new, from the fertile brains of priests, who, after exhausting the common run of Muscovitic titles, such as our Smiths and Joneses, were compelled to fall back upon some personal characteristics of the new claimant for civilized nomenclature. Thus we have to-day on the Seal Islands a "Stepan Bayloglazov," or, "Son of a White Eye," "Oseep Baizyahzeekov," or "Son of Man without a Tongue." A number of the old Russian governors and admirals of the imperial navy are represented here by their family names, though I do not think, from my full acquaintance with the namesakes, that the distinguished owners in the first place had anything to do with their physical embodiment on the Pribylov Islands.

these involuntary sentinels of the night have often startled the whole village by shouting at the top of their voices the pleasant and electric announcement of the "ship's light!" or they have frozen it with superstitious horror at daybreak by then reciting some ghostly vision that had appeared to them.

The urchins play marbles, spin tops, and fly kites, intermittently, with all the feverish energy displayed by such youth of our own surroundings; they frolic at base-ball, and use "shinny" sticks with great volubility and activity. The girls are, however, much more repressed, and, though they have a few games, and play quietly with quaintly dressed dolls, yet they do not appear to be possessed of that usual feminine animation so conspicuously marked in our home-life.

The attachment which the natives have for their respective islands was well shown to me in 1874. Then a number of St. George people were taken over to St. Paul, temporarily, to do the killing incidental to a reduction of the quota of twenty-five thousand for their island and a corresponding increase at St. Paul. They became homesick immediately, and were never tired of informing the St. Paul natives that St. George was a far handsomer and more enjoyable island to live upon; that walking over the long sand reaches of "Pavel" made their legs grievously weary, and that the whole effect of this change of residence was "ochen scootchnie." Naturally the ire of the St. Paul people rose at once, and they retorted in kind, indicating the rocky surface of St. George and its great inferiority as a seal-island. I was surprised at the genuine feeling on both sides, because, as far as I could judge from a residence on each island, it was a clear case of tweedle-dee and tweedle-dum between them as to opportunities and climate necessary for a pleasurable existence. The natives themselves are of one and common stock, though the number of Creoles on St. George is relatively much larger than on St. Paul. Consequently the tone of the St. George village is rather more sprightly and vivacious.

The question is naturally asked, How do these people employ themselves during the long nine months of every year after the close of the sealing season and until it begins again, when they have little or absolutely nothing to do? It may be answered that they simply vegetate, or, in other words, are entirely idle, mentally and physically, during most of this period. But, to their credit, let it be said that mischief does not employ their idle hands. They are passive killers of time, drinking tea and sleeping, with a few disagreeable exceptions, such as the gamblers. There are a half-dozen of these characters at St. Paul, and perhaps as many at St. George, who spend whole nights at their sittings, even during the sealing season, playing games of cards taught by Russians and persons who have been on the island since the transfer of the territory ; but the majority of the men, women, and children, not being compelled to exert themselves to obtain any of the chief or even the least of the necessaries of life, such as tea and hard bread, sleep the greater portion of the time, when not busy in eating and in the daily observances of that routine belonging to the Greek Catholic Church. The teachings, pomp, and circumstance of the religious observances of this faith alone preserve these people from absolute stagnation. In obedience to its promptings they gladly attend church very regularly. They also make and receive calls on their saints' days, and such days are very numerous. The natives add to these entertainments of their saints' day and birth-festivals, or "Emannimiks," the music of accordeons and violins. Upon the former and its variation, the concertina, they play a number of airs, and are real fond of the noise. A great many of the women in particular can render indifferently a limited selection of tunes, many of which are the old battle-songs, so popular during the rebellion, woven into weird Russian waltzes and love-ditties, which they have jointly gathered from their former masters and our soldiers, who were quartered here in 1869. From the Russians and the troops also they have learned to dance various figures, and have been taught to waltz. These dances, however, the old folks do not enjoy very much. They will come in and sit around and look at the young performers with stolid indifference ; but if they manage to get a strong current of tea setting in their direction, nicely sugared and toned up, they revive and join in the mirth. In old times they never danced here unless they were drunk, and it was the principal occupation of the amiable and mischievous treasury agents and others in those early days to stimulate this beery fun.

Seal-meat is their staple food, and in the village of St. Paul they consume on an average fully five hundred pounds a day the year round, and they are, by the permission of the Secretary of the Treasury, allowed occasionally to kill five thousand or six thousand seal-pups, or an average of twenty-two to thirty young "kotickie " for each man, woman, and child in the settlements. The pups will

16

dress ten pounds each. This shows an average consumption of
nearly six hundred pounds of seal-meat by each person, large and
small, during the year. To this diet the natives add a great deal
of butter and many sweet crackers. They are passionately fond of
butter. No epicure at home or butter-taster in Goshen knows or
appreciates that article better than these people do. If they could
get all that they desire, they would consume one thousand pounds
of butter and five hundred pounds of sweet crackers every week,
and indefinite quantities of sugar. The sweetest of all sweet teeth
are found in the jaw of the ordinary Aleut. But it is of course un-
wise to allow them full swing in this matter, for they would turn
their stomachs into fermenting-tanks if they had free access to an
unlimited supply of saccharine food. The company issues them
two hundred pounds a week. If unable to get sweet crackers,
they will eat about three hundred pounds of hard or pilot bread
every week, and, in addition to this, nearly seven hundred pounds
of flour at the same time. Of tobacco they are allowed fifty pounds
per week; candles, seventy-five pounds; rice, fifty pounds. They
burn, strange as it may seem, kerosene-oil here to the exclusion of
that seal-fat which literally overruns the island. They ignite and
consume over six hundred gallons of kerosene-oil a year in the village of St. Paul alone. They do not fancy vinegar very much; perhaps fifty gallons a year are used up there. Mustard and pepper are
sparingly used, one to one pound and a half a week for the whole
village. Beans they peremptorily reject; for some reason or other
they cannot be induced to use them. Those who go about the vessels contract a taste for split-pea soup, and a few of them are sold
in the village-store. Salt meat, beef or pork, they will take reluctantly, if it is given to and pressed upon them; but they will never
buy it. I remember, in this connection, seeing two barrels of prime
salt pork and a barrel of prime mess salt beef opened in the company's store shortly after my arrival in 1872, and, though the people of the village were invited to help themselves, I think I am right
in saying these three barrels were not emptied when I left the isl-
and in 1873. They use a very little coffee during the year—not more
than one hundred pounds—but of tea a great deal. I do not know
exactly—I cannot find among my notes a record as to that article
—but I can say that they each drink not less than a gallon of tea
per diem. The amount of this beverage which they sip from the
time they rise in the morning until they go to bed late at night is

astounding. Their "samovars," and latterly the regular tea-kettles of our American make, are bubbling and boiling from the moment the housewife bestirs herself at daybreak until the fire goes out when she sleeps. It should be stated in this connection that they are supplied with a regular allowance of coal every year by the company, *gratis*, each family being entitled to a certain amount, which alone, if economically used, keeps them warm all winter in their new houses ; but for those who are extravagant, and are itching to spend their extra wages, an extra supply is always kept in the storehouses of the company for sale. Their appreciation of and desire to possess all the canned fruit that is landed from the steamer is marked to a great degree. If they had the opportunity, I doubt whether a single family on that island to-day would hesitate to bankrupt itself in purchasing this commodity. Potatoes they sometimes demand, as well as onions, and perhaps if these vegetables could be brought here and kept to an advantage the people would soon become very fond of them. Most of these articles of food mentioned heretofore are purchased by the natives in the company's store at either island. This food and the wearing apparel, crockery, etc., which the company bring up here for the use of the people, is sold to them at the exact cost price of the same, plus the expenses of transportation, and many times within my knowledge they have bought goods here at these stores at less rates than they would have been subjected to in San Francisco. The object of the company is not, under any circumstances, to make a single cent of profit out of the sale of these goods to the natives. They aim only to clear the cost and no more. Instructions to this effect are given to its agents, while those of the Government are called upon to take notice of the fact.

The store at St. Paul, as well as that at St. George, has its regular annual "opening" after the arrival of the steamer in the spring, to which the natives seem to pay absorbed attention. They crowd the buildings day and night, eagerly looking for all the novelties in food and apparel. These slouchy men and shawl-hooded women, who pack the area before the counters, appear to feel as deep an interest in the process of shopping as the most enthusiastic votaries of that business do in our own streets. It certainly seems to give them the greatest satisfaction of their lives on the Pribylov Islands.

With regard to ourselves up here in so far as a purely physical existence goes, the American method of living on and in the climate of the Pribylov Islands is highly conducive to strength and health. Tea and coffee, seasoned with condensed milk and lump sugar ; hot biscuits, cakes and waffles ; potatoes, served in every method of cookery ; salt salmon, codfish, and corned beef ; mess pork, and, once a week, a fresh roast of beef or steaks ; all the canned vegetables and fruits; all the potted sauces, jams and jellies ; pies, puddings and pastries ; and the exhaustive list of purely seafaring dishes, such as pea and bean, barley and rice soups, curries and maccaroni; these constitute the staples and many of the luxuries with which the agents of the Alaska Commercial Company prolong their existence while living here in the discharge of their duties, and to which they welcome their guests for discussion and glad digestion.

A piano on St. Paul, in the company house ; an assorted library, embracing over one thousand volumes, selected from standard authors in fiction, science, and history, together with many other unexpected adjuncts of high comfort for body and soul, will be found on these islands, wholly unlooked for by those who first set foot upon them. A small Russian printed library has also been given by the company to the natives on each island for their special entertainment. The rising generation of sealers, however, if they read at all, will read our own typography.

Before leaving the consideration of these people, who are so intimately associated with and blended into the business on these islands, it may be well to clearly define the relation existing between them, the Government, and the company leasing the islands. When Congress granted to the Alaska Commercial Company of San Francisco the exclusive right of taking a certain number of fur-seals every year, for a period of twenty years on these islands, it did so with several reservations and conditions, which were confided in their detail to the Secretary of the Treasury. This officer and the president of the Alaska Commercial Company agreed upon a code of regulations which should govern their joint action in regard to the natives. It was a simple agreement that these people should have a certain amount of dried salmon furnished them for food every year, a certain amount of fuel, a school-house, and the right to go to and come from the islands as they chose ; and also the right to work or not, understanding that in case they did not work, their places would and could be supplied by other people who would work.

The company, however, has gone far beyond this exaction of the Government; it has added an inexpressible boon of comfort, in the formation of those dwellings now occupied by the natives, which was not expressed nor thought of at the time of the granting of the lease. An enlightened business-policy suggested to the company that it would be much better for the natives, and much better for company too, if these people were taken out of their filthy, unwholesome hovels, put into habitable dwellings, and taught to live cleanly, for the simple reason that by so doing the natives, living in this improved condition, would be able physically and mentally, every season when the sealing work began, to come out from their long inanition and go to work at once with vigor and energetic persistency. The sequel has proved the wisdom of the company.

Before this action on their part, it was physically impossible for the inhabitants of St. Paul or St. George Islands to take the lawful quota of one hundred thousand seal-skins annually in less than three or four working months. They take them in less than thirty working days now with the same number of men. What is the gain? Simply this, and it is everything : the fur-seal skin, from the 14th of June, when it first arrives, as a rule, up to the 1st of August, is in prime condition ; from that latter date until the middle of October it is rapidly deteriorating, to slowly appreciate again in value as it sheds and renews its coat; so much so that it is practically worthless in the markets of the world. Hence, the catch taken by the Alaska Commercial Company every year is a prime one, first to last—there are no low-grade "stagey" skins in it ; but under the old regimen, three-fourths of the skins were taken in August, in September and even in October, and were not worth their transportation to London. Comment on this is unnecessary ; it is the contrast made between a prescient business-policy, and one that was as shiftless and improvident as language can well devise.*

* Living as the Seal-islanders do, and doing what they do, the seal's life is naturally their great study and objective point. It nourishes and sustains them. Without it they say they could not live, and they tell the truth. Hence, their attention to the few simple requirements of the law, so wise in its provisions, is not forced or constrained, but is continuous. Self-interest in this respect appeals to them keenly and eloquently. They know everything that is done and everything that is said by anybody and by everybody in their little community. Every seal-drive that is made, and every skin that is taken, is recorded and accounted for by them to their chiefs and their church, when

The company found so much difficulty in getting the youth of the villages to attend their schools, taught by our own people, especially brought up there and hired by the company, that they have adopted the plan of bringing one or two of the brightest boys down every year and putting them into our schools, so that they may grow up here and be educated, in order to return and serve as teachers there. This policy is warranted by the success which attended an experiment made at the time when I was up there first, whereby a son of the chief was carried down and over to Rutland, Vt., for his education, remained there four years, then returned and took charge of the school on St. Paul, which he has had until recently, with the happiest results in increased attendance and attention from the children. But, of course, so long as the Russian Church service is conducted in the Russian language, we will find on the islands more Russian-speaking people than our own. The non-attendance at school was not and is not to be ascribed to indisposition on the part of the children and parents. One of the oldest and most intelligent of the natives told me, explanatory of their feeling and consequent action, that he did not, nor did his neighbors, have any objection to the attendance of their children on our English school ; but, if their boys and young men neglected their Russian lessons they knew not who were going to take their places, when they died, in his church, at the christenings, and at their burial. To any one familiar with the teachings of the Greek Catholic faith, the objection of old Philip Volkov seems reasonable. I hope, therefore, that, in the course of time, the Russian Church service may be voiced in English ; not that I want to substitute any other religion for it— far from it ; in my opinion it is the best one we could have for these people—but until this substitution of our language for the Russian is done, no very satisfactory work, in my opinion, will be accomplished in the way of an English education on the Seal Islands.

The Alaska Commercial Company deserves and will receive a brief but comprehensive notice at this point. In order that we may

---

they make up their tithing-roll at the close of each day's labor. Nothing can come to the islands, by day or by night, without being seen by them and spoken of. I regard the presence of these people on the islands at the transfer, and their subsequent retention and entailment in connection with the seal-business, as an exceedingly good piece of fortune, alike advantageous to the Government, to the company, and to themselves.

follow it to these islands, and clearly and correctly appreciate the circumstance which gave it footing and finally the control of the business, I will pass back and review a chain of evidence adduced in this direction from the time of our first occupation, in 1867, of the territory of Alaska.

It will be remembered by many people, that when we were ratifying the negotiation between our Government and that of Russia, it became painfully apparent that nobody in this country knew anything about the subject of Russian America. Every school-boy knew where it was located, but no professor or merchant, however wise or shrewd, knew what was in it. Accordingly, immediately after the purchase was made and the formal transfer effected, a large number of energetic and speculative men, some coming from New England even, but most of them residents of the Pacific coast, turned their attention to Alaska. They went up to Sitka in a little fleet of sail and steam vessels, but among their number it appears there were only two of our citizens who knew of or had the faintest appreciation as to the value of the Seal Islands. One of these, Mr. H. M. Hutchinson, a native of New Hampshire, and the other, a Captain Ebenezer Morgan, a native of Connecticut, turned their faces in 1868 toward them ; also an ex-captain of the Russian-American Company, Gustav Niebaum, who became a citizen immediately after the transfer, knowing of their value, chartered a small vessel, and hastened so as to land there a few days even before Captain Morgan arrived in the *Peru*, a whaling ship.

Mr. Hutchinson gathered his information at Sitka—Captain Morgan had gained his years before by experience on the South Sea sealing grounds. Mr. Hutchinson represented a company of San Francisco or California capitalists when he landed on St. Paul ; Captain Morgan represented another company of New London capitalists and whaling merchants. They arrived almost simultaneously, Morgan a few days or weeks anterior to Hutchinson. He had quietly enough commenced to survey and pre-empt the rookeries on the islands, or, in other words, the work of putting stakes down and recording the fact of claiming the ground, as miners do in the mountains ; but later agreed to co-operate with Mr. Hutchinson. These two parties passed that season of 1868 in exclusive control of those islands, and they took an immense number of seals. They took so many that it occurred to Mr. Hutchinson unless something was done to check and protect these wonderful rookeries, which he

saw here for the first time, and which filled him with amazement,
that they would be wiped out by the end of another season ; although he was the gainer then, and would be perhaps at the end,
if they should be thus eliminated, yet he could not forbear saying to himself that it was wrong and should not be.   To this Captain Morgan also assented, and Captain Niebaum joined with them
cordially.   In the fall of 1868 Mr. Hutchinson and Captain Morgan, by their personal efforts, interested and aroused the Treasury Department and Congress, so that a special resolution was
enacted declaring the Seal Islands a governmental reservation,
and prohibiting any and all parties from taking seals thereon
until further action by Congress.   In 1869, seals were taken on
those islands, under the direction of the Treasury Department,
for the subsistence of the natives only ; and in 1870 Congress
passed the present law, for the protection of the fur-bearing animals
on those islands, and under its provisions, and in accordance there-
with, after an animated and bitter struggle in competition, the
Alaska Commercial Company, of which Mr. Hutchinson was a prime
organizer, secured the award and received the franchise which it
now enjoys and will enjoy for some time yet.   The company is an
American corporation, with a charter, rules, and regulations.   They
employ a fleet of vessels, sail and steam : four steamers, a. dozen or
fifteen ships, barks, and sloops.   Their principal occupation and
attention is given naturally to the Seal Islands, though they have
station sscattered over the Aleutian Islands and that portion of
Alaska west and north of Kadiak.   No post of theirs is less than
five hundred or six hundred miles from Sitka.

   Outside of the Seal Islands all trade in this territory of Alaska
is entirely open to the public.   There is no need of protecting the
fur-bearing animals elsewhere, unless it may be by a few wholesome general restrictions in regard to the sea-otter chase.   The
country itself protects the animals on the mainland and other
islands by its rugged, forbidding, and inhospitable exterior.

   The treasury officials on the Seal Islands are charged with the
careful observance of every act of the company ; a copy of the lease
and its covenant is conspicuously posted in their office ; is translated into Russian, and is familiar to all the natives.   The company
directs its own labor, in accordance with the law, as it sees fit ; se-
lects its time of working, etc.   The natives themselves work under
the direction of their own chosen foremen, or " toyones."   These

chiefs call out the men at the break of every working-day, divide them into detachments according to the nature of the service, and order their working.   All communications with the laborers on the sealing-ground and the company passes through their hands, those chiefs having every day an understanding with the agent of the company as to his wishes, and they govern themselves thereby.

The company pays forty cents for the labor of taking each skin. The natives take the skins on the ground, each man tallying his work and giving the result at the close of the day to his chief or foreman.   When the skins are brought up and counted into the salt-houses, where the agent of the company receives them from the hands of the natives, the two tallies usually correspond very closely, if they are not entirely alike.   When the quota of skins is taken, at the close of two, three, or four weeks of labor, as the case may be, the total sum for the entire catch is paid over in a lump to the chiefs, and these men divide it among the laborers according to their standing as workmen, which they themselves have exhibited on their special tally-sticks.   For instance, at the annual divisions or "catch" settlement, made by the natives on St. Paul Island among themselves, in 1872, when I was present, the proceeds of their work for that season in taking and skinning seventy-five thousand seals, at forty cents per skin, with extra work connected with it, making the sum of $30,637.37, was divided among them in this way : There were seventy-four shares made up, representing seventy-four men, though in fact only fifty-six men worked, but they wished to give a certain proportion to their church, a certain proportion to their priest, and a certain proportion to their widows; so they water their stock, commercially speaking.*

It will be remembered that at the time the question of leasing the islands was before Congress much opposition to the proposal

---

*37 first-class shares, at.........................$451 22 each.
23 second-class shares, at...................... 406 08 each.
4 third-class shares, at........................ 360 97 each.
10 fourth-class shares, at...................... 315 85 each.

These shares do not represent more than fifty-six able-bodied men.

In August, 1873, while on St. George Island, I was present at a similar division, under similar circumstances, which caused them to divide among themselves the proceeds of their work in taking and skinning twenty-five

was made, on several grounds, by two classes, one of which argued against a "monopoly," the other urging that the Government itself would realize more by taking the whole management of the business into its own hands. At that time far away from Washington, in the Rocky Mountains, I do not know what arguments were used in the committee-rooms, or who made them; but, since my careful and prolonged study of the subject on the ground itself, and of the trade and its conditions, I am now satisfied that the act of June, 1870, directing the Secretary of the Treasury to lease the seal-islands of Alaska to the highest bidder, under the existing conditions and qualifications, did the best and the only correct and profitable thing that could have been done in the matter, both with regard to the preservation of the seal-life in its original integrity, and the pecuniary advantage of the treasury itself. To make this statement perfectly clear, the following facts, by way of illustration, should be presented:

First. When the Government took possession of these interests in 1868 and 1869, the gross value of a seal-skin laid down in the best market, at London, was less in some instances and in others but slightly above the present tax and royalty paid upon it by the Alaska Commercial Company.

Second. Through the action of the intelligent business-men who took the contract from the Government in stimulating and encouraging the dressers of the raw material, and in taking sedulous

---

thousand seals, at forty cents a skin, $10,000. They made the following subdivision:

	Per share.
17 shares each, 961 skins	$384 40
2 shares each, 935 skins	374 00
3 shares each, 821 skins	328 40
1 share each, 820 skins	328 00
3 shares each, 770 skins	308 00
3 shares each, 400 skins	160 00

These twenty-nine shares referred to, as stated above, represent only twenty-five able-bodied men; two of them were women. This method of division as above given is the result of their own choice. It is an impossible thing for the company to decide their relative merits as workmen on the ground, so they have wisely turned its entire discussion over to them. Whatever they do they must agree to—whatever the company might do they possibly and probably would never clearly understand, and hence dissatisfaction and suspicion would inevitably arise. As it is, the whole subject is most satisfactorily settled.

care that nothing but good skins should leave the island, and in combination with leaders of fashion abroad, the demand for the fur, by this manipulation and management, has been wonderfully increased.

Third. As matters now stand, the greatest and best interests of the lessees are identical with those of the Government; what injures one instantly injures the other. In other words, both strive to guard against anything that shall interfere with the preservation of the seal-life in its original integrity, and both having it to their interest, if possible, to increase that life; if the lessees had it in their power, which they certainly have not, to ruin these interests by a few seasons of rapacity, they are so bonded and so environed that prudence prevents it.

Fourth. The frequent changes in the office of the Secretary of the Treasury, who has very properly the absolute control of the business as it stands, do not permit upon his part that close, careful scrutiny which is exercised by the lessees, who, unlike him, have but their one purpose to carry out. The character of the leading men among them is enough to assure the public that the business is in responsible hands, and in the care of persons who will use every effort for its preservation and its perpetuation, as it is so plainly their best end to serve. Another great obstacle to the success of the business, if controlled entirely by the Government, would be encountered in disposing of the skins after they had been brought down from the islands. It would not do to sell them up there to the highest bidder, since that would license the sailing of a thousand ships to be present at the sale. The rattling of their anchor-chains and the scraping of their keels upon the beaches of the two little islands would alone drive every seal away and over to the Russian grounds in a remarkably short space of time. The Government would therefore need to offer them at public auction in this country: that would be simple history repeating itself—the Government would be at the mercy of any well-organized combination of buyers. Its agents conducting the sale could not counteract the effect of such a combination as can the agents of a private corporation, who may look after their interest in all the markets of the world in their own time and in their own way, according to the exigencies of the season and the demand, and who are supplied with money which they can use, without public scandal, in the manipulation of the market. On this ground I feel confident in stating that the Treasury of the United States receives more money, net, under the system now in operation than it would by taking the exclusive control of the business. Were any capable government officer supplied with, say, $100,000, to expend in "working the market," and intrusted with the disposal of one hundred thousand seal-skins wherever he could do so to the best advantage of the Government, and were this agent a man of first-class ability and energy, I think it quite likely that the same success might attend his labor in the London market that distinguishes the management of the Alaska Commercial Company. But imagine the cry of fraud and embezzlement that would be raised against him, however honest he might be! This alone would bring the whole business into positive disrepute, and make it a national scandal. As matters are now conducted there is no room for scandal—not one single transaction on the islands but what is as clear to investigation and accountability as the light of the noon-day sun ; what is done is known to everybody, and the tax now laid by the Government upon, and paid into the treasury every year by the Alaska Commercial Company yields alone a handsome rate of interest on the entire purchase-money expended for the ownership of all Alaska.

It is frequently urged with great persistency, by misinformed and malicious authority, that the lessees can and do take thousands of skins in excess of the law, and this catch in excess is shipped *sub rosa* to Japan from the Pribylov Islands. To show the folly of such a move on the part of the Company, if even it were possible, I will briefly recapitulate the conditions under which the skins are taken. The natives of St. Paul and St. George do themselves, in the manner I have indicated, all the driving and skinning of the seals for the company. No others are permitted or asked to land upon the islands to do this work, so long as the inhabitants of the islands are equal to it. They have been equal to it and they are more than equal to it. Every skin taken by the natives is counted by themselves, as they get forty cents per pelt for that labor, and, at the expiration of each day's work in the field, the natives know exactly how many skins have been taken by them, how many of these skins have been rejected by the company's agent because they were carelessly cut and damaged in skinning—usually about three-fourths of one per cent. of the whole catch—and they have it recorded every evening by those among them who are charged with the duty. Thus, were one hundred and one thousand skins taken, instead of one hundred thousand allowed by law, the natives would know it as quickly as it was done, and they would, on the strength of their record and their tally, demand the full amount of their compensation for the extra labor ; and were any ship to approach the islands, at any hour, these people would know it at once, and would be aware of any shipment of skins that might be attempted. It would then be the common talk among the three hundred and ninety-eight inhabitants of the two islands, and it would be a matter of record, open to any person who might come upon the ground charged with investigation.

Furthermore, these natives are constantly going to and from Oonalashka, visiting their relations in the Aleutian settlements, hunting for wives, etc. On the mainland they have intimate intercourse with bitter enemies of the company, with whom they would not hesitate to talk over the whole state of affairs on the islands, as they always do ; for they know nothing else and think of nothing else and dream of nothing else. Therefore, should anything be done contrary to the law, the act could and would be reported by these people. The Government, on its part, through its four agents stationed on these islands, counts these skins into the ship, and one of their number goes down to San Francisco upon her. There the collector of the port details experts of his own, who again count them all out of the hold, and upon that record the tax is paid and the certificate signed by the Government.

It will therefore at once be seen, by examining the state of affairs on the islands, and the conditions upon which the lease is granted, that the most scrupulous care in fulfilling the terms of the contract is compassed, and that this strict fulfilment is the most profitable course for the lessees to pursue ; and that it would be downright folly in them to deviate from the letter of the law, and thus lay themselves open at any day to discovery, the loss of their contract, and forfeiture of their bonds. Their action can be investigated at any time, any moment, by Congress; of which they are fully aware. They cannot bribe these three hundred and ninety-eight people on the islands to secrecy, any more successfully than they could conceal their action from them on the sealing fields ; and any man of average ability could go, and can go, among these natives and inform himself as to the most minute details of the catch, from the time the lease was granted up to the present hour, should he have reason to suspect the honesty of the Treasury agents. The road to and from the islands is not a difficult one, though it is travelled only once a year.

CHAPTER X.

AMPHIBIAN MILLIONS.

Difference between a Hair-seal and a Fur-seal.—The Fur-seal the most Intelligent of all Amphibians.—Its singularly Free Progression on Land.—Its Power in the Water.—The Old Males the First Arrivals in the Spring.—Their Desperate Battles one with Another for Position on the Breeding Grounds.—Subsequent Arrival of the Females.—Followed by the "Bachelors."—Wonderful Strength and Desperate Courage of the Old Males.—Indifference of the Females.—Noise of the Rookeries Sounds like the Roar of Niagara.—Old Males fast from May to August, inclusive; neither Eat nor Drink, nor Leave their Stations in all that Time.—Graceful Females.—Frolicsome "Pups."—They have to Learn to Swim!—How they Learn.—Astonishing Vitality of the Fur-seal.—"Podding" of the Pups.—Beautiful Eyes of the Fur-seal.—How the "Holluschickie," or Bachelor Seals, Pass the Time.—They are the only ones Killed for Fur.—They Herd alone by Themselves in spite of their Inclination; Obliged to.—They are the Champion Swimmers of the Sea.—A Review of the Vast Breeding Rookeries.—Natives Gathering a Drove.—Driving the Seals to the Slaughtering Fields.—No Chasing—no Hunting of Seals.—The Killing Gang at Work: Skinning, Salting, and Shipping the Pelts.—All Sent Direct to London.—Reasons Why.—How the Skins are Prepared for Sacks, Muffs, etc.

"The web-footed seals forsake the stormy swell,
And, sleeping in herds, exhale nauseous smell."—HOMER.

A VIVID realism of the fact that often truth is far stranger than fiction is strikingly illustrated in the life-history of the fur-seal: as it is the one overshadowing and superlatively interesting subject of this discussion, I shall present all its multitudinous details, even at the risk of being thought tedious. That aggregate of animal life shadowed every summer out upon the breeding grounds of the Seal Islands is so vast, so anomalous, so interesting, and so valuable, that it deserves the fullest mention; and even when I shall have done, it will be but feebly expressed.

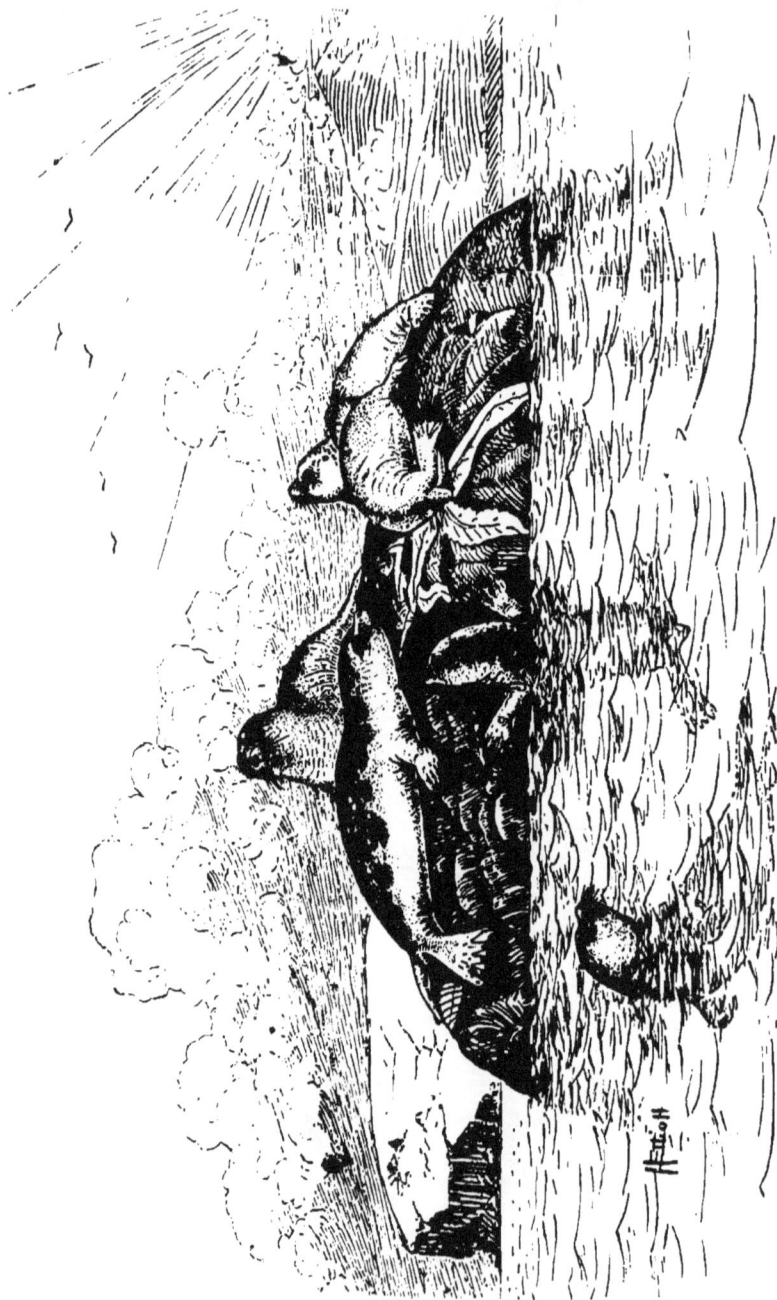

THE HARBOR SEAL

Great as it is, yet a short schedule * embraces the titles of all the pinnipeds found in, on, and around the island-group. Of this list the hair-seal† is the animal which has done so much to found that erroneous popular and scientific opinion as to what a fur-seal appears like. *Phoca vitulina* has, in this manner, given to the people of the world a false idea of its relatives. It is so commonly distributed all over the littoral salt waters of the earth, seen in the harbors of nearly every marine port, or basking along the loneliest and least inhabited of desolate coasts far to the north, that everybody has noticed it, if not in life, then in its stuffed skins at the museums, sometimes very grotesquely mounted. This copy, set everywhere before the eye of the naturalist, has rendered it so difficult for him to correctly discriminate between the *Phocidæ* and the *Otariidæ*, that the synonymy of the *Pinnipedia* has been expanded until it is replete with meaningless description and surmise.

Although the hair-seal belongs to the great group of pinnipeds, yet it does not have even a generic affinity with those seals with which it has been so persistently grouped, namely, the fur-seal and the sea-lion. It no more resembles them, than does the raccoon a black or grizzly bear.

I shall not enter into a detailed description of this seal ; it is wholly superfluous, for excellent, and, I believe, trustworthy accounts have been repeatedly published by writers who have treated of the subject as it was spread before their eyes on the coasts of

---

* The seal-life on the Pribylov Islands may be classified under the following heads, namely : (1) The fur-seal, *Callorhinus ursinus*, the " kautickie " of the Russians ; (2) the sea-lion, *Eumetopias stelleri*, the "seevitchie " of the Russians; (3) the hair-seal, *Phoca vitulina*, the " nearhpahsky " of the Russians ; (4) the walrus, *Odobænus obesus*, the "morsjee" of the Russians.

† The inconsequential numbers of the hair-seal around and on the Pribylov Islands seem to be characteristic of all Alaskan waters and the northwest coast ; also, the phocidæ are equally scant on the Asiatic littoral margins. Only the following four species are known to exist throughout the entire extent of that vast marine area, viz. :

Phoca vitulina—Everywhere between Bering Straits and California.

Phoca fœtida—Plover Bay, Norton's Sound, Kuskokvim mouth, and Bristol Bay, of Bering Sea ; Cape Seartze Kammin, Arctic Ocean, to Point Barrow.

Erignathus barbatus—Kamchatkan coast, Norton's Sound, Kuskokvim mouth and Bristol Bay, of Bering Sea.

Histriophoca equestris—Yukon mouth and coast south to Bristol Bay, of Bering Sea.

Labrador, Newfoundland, and Greenland ; to say nothing of the re-
searches and notes made by European scientists.   It differs completely in shape and habit from its congeners on these islands.
Here, where I have studied its biology, it seldom comes up from
the water more than a few rods at the farthest ; generally hauling
and resting at the margin of the surf-wash.   It takes up no position
on land to hold and protect a family or harem, preferring the detached water-worn rocks, especially those on the lonely north shore
of St. Paul, although I have seen it resting at "Gorbatch,* near the
sea-margin of the great seal-rookery of that name, on the Reef
Point of St. Paul ; its cylindrical, supine, gray and white body
marked in strong contrast with the erect, black, and ochre-colored
forms of the *Callorhinus*, which swarmed round about it.   On such
small spots of rock, wet and isolated from the mainland, and in secluded places of the north shore, the "nearhpah" brings forth its
young, a single pup, perfectly white, covered with long woolly hair,
and weighing from three to seven pounds.   This pup grows rapidly, and after the lapse of four or five months it tips the scales at
fifty pounds ; by that time it has shed its infant coat and donned
the adult soft steel-gray hair over the head, limbs, and abdomen,
with its back most richly mottled and barred lengthwise, by dark
brown and brown-black streaks and blotches, suffused at their edges
into the light steel-gray ground of the body.   When they appear in
the spring following, this bright gray tone to their color has ripened into a dingy ochre, and the mottling spread well over the head
and down on the upper side or back of the flippers, but fades out
as it progresses.   It has no appreciable fur or under-wool.   There
is no noteworthy difference as to color or size between the sexes.
So far as I have observed, they are not polygamous.   They are exceedingly timid and wary at all times, and in this manner and
method they are diametrically opposed, not by shape alone, but
by habit and disposition, to the fashion of the fur-seal in especial,
and the sea-lion.   Their skin is of little value, comparatively, but
their chief merit, according to the natives, is the relative greater
juiciness and sweetness of their flesh, over even the best steaks of
sea-lion or fur-seal pup meat.

One common point of agreement among all authors was, by my
observations of fact, so strikingly refuted, that I will here correct a
prevalent error made by naturalists who, comparing the hair-seal
with the fur-seal, state that in consequence of the peculiar struct-

ure of their limbs, their progression on land is "mainly accomplished by a wriggling, serpentine motion of the body, slightly assisted by the extremities." This is not so in any respect; for, whenever I have purposely surprised these animals, a few rods from the beach-margin, they would awake and excitedly scramble, or rather spasmodically exert themselves, to reach the water instantly, by striking out quickly with both fore-feet simultaneously, lifting in this way alone, and dragging the whole body forward, without any "wriggling motion" whatever to their back or posterior parts, moving from six inches to a foot in advance every time their fore-feet were projected forward, and the body drawn along according to the violence of the effort and the character of the ground; the body of the seal then falls flat upon its stomach, and the fore-feet or flippers are free again for another similar motion. This action of *Phoca* is effected so continuously and so rapidly, that in attempting to head off a young "nearhpah" from the water, at English Bay, I was obliged to leave a brisk walk and take to a dog trot to do it. The hind-feet are not used when exerted in this rapid movement at all; they are dragged along in the wake of the body, perfectly limp and motionless. But they do use those posterior parts, however, when leisurely climbing up and over rocks undisturbed, or playing one with another; still it is always a weak, trembling terrestrial effort, and particularly impotent and clumsy. In their swift swimming the hind-feet of *Phocidæ* evidently do all the work; the reverse is a remarkable characteristic of the *Otariidæ*.

These remarks of mine, it should be borne in mind, apply directly to the *Phoca vitulina*, and I presume indirectly with equal force to all the rest of its more important generic kindred, be they as large as the big maklok, *Erignathus barbata*, or less.

This hair-seal is found around these islands at all seasons of the year, but in very small numbers. I have never seen more than twenty-five or thirty at any one time, and I am told that its occidental distribution, although everywhere found, above and below, from the arctic to the tropics, and especially general over the North Pacific coast, nowhere exhibits any great number at any one place; but we know that it and its immediate kindred form a vast majority of the multitudinous seal-life peculiar to our North Atlantic shores, ice-floes, and contiguous waters. The scarcity of this species, and of all its generic allies, in the waters of the Pacific, is notable as compared with those of the circumpolar Atlantic,

where these hair-seals are the seals of commerce : they are found in
such immense numbers between Greenland and Labrador, and
thence to the eastward at certain seasons of every year, that employment is given to a fleet of about sixty sailing and steam vessels, which annually goes forth from St. John, Newfoundland, and
elsewhere, fitted for seal-fishing : taking in all this cruising over
three hundred thousand of these animals each season. The principal object of value, however, is the oil rendered from them : the
skins have a very small commercial importance.

The fur-seal, *Callorhinus ursinus*, which repairs to these islands
to breed and to shed its hair and fur, in numbers that seem almost
fabulous, is the highest organized of all the *Pinnipedia*, and, indeed,
for that matter, when land and water are weighed in the account
together, there is no other animal known to man which may be truly
classed as its superior, from a purely physical point of view.
Certainly there are few, if any, creatures in the animal kingdom
that can be said to exhibit a higher order of instinct, approaching
even our intelligence.

I wish to draw attention to a specimen of the finest of this race
—a male in the flush and prime of his first maturity, six or seven
years old, and full grown. When it comes up from the sea early
in the spring, out to its station for the breeding season, we have an
animal before us that will measure six and a half to seven and a
quarter feet in length from tip of nose to the end of its abbreviated
abortive tail. It will weigh at least four hundred pounds, and I
have seen older specimens much more corpulent, which, in my best
judgment, could not be less than six hundred pounds in weight.*

---

* Those extremely heavy adult males which arrive first in the season and
take their stations on the rookeries, are so fat that they do not exhibit a
wrinkle or a fold of the skins enveloping their blubber-lined bodies. Most of
this fatty deposit is found around the shoulders and the neck, though a warm
coat of blubber covers all the other portions of the body save the flippers. This
blubber-thickening of the neck and chest is characteristic of the adult males
only, which are, by its provisions, enabled to sustain the extraordinary protracted fasting periods incident to their habit of life and reproduction.

When those superlatively fleshy bulls first arrive, a curious body-tremor
seems to attend every movement which the animals make on land ; their fat
appears to ripple backward and forward under their hides, like waves. As
they alternate with their flippers in walking, the whole form of the "see-
catchie" fairly shakes as a bowl full of jelly does when agitated on the table
before us.

GROUP OF FUR SEALS

Young Females   Old Male   Young Male   Mother Seal   Old Male " Roaring"
   2 years   18 years   6 years   and Pup nursing   Young Males
                                                                            2 years

The head of this animal now before us appears to be disproportionately small in comparison with an immensely thick neck and shoulders ; but, as we come to examine it, we will find it is mostly all occupied by the brain. The light frame-work of its skull supports an expressive pair of large bluish-hazel eyes, alternately burning with revengeful, passionate light, then suddenly changing to the tones of tenderness and good-nature. It has a muzzle and jaws of about the same size and form observed in any full-blooded Newfoundland dog, with this difference, that the lips are not flabby and overhanging ; they are as firmly lined and pressed against one another as our own. The upper lips support a yellowish-white and gray mustache, composed of long, stiff bristles, which, when not torn out and broken off in combat, sweeps down and over the shoulders as a luxuriant plume. Look at it as it comes leisurely swimming on toward the land ; see how high above the water it carries its head, and how deliberately it surveys the beach, after having stepped upon it (for it may be truly said to step with its fore-flippers, as they regularly alternate when it moves up), carrying the head well above them, erect and graceful, at least three feet from the ground. The fore-feet, or flippers, are a pair of dark bluish-black hands, about eight or ten inches broad at their junction with the body, and the metacarpal joint, running out to an ovate point at their extremity, some fifteen to eighteen inches from this union—all the rest of the forearm, the ulna, radius, and humerus being concealed under the skin and thick blubber-folds of the main body and neck, hidden entirely at this season, when it is so fat. But six weeks to three months after this time of landing, when that superfluous fat and flesh is consumed by self-absorption, then those bones will show plainly under its shrunken skin. On the upper side of these flippers the hair of the body straggles down finer and fainter as it comes below to a point close by, and slightly beyond that spot of junction where the phalanges and the metacarpal bones unite, similar to that point on our own hand where our knuckles are placed ; and here the hair ends, leaving the rest of the skin to the end of the flipper bare and wrinkled in places at the margin of the inner side ; showing, also, five small pits, containing abortive nails, which are situated immediately over the union of the phalanges with their cartilaginous continuations to the end of the flipper.

On the under side of the flipper the skin is entirely bare from its outer extremity up to the body-connection. It is sensibly tougher and thicker than elsewhere on the body ; it is deeply and regularly wrinkled with seams and furrows, which cross one another so as to leave a kind of sharp diamond-cut pattern. When they are placed by the animal upon the smoothest rocks, shining and slippery from algoid growths and the sea-polish of restless waters, they seldom fail to adhere.

When we observe this seal moving out on the land, we notice that, though it handles its fore-feet in a most creditable manner, it brings up its rear in quite a different style : for, after every second step ahead with the anterior limbs it will arch its spine, and in arching, it drags and lifts up, and together forward, the hind-feet, to a fit position under its body, giving it in this manner fresh leverage for another movement forward by the fore-feet, in which the spine is again straightened out, and then a fresh hitch is taken up on the posteriors once more, and so on as the seal progresses. This is the leisurely and natural movement on land, when not disturbed, the body all the time being carried clear of and never touching the ground ; but if the creature is frightened, this method of progression is radically changed. It launches into a lope and actually gallops so fast that the best powers of a man in running are taxed to head it off. Still, it must be remembered that it cannot run far before it sinks, trembling, gasping, breathless, to the earth. Thirty or forty yards of such speed marks the utmost limit of its endurance.

The radical difference in the form and action of the hind-feet cannot fail to strike the eye at once. They are one-seventh longer than the fore-hands and very much lighter and more slender ; they resemble, in broad terms, a pair of black-kid gloves, flattened out and shrivelled, as they lie in their box.

There is no suggestion of fingers on the fore-hands ; but the hind-feet seem to be toes run into ribbons, for they literally flap about involuntarily from that point where the cartilaginous processes unite with the phalangeal bones. The hind-feet are also merged in the body at their junction with it, like those anterior. Nothing can be seen of the leg above the tarsal joint.

The shape of the hind flipper is strikingly like that of a human foot, provided the latter were drawn out to a length of twenty or twenty-two inches, the instep flattened down and the toes run out into thin, membranous, oval-tipped points, only skin-thick, leaving three strong cylindrical, grayish, horn-colored nails, half an inch long each, back six inches from these skinny toe-ends, without any sign of nails to mention on the outer big and little toes.

On the upper side of this hind-foot the body-hair comes down to that point where the metatarsus and phalangeal bones join and fade out. From that junction the phalanges, about six inches down to the nails above mentioned, are entirely bare and stand ribbed up in bold relief on the membrane which unites them, as the web to a duck's foot. The nails just referred to mark the ends of the phalangeal bones and their union in turn with the cartilaginous processes, which run rapidly tapering and flattening out to the ends of the thin toe-points. Now, as we are looking at this fur-seal's motion and progression, that which seems most odd is the gingerly manner (if I may be allowed to use the expression) in which it carries these hind flippers. They are held out at right angles from the body directly opposite the pelvis, the toe-ends or flaps slightly waving, curled, and drooping over, supported daintily, as it were, above the earth, the animal only suffering its weight behind to fall upon its heels, which are themselves opposed to each other, scarcely five inches apart.

We shall, as we see this seal again later in the season, have to notice a different mode of progression and bearing, both when it is lording over its harem or when it grows shy and restless at the end of the breeding season, then faint, emaciated, and dejected. But we will now proceed to observe him in the order of his arrival and that of his family. His behavior during the long period of fasting and unceasing activity and vigilance, and other cares which devolve upon him as the most eminent of all polygamists in the brute world, I shall carefully relate, and to fully comprehend the method of this exceedingly interesting animal it will be frequently necessary for the reader to refer to my sketch-maps of its breeding grounds or rookeries, and the islands.

The adult males are the first examples of the *Callorhinus* to arrive in the spring on the seal-ground, which has been deserted by all of them since the close of the preceding year. *

* The distances at sea, away from the Pribylov Islands, in which fur-seals are found during the breeding season, are very considerable. Scattered records have been made of seeing large bands of them during August as far down the northwest coast as they probably range at any season of the year, viz., well out at sea in the latitude of Cape Flattery, 47° to 49° south latitude. In the winter and spring, up to middle of June, all classes are found here spread out

Between May 1st and 5th, usually, a few males will be found scattered over the rookeries pretty close to the water. They are at this time quite shy and sensitive, seeming not yet satisfied with the land, and a great many spend day after day idly swimming out among the breakers a little distance from the shore before they come to it, perhaps somewhat reluctant at first to enter upon the assiduous duties and the grave responsibilities before them of fighting for and maintaining their positions in the rookeries.

The first arrivals are not always the oldest bulls, but may be said to be the finest and most ambitious of their class. They are full grown and able to hold their places on the rookeries or the breeding flats, which they immediately take up after coming ashore. Their method of landing is to come collectively to those breeding grounds where they passed the prior season ; but I am not able to say authoritatively, nor do I believe it, strongly as it has been urged by many careful men who were with me on the islands, that these animals come back to and take up the same position on their breeding grounds that they individually occupied when there last year. From my knowledge of their action and habit, and from what I have learned of the natives, I should say that very few, if any, of them make such a selection and keep these places year after year. Even did the seal itself intend to come directly from the sea to that spot on the rookery which it left last summer, what could it do if it came to that rookery margin a little later and found that another " see-catch " had occupied its ground? The bull could do nothing. It would either have to die in its tracks, if it persisted in attaining this supposed objective point, or do what undoubtedly it does do— seek the next best locality which it can secure adjacent.

One aged " see-catch " was pointed out to me at the " Gorbatch " section of the Reef rookery, as an animal that was long known to the natives as a regular visitor, close by or on the same rock, every season during the past three years. They called him " Old John," and they said they knew him because he had one of his posterior digits missing, bitten off, perhaps, in a combat. I

over wide areas of ocean. Then by June 15th they will have all departed, the first and the latest, *en route* for the Pribylov Islands. Then, when seen again in this extreme southern range, I presume the unusually early examples of return toward the end of August are squads of the yearlings of both sexes, for this division is always the last to land on and the first to leave the Seal Islands annually.

" OLD JOHN "

A Life Study of an aged Fur Seal-Bull or "Seecatch."—Gorbotch Rookery, July 2, 1872

saw him in 1872, and made careful drawings of him in order that I might recognize his individuality, should he appear again in the following year, and when that time rolled by, I found him not ; he failed to reappear, and the natives acquiesced in his absence. Of course it was impossible to say that he was dead when there were ten thousand rousing, fighting bulls to the right, left, and below us, under our eyes, for we could not approach for inspection. Still, if these animals came each to a certain place in any general fashion, or as a rule, I think there would be no difficulty in recognizing the fact ; the natives certainly would do so ; as it is, they do not. I think it very likely, however, that the older bulls come back to the same common rookery ground where they spent the previous season ; but they are obliged to take up their position on it just as the circumstances attending their arrival will permit, such as finding other seals which have arrived before them, or of being whipped out by stronger rivals from their old stands.

It is entertaining to note, in this connection,' that the Russians themselves, with the object of testing that mooted query, during the later years of their possession of the islands, drove up a number of young males from Lukannon, cut off their ears, and turned them out to sea again. The following season, when the droves came in from the "hauling-grounds" to the slaughtering-fields, quite a number of those cropped seals were in the drives, but instead of being found all at one place—the place from whence they were driven the year before—they were scattered examples of croppies from every point on the island. The same experiment was again made by our people in 1870 (the natives having told them of such prior undertaking), and they went also to Lukannon, drove up one hundred young males, cut off their left ears, and set them free in turn. Of this number, during the summer of 1872, when I was there, the natives found in their driving of seventy-five thousand seals from the different hauling-grounds of St. Paul up to the village killing-grounds, two on Novastoshnah rookery, ten miles north of Lukannon, and two or three from English Bay and Tolstoi rookeries, six miles west by water ; one or two were taken on St. George Island, thirty-six miles to the southeast, and not one from Lukannon was found among those that were driven from there ; probably, had all the young males on the two islands this season been examined, the rest of the croppies that had returned from the perils of the deep, whence they sojourned during the winter, would have been distributed quite equally about the Pribylov hauling-grounds. Although the natives say that they think the cutting off of the animal's ear gives the water such access to its head as to cause its death, yet I noticed that those examples which we had recognized by this auricular mutilation, were normally fat and well developed. Their theory does not appeal to my belief, and it certainly requires confirmation.

These experiments would tend to prove very cogently and conclusively that when the seals approach the islands in the spring they have nothing in their minds but a general instinctive appreciation of the fitness of the land as a whole, and no special fondness or determination to select any one particular spot, not even the place of their birth. A study of my map of the distribution of the seal-life on St. Paul, clearly indicates that the landing of the seals on the respective rookeries is influenced greatly by the direction of the wind at the time of their approach to the islands in the spring and early summer. The prevailing airs, blowing, as they do at that season, from the north and northwest, carry far out to sea the odor of the old rookery flats, together with a fresh scent of the pioneer bulls which have located themselves on these breeding grounds three or four weeks in advance of their kind. The seals come up from the great North Pacific, and hence it will be seen that the rookeries of the south and southeastern shores of St. Paul Island receive nearly all the seal-life, although there are miles of perfectly eligible ground at Nahsayvernia or north shore. To settle this matter beyond all argument, however, I know is an exceedingly difficult task, since the identification of individuals, from one season to another, among the hundreds of thousands, and even millions, that come under your eye on one of these great rookeries, is well-nigh impossible. From the time of the first arrival in May up to the beginning of June, or as late as the middle of that month, if the weather be clear, is an interval in which everything seems quiet. Very few seals are added to the pioneers that have landed, as we have described. About June 1st, however, sometimes a little before, and never much later, the seal-weather—the foggy, humid, oozy damp of summer—sets in ; and with it, as the gray banks roll up and shroud the islands, old bull-seals swarm from the depths by hundreds and thousands, and locate themselves in advantageous positions for the reception of the females, which are generally three weeks or a month later than this date in arrival.

It appears from my survey of these breeding grounds that a well-understood principle exists among the able-bodied males, to-wit : that each one shall remain undisturbed on his own ground, which is usually about six to eight feet square: provided, that at the start, and from that time until the arrival of the females, he is strong enough to hold this ground against all comers ; inasmuch as the crowding in of fresh arrivals often causes a removal of thóso which, though equally able-bodied at first, have exhausted themselves by fighting earlier and constantly, they are finally driven by these fresher animals back farther and higher up on the rookery, and sometimes off altogether.

The labor of locating and maintaining a position on the rookery is real, terrible and serious business for these bulls which come in last, and it is so all the time to those males that occupy the water-line of the breeding grounds. A constantly sustained fight between the new-comers and the occupants goes on morning, noon, and night, without cessation, frequently resulting in death to one, or even both, of the combatants. The " seecatchie " under six years of age, although hovering about the sea-margins of the breeding grounds, do not engage in much fighting there ; it is the six and seven year old males, ambitious and flushed with a full sense of their reproductive ability, that swarm out and do battle with the older males of these places. A young male of this latter class is, however, no match for any fifteen or twenty year old bull, provided that an old "seecatchie" retains his teeth ; for, with these weapons, his relatively harder thews and sinews give him the advantage in almost every instance among the hundreds of combats that I have witnessed. These trials of strength between the old and the young are incessant until the rookeries are mapped out ; since, by common consent, the males of all classes recognize the coming of the females. After their arrival and settlement over the whole extent of the breeding grounds, about July 15th at the latest, very little fighting takes place.

Many of those bulls exhibit wonderful strength and desperate courage. I marked one veteran at Gorbatch, who was the first to take up his position early in May, and that position, as usual, directly on the water-line. This male seal had fought at least forty or fifty desperate battles, and beaten off his assailants every time— perhaps nearly as many different seals each of which had coveted his position—when the fighting season was over (after the cows are mostly all hauled up), I saw him still there, covered with scars and frightfully gashed—raw, festering, and bloody, one eye gouged out —but lording it bravely over his harem of fifteen or twenty females, which were all huddled together around him on the same spot of his first location.

This fighting between the old and adult males (for none others fight) is mostly, or rather entirely, done with the mouth. The opponents seize one another with their teeth, and thus clinching their jaws, nothing but the sheer strength of the one and the other tugging to escape can shake them loose ; then, that effort invariably leaves an ugly wound, for the sharp canines tear out deep gutters in the skin and furrows in the blubber, or shred the flippers into ribbon-strips.

They usually approach each other with comically averted heads, just as though they were ashamed of the rumpus which they are determined to precipitate. When they get near enough to reach one another, they enter upon the repetition of many feints or passes before either one or the other takes the initiative by gripping. The heads are darted out and back as quick as a flash ; their hoarse roaring and shrill, piping whistle never ceases, while their fat bodies writhe and swell with exertion and rage ; furious lights gleam in their eyes, their hair flies in the air, and their blood streams down,—all combined makes a picture so fierce and so strange that, from its unexpected position and its novelty, it is perhaps one of the most extraordinary brutal contests which a man can witness.

In these battles of the seals the parties are always distinct ; the one is offensive, the other, defensive. If the latter proves the weaker, he withdraws from the position occupied, and is never followed by his conqueror, who complacently throws up one of his hind flippers, fans himself, as it were, to cool his fevered wrath and blood from the heat of the conflict, sinks into comparative quiet, only uttering a peculiar chuckle of satisfaction or contempt, with a sharp eye open for another covetous bull or " see-catch." *

That period occupied by the males in taking and holding their positions on a rookery offers a very favorable opportunity to study them in the thousand and one different attitudes and postures

* " See-catch," is the native name for a bull on the rookeries, especially one which is able to maintain its position.

Jealous Eyes Compare Them

OLD BULLS FIGHTING

Fur Seals in Deadly Combat : a Thousand such Conflicts are in simultaneous Action during every Minute of the Breeding Season on the Pribylov Islands

assumed between the two extremes of desperate conflict and deep
sleep—sleep so profound that one can, if he keeps to the leeward,
approach close enough, stepping softly, to pull the whiskers of any
old male taking a nap on a clear place. But after the first touch to
these mustaches the trifler must jump back with electrical celerity,
if he has any regard for the sharp teeth and tremendous shaking
which will surely overtake him if he does not. The younger seals
sleep far more soundly than the old ones, and it is a favorite pas-
time for the natives to surprise them in this manner—favorite, because it is attended with no personal risk. The little beasts, those
amphibious sleepers, rise suddenly, and fairly shrink to the earth,
spitting and coughing out in their terror and confusion.

The neck, chest, and shoulders of a fur-seal bull comprise more
than two-thirds of his whole weight; and in this long, thick neck
and the powerful muscles of the fore-limbs and shoulders is embodied the larger portion of his strength. When on land, with the
fore-hands he does all climbing over rocks and grassy hummocks
back of the rookery, or shuffles his halting way over smooth
parades—the hind-feet are gathered up as useless trappings after
every second step forward, which we have described at the outset
of this chapter. These anterior flippers are also the propelling
power when in water, and exclusive machinery with which they
drive their rapid passage—the hinder ones float behind like the
steering sweep to a whale-boat, and are used evidently as rudders,
or as the tail of a bird is, while its wings sustain and force its rapid
flight.

The covering to its body is composed of two coats, one being a
short, crisp, glistening over-hair, and the other a close, soft, elastic
pelage or fur, which gives a distinctive value to the pelt. I can
call it readily to the mind of my readers when I say to them that
the down and feathers on the breast of a duck lie relatively as the
fur and hair do upon the skin of the seal.

At this season of first "hauling up"* in the spring the prevailing color of the bulls, after they dry off and have been exposed to
the weather, is a dark, dull brown, with a sprinkling in it of lighter
brown-black, and a number of hoary or grizzled gray coats peculiar

* "Hauling up," is a technical term applied to the action of seals when
they land from the surf and haul up or drag themselves over the beach. It
is expressive and appropriate, as are most of the sealing phrases.

to the very old males. On the shoulders of all of them—that is,
the adults—the over-hair is either a gray or rufous-ochre or a
very emphatic "pepper and salt." This is called the "wig." The
body-colors * are most intense and pronounced upon the back of
the head, neck, and spine, fading down on the flanks lighter, to
much lighter ground on the abdomen ; still never white or even a
clean gray, so beautiful and peculiar to them when young, and to
the females. The skin of the muzzle and flippers is a dark bluish-
black, fading in the older examples to a reddish and purplish tint.
The color of the ears and tail is similar to that of the body, perhaps
a trifle lighter. The ears on a bull fur-seal are from one inch to an
inch and a half in length. The pavilions or auricles are tightly
rolled up on themselves, so that they are similar in shape to and
exactly the size of the little finger on the human hand, cut off at
the second phalangeal joint—a trifle more cone-shaped, however—as
they are greater at the base than they are at the tip. They are
haired and furred as the body is.

I think it probable that this animal is able to and does exert the
power of compressing or dilating this scroll-like pavilion to its ear,
just according as it dives deeper or rises in the water, and also I
am quite sure that the hair-seal has this control over its *meatus externus*, from what I have seen of it. I have not been able to verify
it in either case by actual observation ; yet such opportunity as I
have had gives me undoubted proof of the fact that the hearing of
a fur-seal is wonderfully keen and surpassingly acute. If you
make any noise, no matter how slight, an alarm will be given instantly by these insignificant-looking auditors, and the animal,
awaking from profound sleep, assumes, with a single motion, an
erect posture, gives a stare of stupid astonishment, at the same time
breaking out into incessant, surly roaring, growling, and "spitting," if it be an old male.

This spitting, as I call it, is by no means a fair or full expression
of a most characteristic sound or action, so far as I have ob-

* There is also perfect uniformity in the coloration of the breeding coats
of fur-seals, which is strikingly manifest while inspecting the rookeries
late in July, when they are solidly massed thereon. At a quarter-mile distance the whole immense aggregate of animal life seems to be fused into a
huge homogeneous body that is alternately roused up in sections and then
composed, just as a quantity of iron-filings covering the bottom of a saucer
will rise and fall when a magnet is passed over and around the dish.

served, peculiar to fur-seals alone, the bulls in particular. It is the usual prelude to all their combats, and it is their signal of astonishment. It follows somewhat in this way: when the two disputants are nearly within reaching or striking distance, they make a number of feints or false passes, as fencing-masters do, at one another, with the mouth wide open, lifting the lips or snarling so as to exhibit their glistening teeth ; with each pass of the head and neck they expel the air so violently through the larynx as to cause a rapid *choo-choo-choo* sound, like steam-puffs as they escape from the smoke-stack of a locomotive when it starts a heavy train, especially while the driving-wheels slip on the rail.

All of the bulls have the power and frequent inclination to utter four distinct calls or notes. This is not the case with the sea-lion, whose voice is confined to a single bass roar, or that of the walrus, which is limited to a dull grunt, or that of the hair-seal, which is almost inaudible. This volubility of the adult male is decidedly characteristic and prominent. He utters a hoarse, resonant roar, loud and long ; he gives vent to a low, entirely different gurgling growl ; he emits a chuckling, sibilant, piping whistle, of which it is impossible to convey an adequate idea, for it must be heard to be understood, and this spitting or *choo* sound just mentioned. The cow * has but one note—a hollow, prolonged, bla-a-ting call, addressed only to her pup : on all other occasions she is usually silent ; it is something strangely like the cry of a calf or an old sheep. She also makes a spitting sound or snort when suddenly disturbed —a kind of cough, as it were. The pups "blaat" also, with little or no variation, their sound being somewhat weaker and hoarser

---

* Without explanation I may be considered as making use of paradoxical language by using these terms of description, since the inconsistency of talking of "pups," with "cows," and "bulls," and "rookeries," on the breeding grounds of the same, cannot fail to be noticed ; but this nomenclature has been given and used by the American and English whaling and sealing parties for many years, and the characteristic features of the seals themselves so suit the naming that I have felt satisfied to retain the style throughout as rendering my description more intelligible, especially so to those who are engaged in the business or may be hereafter. The Russians are more consistent, but not so "pat." They call the bull "see-catch," a term implying strength, vigor, etc. ; the cow, "matkah," or mother; the pups, "kotickie," or little seals ; the non-breeding males under six and seven years, "holluschickie," or bachelors. The name applied collectively to the fur-seal by them is "morskie-kot," or sea-cat.

after birth than their mother's. They, too, comically spit or cough when aroused suddenly from a nap or driven into a corner, opening their little mouths (like young birds in a nest) when at bay, backed up in some crevice or against grassy tussocks.

Indeed, so similar is that call of the female to the bleating of sheep that a number of the latter, which the Alaska Commercial Company had brought up from San Francisco to St. George Island during the summer of 1873, were constantly attracted to the rookeries, and were running in among the "holluschickie" so much that they neglected better pasturage on the uplands beyond, and a small boy had to be regularly employed to herd them where they would feed to advantage. These transported *Ovidæ*, though they could not possibly find anything in their eyes suggestive of companionship among the seals, had their ears so charmed by those sheep-like accents of the female pinnipeds as to persuade them in spite of their senses of vision and smell.

The sound which arises from these great breeding grounds of the fur-seal, where thousands upon tens of thousands of angry, vigilant bulls are roaring, chuckling, and piping, and multitudes of seal-mothers are calling in hollow, bleating tones to their young, that in turn respond incessantly, is simple defiance to verbal description. It is, at a slight distance, softened into a deep booming, as of a cataract; and I have heard it, with a light, fair wind to the leeward, as far as six miles out from land on the sea; even in the thunder of the surf and the roar of heavy gales, it will rise up and over to your ear for quite a considerable distance away. It is a monitor which the sea-captains anxiously strain their ears for, when they run their dead-reckoning up, and are lying to for the fog to rise, in order that they may get their bearings of the land. Once heard, they hold on to the sound, and feel their way in to anchor. The seal-roar at "Novastoshnah" during the summer of 1872 saved the life of a surgeon,* and six natives belonging to the village, who had pushed out on an egging trip from Northeast Point to Walrus Island. I have sometimes thought, as I have lis-

---

* Dr. Otto Cramer: The suddenness with which fog and wind shut down and sweep over the sea here, even when the day opens most auspiciously for a short boat-voyage, has so alarmed the natives in times past that a visit is now never made by them from island to island, unless on one of the company's vessels. Several bidarrahs have never been heard from, which, in earlier times, attempted to sail, with picked crews of the natives, from one island to the other.

tened all night long to this volume of extraordinary sound, which never ceases with the rising or the setting of the sun throughout the entire period of breeding, that it was fully equal to the churning boom of the waves of Niagara. Night and day, belonging to that season, vibrates with this steady and constant din upon the rookeries.

Fur-seals Scratching Themselves.
[*Off the Black Bluffs, St. Paul's Island.*]

The most casual observer will notice that these seals seem to suffer great inconvenience and positive misery from a comparatively low degree of heat. I have often been surprised to observe that, when the temperature was 46° and 48° Fahr. on land during the summer, they would show everywhere signs of distress, whenever they made any exertion in moving or fighting, evidenced by panting and the elevation of their hind flippers, which they used incessantly as so many fans. With the thermometer again higher, as it is at rare intervals, standing at 55° and 60°, they are then oppressed even when at rest ; and at such times the eye is struck by the kaleidoscopic appearance of a rookery—in any of these rookeries where the seals are spread out in every imaginable position their lithesome bodies can assume, all industriously fan themselves : they use sometimes the fore flippers as ventilators, as it were, by holding them aloft motionless, at the same time fanning briskly with the hinder ones, according as they sit or lie. This wavy motion of fanning or flapping gives a hazy indistinctness to the whole scene, which is difficult to express in language ; but one of the most prominent characteristics of the fur-seal, and perhaps the most unique feature, is this very fanning manner in which they use their flippers, when seen on the breeding grounds at this season. They also, when idle, as it were, off-shore at sea, lie on their sides in the water with only a partial exposure of the body, the head submerged, and then hoist up a fore or hind flipper clear out of the water, at the same time scratching themselves or enjoying a momentary nap ; but in this position there is no fanning. I say " scratching," because the seal, in common with all animals, is preyed upon by vermin, and it has a peculiar species of louse, or parasitic tick, which annoys it.

Speaking of seals as they rest in the water leads me to remark that they seem to sleep as sound and as comfortably, bedded on the waves or rolled by the swell, as they do on the land. They lie on their backs, fold the fore flippers down across the chest, and turn the hind ones up and over, so that the tips rest on their necks and chins, thus exposing simply the nose and the heels of the hind flippers above water, nothing else being seen. In this position, unless it is very rough, the seal sleeps as serenely as did the prototype of that memorable song, who was " rocked in the cradle of the deep."

All the bulls, from the very first, that have been able to hold their positions, have not left them from the moment of their landing for a single instant, night or day ; nor will they do so until the end of the rutting season, which subsides entirely between August 1st and 10th—it begins shortly after the coming of the cows in June. Of necessity, therefore, this causes them to fast, to abstain entirely from food of any kind, or water, for three months at least ; and a few of them actually stay out four months, in total abstinence, before going back into the ocean for the first time after "hauling up" in May. They then return as so many bony shadows of what they were only a few months previously; covered with wounds, abject and spiritless, they laboriously crawl back to the sea to renew a fresh lease of life.

Such physical endurance is remarkable enough alone; but it is simply wonderful when we come to associate this fasting with the unceasing activity, restlessness, and duty devolved upon the bulls as the heads of large families. They do not stagnate like hibernating bears in caves; there is not one torpid breath drawn by them in the whole period of their fast. It is evidently sustained and accomplished by the self-absorption of their own fat, with which they are so liberally supplied when they first come out from the sea and take up their positions on the breeding grounds, and which gradually disappears, until nothing but the staring hide, protruding tendons and bones mark the limit of their abstinence. There must be some remarkable provision made by nature for the entire torpidity of the seals' stomachs and bowels, in consequence of their being empty and unsupplied during this long period, coupled with the intense activity and physical energy of the animals throughout that time, which, however, in spite of the violation of a supposed physiological law, does not seem to affect them, for they come back just as sleek, fat, and ambitious as ever, in the following season. That the seals drink or need fresh water, I doubt; but they cool their mouths incessantly by swimming with them wide open through the waves, laving as it were their hot throats and lips in the flood.*

Between June 12th and 14th, the first of the cow-seals, as a rule, come up from the sea; then that long agony of the waiting bulls is over, and they signalize it by a period of universal, spasmodic, desperate fighting among themselves. Though they have quarrelled all the time from the moment they first landed, and con-

---

* "Do these seals drink?" is a question doubtless often uppermost and suggested to the observer's eye as he watches those animals going to the water from the hauling-grounds and the rookeries; at least it was in mine. I never could detect a *callorhinus* or a *eumetopias* lapping, either in the fresh-water pools and lakes, or in the brackish lagoon, or the sea; but it plunges at times into the rollers with its jaws wide open as it dives, reappearing quickly in the same manner to dip and rise again, many times in rapid succession, as it swims along, the water running in little streams from the corners of the open mouth whenever its head pops above the surface. Whether this action was simply to cool itself, or that of drinking, I am not prepared to assert positively. I think it was to meet both purposes.

18

tinue to do so until the end of the season, in August, yet that fighting which takes place at this date is the bloodiest and most vindictive known to the seal. I presume that the heaviest percentage of mutilation and death among the old males from these brawls, occur in this week of the earliest appearance of the females.

A strong contrast now between the males and females looms up, both in size and shape, which is heightened by an air of exceeding peace and dove-like amiability which the latter class exhibit, in contradistinction to the ferocity and saturnine behavior of the former.* The cows are from four to four and a half feet in length from head to tail, and much more shapely in their proportions than the bulls; there is no wrapping around their necks and shoulders of unsightly masses of blubber; their lithe, elastic forms, from the first to the last of the season, are never altered; they are, however, enabled to keep such shape, because, in the provision of seal economy, they sustain no protracted fasting period; for, soon after the birth of their young they leave it on the ground and go to the sea for food, returning perhaps to-morrow, may be later, or even not for several days in fact, to again suckle and nourish it; having in the meantime sped far off to distant fishing banks, and satiated a hunger which so active and highly organized an animal must experience, when deprived of sustenance for any length of time.

As the females come up wet and dripping from the water, they are at first a dull, dirty-gray color, dark on the back and upper parts, but in a few hours the transformation in their appearance made by drying is wonderful. You would hardly believe that they could be the same animals, for they now fairly glisten with a rich steel and maltese gray lustre on the back of the head, the neck, and

---

* The old males, when grouped together by themselves, indulge in no humor or frolicsome festivities whatsoever. On the contrary, they treat each other with surly indifference. The mature females, however, do not appear to lose their good nature to anything like so marked a degree as do their lords and masters, for they will at all seasons of their presence on the islands be observed, now and then, to suddenly unbend from severe matronly gravity by coyly and amiably tickling and gently teasing one another, as they rest in the harems, or later, when strolling in September. There is no sign given, however, by these seal-mothers of a desire or attempt to fondle or caress their pups; nor do the young appear to sport with any others than the pups themselves, when together. Sometimes a yearling and a five or six months old pup will have a long-continued game between themselves. They are decidedly clannish in this respect—creatures of caste, like Hindoos.

along down the spine, which blends into an almost snow-white over the chest and on the abdomen. But, this beautiful coloring in turn is again altered by exposure to the same weather ; for, after a few days it will gradually change, so that by the lapse of two or three weeks it is a dull, rufous-ochre below, and a cinereous brown and gray mixed above. This color they retain throughout the breeding season, up to the time of shedding their coats in August.

The head and eye of the female are exceedingly beautiful ; the expression is really attractive, gentle, and intelligent ; the large, lustrous, blue-black eyes are humid and soft with the tenderest expression, while the small, well-formed head is poised as gracefully on her neck as can be well imagined ; she is the very picture of benignity and satisfaction, when she is perched up on some convenient rock, and has an opportunity to quietly fan herself, the eyes half-closed and the head thrown back on her gently-swelling shoulders.

The females land on these islands, not from the slightest desire to see their uncouth lords and masters, but from an accurate and instinctive appreciation of the time in which their period of gestation ends. They are in fact driven up to the rookeries by this cause alone ; the young cannot be brought forth in the water, and, in all cases marked by myself, the pups were born soon after landing, some in a few hours, but, most usually, a day or so elapses before delivery. They are noticed and received by the males at the water-line stations with attention ; they are alternately coaxed and urged up on to the rocks, as far as these beach-masters can do so, by chuckling, whistling, and roaring, and then they are immediately under the most jealous supervision ; but, owing to the covetous and ambitious nature of those bulls which occupy these stations to the rear of the water-line and away back, the little cows have a rough-and-tumble time of it, when they begin to arrive in small numbers at first ; for no sooner is the pretty animal fairly established on the station of male number one, who has welcomed her there, than he, perhaps, sees another one of her style in the water from whence she has come, and, in obedience to his polygamous feeling, devotes himself anew to coaxing the later arrival, by that same winning manner so successful in the first case ; then when bull number two, just back, observes bull number one off guard, he reaches out with his long strong neck and picks up the unhappy but passive cow by the scruff of hers, just as a cat does a kitten, and deposits her upon his seraglio ground ; then bulls number three and four, and so on, in the vicinity, seeing this high-handed operation, all assail one another, especially number two, and for a moment have a tremendous fight, perhaps lasting half a minute or so, and during this commotion the little cow is generally moved, or moves, farther back from the water, two or three stations more, where, when all gets quiet again, she usually remains in peace. Her last lord and master, not having that exposure to such diverting temptation as her first, gives her such care that she not only is unable to leave, did she wish, but no other bull can seize upon her : this is only a faint and (I fully appreciate it) wholly inadequate description of the hurly-burly and that method by which the rookeries are filled up, from first to last, when the females arrive—it is only one instance of the many trials and tribulations which both parties on the rookery subject themselves to, before the harems are filled. *

Far back, fifteen or twenty "see-catchie" stations deep from the water-line, and sometimes more, but generally not over an average of ten or fifteen, the cows crowd in at the close of the

---

* When the females first come ashore there is no sign whatever of affection manifested between the sexes. The males are surly and morose, and the females entirely indifferent to such reception. They are, however, subjected to very harsh treatment sometimes in progress of battles between the males for their possession, and a few of them are badly bitten and lacerated every season.

. One of the cows that arrived at Nahspeel, St. Paul's Island, early in June, 1872, was treated to a mutilation in this manner, under my eyes. When she had finally landed on the barren rocks of one of the numerous "seecatchie" at the water-front of this small rookery, and while I was carefully making a sketch of her graceful outlines, a rival bull, adjacent, reached out from his station and seized her with his mouth at the nape of the neck, just as a cat lifts a kitten. At the same instant, almost simultaneously, the old male that was rightfully entitled to her charms, turned, and caught her in his teeth by the skin of her posterior dorsal region. There she was, lifted and suspended in mid-air, between the jaws of the furious rivals, until, in obedience to their powerful struggles, the hide of her back gave way, and, as a ragged flap of the raw skin more than six inches broad and a foot in length was torn up and from her spine, she passed, with a rush, into the possession of the bull which had covetously seized her. She uttered no cry during this barbarous treatment, nor did she, when settled again, turn to her torn and bleeding wound to notice it in any way whatsoever that I could observe.

I may add here that I never saw the seals under such, or any circumstances, lick or nurse their wounds as dogs or cats do; but, when severe inflammation takes place, they seek the water, disappearing promptly from scrutiny

SUNDRY SEAL SKETCHES FROM THE AUTHOR'S PORTFOLIO

season for arriving, which is by July 10th or 14th ; then they are able to go about very much as they please, for the bulls have become so greatly enfeebled by this constant fasting, fighting, and excitement during the past two months, that they are quite content now with only one or two partners, even if they should have no more.

The cows seem to haul up in compact bodies from the water, covering in the whole ground to the rear of the rookeries, never scattering about over the surface of this area ; they have mapped out, from the first, their chosen resting-places, and they will not lie quietly in any position outside of the great mass of their kind. This is due to their intensely gregarious nature, and is especially adapted for their protection. And here I should call attention to the fact that they select this rookery ground with all the skill of civil engineers. It is preferred with special reference to drainage, for it must slope so that the produce of constantly dissolving fogs and rain-clouds shall not lie upon it, since they have a great aversion to, and a firm determination not to rest on water-puddled ground. This is admirably exhibited, and will be understood by a study of my sketch-maps which follow, illustrative of these rookeries and the area and position of the seals upon them. Every one of those breeding grounds rises up gently from the sea, and on no one of them is there anything like a muddy flat.

I found it an exceedingly difficult matter to satisfy myself as to a fair general average number of cows to each bull on the rookery, but, after protracted study, I think it will be nearly correct when I assign to each male a specific ratio of from fifteen to twenty females at the stations nearest the water, and for those back, in order, from that line to the rear, from five to twelve ; but there are many exceptional cases, and many instances where forty-five and fifty females are all under the charge of one male : and then, again, where there are only two or three females : hence this question was and is not entirely satisfactory in its settlement to my mind.

Near Ketavie Point, and just above it to the north, is an odd wash-out of basalt by the surf, which has chiselled, as it were, from the foundation of the island, a lava table, with a single roadway or land passage to it. Upon the summit of this footstool I counted forty-five cows, all under the charge of one old veteran. He had them penned on this table-rock by taking his stand at the gate, as it were, through which they passed up and passed down—a Turkish brute typified.

At the rear of all these rookeries there is invariably a large number of able-bodied males which have come late, and wait patiently, yet in vain, for families ; most of them having had to fight as desperately for the privilege of being there as any of their more fortunately located neighbors, who are nearer the water, and in succession from there to where they are themselves ; but the cows do not like to be in any outside position. They cannot be coaxed out where they are not in close company with their female mates and masses. They lie most quietly and contentedly in the largest harems, and cover the surface of the ground so thickly that there is hardly moving or turning room when they cease to come from the sea. The inaction on the part of those males in the rear during the breeding season only serves to well qualify them for moving into the places which are necessarily vacated by disabled males that are, in the meantime, obliged to leave from virile exhaustion, or incipient wounds. All the surplus able-bodied bulls, which have not been successful in effecting a landing on the rookeries cannot be seen at any one time, however, in the season, on this rear line. Only a portion of their number are in sight ; the others are either loafing at sea, adjacent, or are hauled out in morose squads between the rookeries on the beaches. The cows, during the whole season, do great credit to their amiable expression by their manner and behavior on the rookery. They never fight or quarrel one with another, and never or seldom utter a cry of pain or rage when they are roughly handled by the bulls, which frequently get a cow between them and actually tear the skin from her back with their teeth, cutting deep gashes in it as they snatch her from mouth to mouth. If sand does not get into these wounds it is surprising how rapidly they heal ; and, from the fact that I never could see scars on them anywhere except the fresh ones of this year, they must heal effectually and exhibit no trace the next season.

The cows, like the bulls, vary much in weight, but the extraordinary disparity in the adult size of the sexes is exceedingly striking. Two females taken from the rookery nearest to St. Paul village, right under the bluffs (and almost beneath the eaves of the natives' houses) called "Nah Speel," after they had brought forth their young, were weighed by myself, and their respective returns on the scales were fifty-six and one hundred pounds each ; the former being about three or four years old, and the latter over six— perhaps ten. Both were fat, or rather, in good condition—as good as they ever are. Thus the female is just about one-sixth the size of the male. Among the sea-lions the proportion is just one-half the bulk of the male, while the hair-seals, as I have before stated, are not distinguishable in this respect, as far as I could observe, but my notice was limited to a few specimens only.

The courage with which the fur-seal holds his position as the head and guardian of a family is of the highest order. I have repeatedly tried to drive them from their harem-posts, when they were fairly established on their stations, and have, with very few exceptions, failed. I might use every stone at my command, making all the noise I could. Finally, to put this courage to its fullest test, I have walked up to within twenty feet of an old veteran, toward the extreme end of Tolstoi, who had only four cows in charge, and commenced with my double-barrelled fowling-piece to pepper him all over with fine mustard-seed shot, being kind enough, in spite of my zeal, not to put out his eyes. His bearing, in spite of the noise, smell of powder, and painful irritation which the fine shot must have produced, did not change in the least from the usual attitude of determined, plucky defence (which nearly all of the bulls assume) when he was attacked with showers of stones and noise. He would dart out right and left with his long neck and catch the timid cows that furtively attempted to run after each report of my gun, fling and drag them back to their places under his head ; and then, stretching up to his full height, look me directly and defiantly in the face, roaring and chuckling most vehemently. The cows, however, soon got away from him : they could not endure my racket, in spite of their dread of him. But he still . stood his ground, making little charges on me of ten or fifteen feet in a succession of gallops or lunges, spitting furiously, and then comically retreating, with an indescribable leer and swagger, to the old position, back of which he would not go, fully resolved to hold his own or die in the attempt.

This courage is all the more noteworthy from the fact that, in regard to man, it is invariably of a defensive character. The seal is always on the defensive ; he never retreats, and he will not assail. If he makes you return when you attack him he never follows you much farther than the boundary of his station, and then no aggravation will compel him to take the offensive, so far as I have been able to observe. I was very much impressed by this trait.

It is quite beyond my power—indeed, entirely out of the ques-

tion—to give a fair idea of the thousand and one positions in which
seals compose themselves and rest when on land. They may be said
to assume every possible attitude which a flexible body can be put
into, no matter how characteristic or seemingly forced or constrained. Their joints seem to be double-hinged—in fact, fitted with
ball and socket union of the bones. One favorite position, especially
with the females, is to perch upon a point or edge-top of some
rock, and throw their heads back upon their shoulders, with the
nose held directly up and aloft ; and then, closing their eyes, take
short naps without changing their attitude, now and then softly
lifting one or the other of their long, slender hind flippers, which
they slowly wave with that peculiar fanning motion to which I have
alluded heretofore. Another attitude, and one of the most common, is to curl themselves up just as a dog does on a hearth-rug,
bringing the tail and nose close together. They also stretch out,
laying the head close to the body, and sleep an hour or two without
rising, holding one of the hind flippers up all the time, now and
then gently moving it, the eyes being tightly closed.

I ought, perhaps, to define the anomalous tail of the fur-seal
here. It is just about as important as the caudal appendage to a
bear ; even less significant. It is the very emphasis of abbreviation.
In the old males it is positively only four or five inches in length,
while among the females only two and a half to three inches,
wholly inconspicuous, and not even recognized by the casual
observer : they never wag or move it at all.

I come now to speak of another feature which interested me
nearly, if not quite, as much as any other characteristic of this
creature, and that is their fashion of slumber. The sleep of the
fur-seal, seen on land, from the old male down to the youngest, is
always accompanied by an involuntary, nervous, muscular twitching
and slight shifting of the flippers, together with ever and anon
quivering and uneasy rollings of the body, accompanied by a quick
folding anew of the fore flippers ; all of which may be signs, as it
were, in fact, of their simply having nightmares, or of sporting, in
a visionary way, far off in some dreamland sea. But, it may be
that as an old nurse said in reference to the smiles on a sleeping
child's face, they are disturbed by their intestinal parasites. I have
studied hundreds of such somnolent examples. Stealing softly up
so closely that I could lay my hand upon them from the point
where I was sitting, did I wish to, and watching the sleeping seals,

I have always found their sleep to be of this nervous description. The respiration is short and rapid, but with no breathing (unless the ear is brought very close). The quivering, heaving of the flanks only indicates the action of the lungs. I have frequently thought that I had succeeded in finding a snoring seal, especially among the pups; but a close examination always gave some abnormal reason for it—generally a slight distemper; never anything more severe, however, than some trifle by which the nostrils were stopped to a greater or less degree.

The cows on the rookeries sleep a great deal, but the bulls have the veriest cat-naps that can be imagined. I never could time the slumber of any old male on the breeding grounds which lasted, without interruption, longer than five minutes, day or night. While away from these places, however, I have known them to lie sleeping in the manner I have described, broken by such fitful, nervous, dreamy starts, yet without opening the eyes, for an hour or so at a time.

With an exception of the pups, the fur-seal seems to have very little rest, awake or sleeping. Perpetual motion is well-nigh incarnate with its being. I naturally enough, when beginning my investigation of these seal-rookeries, expected to find the animals subdued at night, or early morning, on those breeding grounds; but a few consecutive nocturnal watches satisfied me that the family organization and noise was as active at one time as at another, throughout the whole twenty-four hours. If, however, the day preceding had chanced to be abnormally warm, I never failed then to find the rookeries much more noisy and active during the night than they were by daylight. The seals, as a rule, come and go to and from the sea, fight, roar, and vocalize as much during midnight moments as they do at noonday times. An aged native endeavored to satisfy me that the "seecatchie" could see much better by twilight and night than by daylight. I am not prepared to prove to the contrary, but I think that the fact of his not being able to see so well himself at that hour of darkness was a true cause of most of his belief in the improved nocturnal vision of the seals.*

* This old Aleut, Philip Vollkov, passed to his final rest—"un konchielsah"—in the winter of 1878-79. He was one of the real characters of St. Paul. He was esteemed by the whites on account of his relative intelligence, and beloved by the natives, who called him their "wise man," and who exulted in his piety. Philip, like the other people there of his kind,

As I have said before, the females, soon after landing, are de-livered of their young. Immediately after the birth of the pup (twins are rare, if ever) the little creature finds its voice—a weak, husky blaat—and begins to paddle about with its eyes wide open from the start, in a confused sort of way for a few minutes, until the mother turns round to notice her offspring and give it atten-tion, and still later, to suckle it ; and for this purpose she is sup-plied with four small, brown nipples, almost wholly concealed in the fur, and which are placed about eight inches apart, lengthwise with the body, on the abdomen, between the fore and hind flippers, with about four inches of space between them transversely. These nipples are seldom visible, and then faintly seen through the hair and fur. The milk is abundant, rich, and creamy. The pups nurse very heartily, almost gorging themselves ; so much so, that they often have to yield up the excess of what they have taken down, mewling and puking in a most orthodox manner.

The pup at birth, and for the next three months, is of a jet-black color, hair and flippers, save a tiny white patch just back of each forearm. It weighs from three to four pounds, and is twelve to fourteen inches long. It does not seem to nurse more than once every two or three days ; but in this I am very likely mistaken, for it may have received attention from its mother in the night, or other times in the day when I was unable to keep up my watch over the individual which I had marked for this supervision.

The apathy with which the young are treated by the old on the breeding grounds, especially by the mothers, was very strange to me, and I was considerably surprised at it. I have never seen a seal-mother caress or fondle her offspring ; and should it stray to a short distance from the harem, I could step to and pick it up, and even kill it before the mother's eye, without causing her the slightest concern, as far as all outward signs and manifestations should indi-cate. The same indifference is also exhibited by the male to all that may take place of this character outside of the boundary of his seraglio ; but the moment the pups are inside the limits of his harem-ground he is a jealous and a fearless protector, vigilant and determined. But if the little animals are careless enough to pass

---

was not much comfort to me when I asked questions as to the seals. He usually answered important inquiries by crossing himself and replying, " God knows." There was no appeal from this.

beyond this boundary, then I can go up to them and carry them off before the eye of the old Turk without receiving from him the slightest attention in their behalf—a curious guardian, forsooth!

It is surprising to me how few of these young pups get crushed to death while the ponderous bulls are floundering over them, engaged in fighting and quarrelling among themselves. I have seen two bulls dash at each other with all the energy of furious rage, meeting right in the midst of a small "pod" of forty or fifty pups, tramp over them with all their crushing weight, and bowling them out right and left in every direction by the impetus of their movements, without injuring a single one, as far as I could see. Still, when we come to consider the fact that, despite the great weight of the old males, their broad, flat flippers and yielding bodies may press down heavily on these little fellows without actually breaking bones or mashing them out of shape, it does seem questionable whether more than one per cent. of all the pups born each season on these great rookeries of the Pribylov Islands are destroyed in this manner on the breeding grounds.*

The vitality of a fur-seal is simply astonishing. Its physical organization passes beyond the fabled nine lives of the cat. As a slight illustration of its tenure of life, I will mention the fact that one morning Philip came to me with a pup in his arms, which had just been born and was still womb-moist,· saying that the mother had been killed at Tolstoi by accident, and he supposed that I would like to have a "choochil." I took it up into my laboratory, and, finding that it could walk about and make a great noise, I attempted to feed it, with the idea of having a comfortable subject to my pencil for life-study of the young in varied attitudes of sleep and motion. It refused everything that I could summon to its attention as food, and, alternately sleeping and walking in its clumsy fashion about the floor; it actually lived nine days, spending the half of every one in floundering over the floor, accompanying all movements with a persistent, hoarse, blaating cry, and I do not believe it ever had a single drop of its mother's milk.

In a pup the head is the only disproportionate feature at birth

---

* The only danger which these little fellows are subject to up here is being caught by an October gale down at the surf-margin, when they have not fairly learned to swim. Large numbers have been destroyed by sudden "nips" of this character.

when it is compared with an adult form, the neck being also relatively shorter and thicker. The eye is large, round, and full ; but,
almost a "navy blue" at times, it soon changes into the blue-black
of adolescence.

The females appear to go to and come from the water feeding
and bathing quite frequnetly after bearing their young and an immediate subsequent coitus with the male : they usually return to the
spot or its immediate neighborhood, where they leave their pups,
crying out for them and recognizing the individual replies ; though
ten thousand around, all together, should blaat at once, they
quickly single out their own and nurse them. It would certainly
be a very unfortunate matter if the mothers could not identify their
young by sound, since these pups get together like a great swarm
of bees, and spread out upon the ground in what the sealers call
" pods," or clustered groups, while they are young and not very
large ; thus, from the middle or end of September until they leave
the islands for the dangers of the great Pacific in the winter, along
by the first of November, they gather in this manner, sleeping and
frolicking by tens of thousands, bunched together at various places
all over the islands contiguous to the breeding grounds, and right
on them. A mother comes up from the sea, whither she has been
to wash, and perhaps to feed, for the last day or two, feeling her
way along to about where she thinks her pup should be—at least,
where she left it last ; but perhaps she misses it, and finds instead
a swarm of pups in which it has been incorporated, owing to its
great fondness for society. The mother, without first entering into
a crowd of thousands, calls just as a sheep does for a lamb,
and out of all the din she—if not at first, at the end of a few trials
—recognizes the voice of her offspring, and then advances, striking
out right and left toward the position from which it replies ; but if
the pup happens at this time to be asleep, it gives, of course, no
response, even though it were close by. In the event of such silence
the cow, after calling for a time without being answered, curls herself up and takes a nap or lazily basks, to be usually more successful, or wholly so, when she calls again.

The pups themselves do not know their own mothers, a fact
which I ascertained by careful observation ; but they are so constituted that they incessantly cry out at short intervals during the
whole time they are awake, and in this way the mother can pick out
from the monotonous blaating of thousands of pups her own, and

she will not permit any other one to suckle her. But the "kotickie" themselves attempt to nose around every seal-mother that comes in contact with them. I have repeatedly watched young pups as they made advances to nurse from another pup's mother, the result invariably being that, while the "matkah" would permit her own off-spring to suckle freely, yet when these little strangers touched her nipples she would either move abruptly away or else turn quickly down upon her stomach, so that the maternal fountains were inaccessible to alien and hungry "kotickie." I have witnessed so many examples of the females turning pups away to suckle only some particular other one, that I feel sure I am entirely right in saying that the seal-mothers know their own young, and that they will not permit any others except them to nurse. I believe that this maternal recognition is due chiefly to the mother's scent and hearing.

Between the end of July and August 5th or 8th of every year the rookeries are completely changed in appearance. The systematic and regular disposition of the families or harems over the whole extent of breeding-ground has disappeared. All that clock-work order which has heretofore existed seems to be broken up. The breeding season closed, those bulls which have held their positions since May 1st leave, most of them thin in flesh and weak, and of their number a very large proportion do not come out again on land during the season; but such as are seen at the end of October and November are in good shape. They have a new coat of rich, dark, gray-brown hair and fur, with gray or grayish-ochre "wigs" of longer hair over the shoulders, forming a fresh, strong contrast to the dull, rusty, brown and umber dress in which they appeared to us during the summer, and which they had begun to shed about August 1st, in common with the females and the "holluschickie." After these males leave at the end of their season's work, and of the rutting for the year, those of them that happen to return to land in any event do not come back until the end of September and do not haul up on the rookery grounds again. As a rule, they prefer to herd altogether, like younger males, upon the sand-beaches and rocky points close to the water.

The cows and pups, together with those bulls which we have noticed in waiting at the rear of the rookeries, and which have been in retirement throughout the whole of the breeding season, now take possession, in a very disorderly manner, of these rookeries;

also, a large number of young, three, four, and five year old males, come, which have been prevented by the menacing threats of stronger bulls from an earlier landing among the females during the breeding season.

Before the middle of August three-fourths, at least, of the cows at this date are off in the water, only coming ashore at irregular intervals to nurse and look after their pups a short time. They presented to my eye, from the summits of the bluffs round about, a picture more suggestive of entire comfort and enjoyment than any-thing I have ever seen presented by animal life. Here, just out and beyond the breaking of the rollers, they idly lie on the rocks or sand-beaches, ever and anon turning over and over, scratching their backs and sides with their fore and hind flippers. The seals on the breeding ground appear to get very lousy.*

Frequent winds and showers will drive and spatter sand into their fur and eyes, often making the latter quite sore. This occurs when they are obliged to leave the rocky rookeries and follow their pups out over the sand-ridges and flats, to which they always have a natural aversion. On the hauling-grounds they pack the soil under their feet so hard and tightly in many places that it holds water in shallow surface-depressions, just like so many rock-basins. Out of and into these puddles the pups and the females flounder

---

* The fur-seal spends a great deal of time, both at sea and on land, in scratching its hide ; for it is annoyed by a species of louse, a *pediculus*, to just about the same degree and in the same manner that our dogs are by fleas. To scratch, it sits upon its haunches, and scrapes away with the toe-nails of first one and then the other of its hind flippers, by which action it reaches readily all portions of its head, neck, chest and shoulders, and with either one or the other of its fore flippers it rubs down its spinal region back of the shoulders to the tail. By that division of labor with its feet it can promptly reduce, with every sign of comfort, any lousy irritation wheresoever on its body. This *pediculus* peculiar to the fur-seal attaches itself almost exclusively to the pec-toral regions ; a few also are generally found at the bases of the auricular pavilions.

When the fur-seal is engaged in this exercise it cocks its head and wears exactly the same expression that our common house-dog does while subjugat-ing and eradicating fleas ; the eyes are partly or wholly closed ; the tongue lolls out ; and the whole demeanor is one of quiet but intense satisfaction.

The fur-seal appears also to scratch itself in the water with the same facil-ity and unction so marked on land, only it varies the action by using its fore-hands principally in its pelagic exercise, while its hind-feet do most of the terrestrial scraping.

and patter incessantly, until evaporation slowly abates the nuisance for a time only, inasmuch as the next day, perhaps, brings more rain and the dirty pools are replenished.

The pups sometimes get so thoroughly plastered in these muddy, slimy puddles, that the hair falls off in patches, giving them, at first sight, the appearance of being troubled with scrofula or some other plague : from my investigations directed to this point, I became satisfied that they were not permanently injured, though evidently very much annoyed. With reference to this suggestion as to sickness or distemper among these seals, I gave the subject direct and continued attention, and in no' one of the rookeries could I discover a single seal, no matter how old or young, which appeared to be suffering in the least from any physical disorder other than that which they themselves had inflicted, one upon the other, by fighting. The third season, passing directly under my observation, failed to reward my search with any manifestation of disease among the seals which congregate in such mighty numbers on those rookeries of St. Paul and St. George. That remarkable freedom from all such complaints enjoyed by these animals is noteworthy, 'and a most trenchant and penetrating cross-questioning of the natives also failed to give me any history or evidence of an epidemic in the past.

The observer will, however, notice every summer, gathered in melancholy squads of a dozen to one hundred or so (scattered along the coast where the healthy seals never go), those sick and disabled bulls which have, in the earlier part of the season, been either internally injured or dreadfully scarred by the teeth of their opponents in fighting. Sand is blown by strong wind into their fresh wounds, causing inflammation and sloughing which very often finishes the life of a victim. The sailors term these invalid gatherings "hospitals," a phrase which, like the most of their homely expressions, is quite appropriate.

Early in August, usually by the 8th or 10th, I noticed one of the remarkable movements of the season. I refer to the pup's first essay in swimming. Is it not odd—paradoxical—that the young seal, from the moment of his birth until he is a month or six weeks old, is utterly unable to swim? If he is seized by the nape of the neck and pitched out a rod into the water from shore, his bullet-like head will drop instantly below the surface, and his attenuated posterior extremities flap impotently on it.

Suffocation is the question of only a few minutes, the stupid little creature not knowing how to raise his immersed head and gain the air again. After they have attained the age indicated above, their instinct drives them down to the margin of the surf, where an alternate ebbing and flowing of its wash, covers and uncovers the rocky or sandy beaches. They first smell and then touch the moist pools, and flounder in the upper wash of the surf, which leaves them as suddenly high and dry as it immersed them at first. After this beginning they make slow and clumsy progress in learning the knack of swimming. For a week or two, when over-head in depth, they continue to flounder about in the most awkward manner, thrashing the water as little dogs do with their fore feet, making no attempt whatever to use the hinder ones. Look at that pup now, launched out for the first time beyond his depth ; see how he struggles—his mouth wide open, and his eyes fairly popping. He turns instantly to the beach, ere he has fairly struck out from the point whence he launched in, and, as the receding swell which at first carried him off his feet and out, now returning, leaves him high and dry, for a few minutes he seems so weary that he weakly crawls up, out beyond its swift returning wash, and coils himself immediately to take a recuperative nap. He sleeps a few minutes, perhaps half an hour, then awakes as bright as a dollar, apparently rested, and at his swimming lesson he goes again. By repeated and persistent attempts, this young seal gradually becomes familiar with the water and acquainted with his own power over that element, which is to be his real home and his whole support. Once boldly swimming, the pup fairly revels in a new happiness. He and his brethren have now begun to haul and swarm along the entire length of St. Paul coast, from Northeast Point down and around to Zapadnie, lining the alternating sand-beaches and rocky shingle with their chunky, black forms. How they do delight in it ! They play with a zest, and chatter like our own children in the kindergartens—swimming in endless evolutions, twisting, turning, or diving—and when exhausted, drawing their plump, round bodies up again on the beach. Shaking themselves dry as young dogs would do, they now either go to sleep on the spot or have a lazy terrestrial frolic among themselves.

Why an erroneous impression ever got into the mind of any man as to this matter of a pup's learning to swim, I confess that I am wholly unable to imagine. I have not seen any "driving" of the young pups into the water by the old ones, in order to teach them this process, as certain authors have positively affirmed. There is not the slightest supervision by the mother or father of the pup, from the first moment of his birth, in this respect, until he leaves for the North Pacific, full-fledged with amphibious power. At the close of the breeding season, every year, the pups are restlessly and constantly shifting back and forth over the rookery ground of their birth, in large squads, sometimes numbering thousands upon thousands. In the course of this change of position they all sooner or later come in contact with the sea; they then blunder into the water for the first time, in a most awkward, ungainly manner, and get out as quick as they can; but so far from showing any fear or dislike of this, their most natural element, as soon as they rest from their exertion they are immediately ready for a new trial, and keep at it, provided the sea is not too stormy or rough. During all this period of self-tuition they seem thoroughly to enjoy the exercise, in spite of their repeated and inevitable discomfitures at the beginning.

That "podding" of these young pups in the rear of the great rookeries of St. Paul, is one of the most striking and interesting phases of this remarkable exhibition of highly-organized life. When they first bunch together they are all black, for they have not begun to shed the natal coat; they shine with an unctuous, greasy reflection, and grouped in small armies or great regiments on the sand-dune tracts at Northeast Point, they present a most extraordinary and fascinating sight. Although the appearance of the "holluschickie" at English Bay fairly overwhelms the observer with an impression of its countless multitudes, yet I am free to declare that at no one point in this evolution of the seal-life, during its reproductive season, have I been so deeply impressed by a sense of overwhelming enumeration, as I have when, standing on the summit of Cross Hill, I looked down to the southward and westward over a reach of six miles of alternate grass and sand-dune stretches, mirrored upon which were hundreds of thousands of these little black pups, spread in sleep and sport within this restricted field of vision. They appeared as countless as the grains of that sand upon which they rested!

By September 15th, all the pups born during the year have become familiar with the water; they have all learned to swim, and are now nearly all down by the water's edge, skirting in

19

large masses the rocks and beaches hitherto unoccupied by seals of
any class this year. Now they are about five or six times their
original weight, or, in other words, they are thirty to forty pounds
avoirdupois, as plump and fat as butter-balls, and they begin to
take on their second coat, shedding their black pup-hair completely.
This second coat does not vary in color, at this age, between the
sexes. They effect such transformation in dress very slowly, and
cannot, as a rule, be said to have ceased their moulting until the
middle or 20th of October.

That second coat, or sea-going jacket, of the pup, is a uniform,
dense, light gray over-hair, with an under-fur which is slightly grayish in some, but is, in most cases, of a soft light brown hue. The over-
hair is fine, close and elastic, from two-thirds of an inch to an
inch in length, while the fur is not quite half an inch long. Thus
the coarser hair shingles over and conceals the soft under-wool
completely, giving the color by which, after the second year, the
sex of the animal is recognized. A pronounced difference between
the sexes is not effected, however, by color alone until the third
year of the animal's life. This over-hair of the pup's new jacket
on its back, neck, and head, is a dark chinchilla-gray, blending into
stone-white, just tinged with a grayish tint on the abdomen and
chest. The upper lip, upon which the whiskers or mustaches take
root, is covered with hair of a lighter gray than that of the body.
This mustache consists of fifteen or twenty longer or shorter
bristles, from half an inch to three inches in length, some brownish,
horn-colored, and others whitish-gray and translucent, on each side
and back and below the nostrils, leaving the muzzle quite prominent and hairless. The nasal openings and their surroundings are,
as I have before said when speaking of this feature, hairless and
similar to those of a dog.*

---

* It has been suggested to me that the exquisite power of scent possessed
by these animals enables them to reach the breeding grounds at about the place
where they left them the season previously: surely the nose of the fur-seal is
endowed to a superlative degree with those organs of smell, and its range of
appreciation in this respect must be very great.

I noticed in all sleeping and waking seals that the nasal apertures were
never widely expanded; and that they were at intervals rapidly opened and
closed with inhalation and exhalation of each breath; the nostrils of the fur-
seal are, as a rule, well opened when the animal is out of water, and remain
so while it is on land.

The most attractive feature about the fur-seal pup, and that which holds this place as it grows on and older, is the eye. That organ is exceedingly clear, dark, and liquid, with which, for beauty and amiability, together with real intelligence of expression, those of no other animal that I have ever seen, or have ever read of, can be compared; indeed, there are few eyes in the orbits of men and women which suggest more pleasantly the ancient thought of their being "windows to the soul." The lids to that eye are fringed with long, perfect lashes, and the slightest irritation in the way of dust or sand, or other foreign substances, seems to cause them exquisite annoyance, accompanied by immoderate weeping. This involuntary tearfulness so moved Steller that he ascribed it to the processes of a mind, and declared that seal-mothers actually "shed tears"!

I do not think a seal's range of vision on land, or out of the water, is very great. I have frequently experimented with adult fur-seals, by allowing them to catch sight of my person, so as to distinguish it as of foreign character, three and four hundred paces off, taking the precaution of standing quietly to the leeward when the wind was blowing strong, and then walking unconcernedly up to them. I have invariably noticed that they would allow me to approach quite close before recognizing my strangeness; then, as it occurred to them, they at once made a lively noise, a medley of coughing, spitting, snorting, and blaating, and plunged in spasmodic lopes and shambled to get away from my immediate neighborhood. As to the pups, they all stupidly stare at the form of a human being until it is fairly on them, when they also repeat in miniature these vocal gymnastics and physical efforts of the older ones, to retreat or withdraw a few rods, sometimes only a few feet, from the spot upon which you have cornered them, after which they instantly resume their previous occupation of either sleeping or playing, as though nothing had happened. Perhaps it is safe to say that the greatest activity displayed by any one of the five senses of the seal is evidenced in its power of scent. This faculty is all that can be desired in the line of alertness. I never failed to awaken an adult seal from the soundest sleep, when from a half to a quarter of a mile distant, no matter how softly I proceeded, if I got to the windward, though they sometimes took alarm when I was a mile off.

They leave evidences of their being on these great reproductive fields, chiefly at the rookeries, in the hundreds of dead carcasses

which mark the last of those animals that had been rendered infirm, sick, and killed by fighting among themselves in the early
part of the season, or of those which have crawled far away from
the scene of battle to die from death-wounds received in bitter
struggles for a harem. On the rookeries, wherever these lifeless
bodies rest, the living, old and young, clamber and patter backward
and forward over and on the putrid remains : thus such constant
stirring up of decayed matter, gives rise to an exceedingly disagreeable and far-reaching "funk." This has been, by all writers who
have dwelt on the subject, referred to as the smell which those animals emit for another reason—erroneously called the "rutting
odor." If these creatures have any odor peculiar to them when in
this condition, I will frankly confess that I am unable to distinguish
it from the fumes which are constantly being stirred up and arising
out of those putrescent carcasses so disturbed, as well as from
the bodies of the few pups which have been killed accidentally by
heavy bulls fighting over them, charging back and forth against one
another, so much of the time.

They have, however, a very characteristic and peculiar smell
when they are driven and get heated ; their breath-exhalations
possess a disagreeable, faint, sickly odor, and when I have walked
within its influence at the rear of a seal-drive, I could almost fancy,
as it entered my nostrils, that I stood beneath an ailantus-tree in
full bloom ; but this odor can by no means be confounded with
what is universally ascribed to another cause. It is also noteworthy
that if your finger is touched ever so lightly to a little fur-seal
blubber, it will smell very much like that which I have appreciated
and described as peculiar to their breath, which arises from them
when they are driven, only it is a little stronger. Both the young
and old fur-seals have this same breath-taint at all seasons of the
year.

With the precision of clock-work and the regularity of the pre-
cession of the seasons, fur-seals have adopted and enforced the
following method of life on these islands of Pribylov. In this system millions of those highly organized animals sustain themselves.

First.—The earliest bulls land in a negligent, indolent way, at
the opening of the season, soon after the rocks at the water's edge
are free from ice, frozen snow, etc. This is, as a rule, about the
1st to the 5th of every May. They land from the beginning to the
end of the season in perfect confidence and without fear ; they are

very fat, and will weigh on an average five hundred pounds each ; some stay at the water's edge, some go to the tier back of them again, and so forth, until the whole rookery is mapped out by them, weeks in advance of the arrival of the first female.

Second.—That by the 10th or 12th of June, all the male stations on the rookeries have been mapped out and fought for, and held in waiting by the "see-catchie." These males are, as a rule, bulls rarely ever under six years of age ; most of them are over that age, being sometimes three, and occasionally doubtless four or five times as old.

Third.—That the cows make their first appearance, as a class, on or after the 12th or 15th of June, in very small numbers, but rapidly after the 23d and 25th of this month, every year, they begin to flock up in such numbers as to fill the harems very perceptibly, and by the 8th or 10th of July they have all come, as a rule—a few stragglers excepted. The average weight of the females now will not be much more than eighty to ninety pounds each.

Fourth.—That the breeding season is at its height from the 10th to the 15th of July every year, and that it subsides entirely at the end of this month and early in August ; also, that its method and system are confined entirely to the land, never effected in the sea.

Fifth.—That the females bear their first young when they are three years old, and that the period of gestation is nearly twelve months, lacking a few days only of that lapse of time.

Sixth.—That the females bear a single pup each, and that this is born soon after landing. No exception to this rule as ever been witnessed or recorded.

Seventh.—That the "see-catchie" which have held the harems from the beginning to the end of the season, leave for the water in a desultory and straggling manner at its close, greatly emaciated, and do not return, if they do at all, until six or seven weeks have elapsed, when the regular systematic distribution of the families over the rookeries is at an end for this season. A general medley of young males now are free : they come out of the water, and wander over all these rookeries, together with many old males, which have not been on seraglio duty, and great numbers of the females. An immense majority over all others present are pups, since only about twenty-five per cent. of the mother-seals are out of the water now at any one time.

Eighth.—That the rookeries lose their compactness and definite boundaries of true breeding limit and expansion by the 25th to the 28th of July every year; then, after this date, the pups begin to haul back, and to the right and left, in small squads at first, but as the season goes on, by the 18th of August, they depart without reference to their mothers; and when thus scattered, the males, females and young swarm over more than three and four times the area occupied by them when breeding and born on the rookeries. The system of family arrangement and uniform compactness of the breeding classes breaks up at this date.

Ninth.—That by the 8th or 10th of August the pups born nearest the water first begin to learn to swim; and that by the 15th or 20th of September they are all familiar, more or less, with the exercise.

Tenth.—That by the middle of September the rookeries are entirely broken up; confused, straggling bands of females are seen among bachelors, pups, and small squads of old males, crossing and recrossing the ground in an aimless, listless manner. The season is now over.

Eleventh.—That many of the seals do not leave these grounds of St. Paul and St. George before the end of December, and some remain even as late as the 12th of January; but that by the end of October and the beginning of November every year, all the fur-seals of mature age—five and six years, and upward—have left the islands. The younger males go with the others; many of the pups still range about the islands, but are not hauled to any great extent on the beaches or the flats. They seem to prefer the rocky shore-margin, and to lie as high up as they can get on such bluffy rookeries as Tolstoi and the Reef. By the end of this month, November, they are, as a rule, all gone.

I now call the attention of the reader to another very remarkable feature in the economy of the seal-life on these islands. The great herds of "holluschickie,"* numbering from one-third to one-half, perhaps, of the whole aggregate of near five million seals known to the Pribylov group, are never allowed by the old "see-catchie" (which threaten frightful mutilation or death) to put their flippers on or near the rookeries.

By reference to my map, it will be observed that I have located

---

* The Russian term "holluschickie" or "bachelors" is very appropriate, and is usually employed.

a large extent of ground—markedly so on St. Paul—as that occupied by the seals' "hauling-ground"; this area, in fact, represents those portions of the island upon which the "holluschickie" roam in heavy squadrons, wearing away and polishing the surface of the soil, stripping every foot, which is indicated on the chart as such, of its vegetation and mosses, leaving a margin as sharply defined on those bluffy uplands and sandy flats as it is on the map itself.

The reason that so much more land is covered by the "holluschickie" than by the breeding seals—ten times as much at least—is due to the fact that, though not as numerous, perhaps, as the breeding seals, yet they are tied down to nothing, so to speak—are wholly irresponsible, and roam hither and thither as caprice and the weather may dictate. Thus they wear off and rub down a much larger area than the rookery seals occupy ; wandering aimlessly, and going back, in some instances, notably at English Bay, from one-half to a whole mile inland, not travelling in desultory files along winding, straggling paths, but sweeping in solid platoons, they obliterate every spear of grass and rub down nearly every hummock in their restless marching.

All the male seals, under six years of age, are compelled to herd apart by themselves and away from the breeding grounds, in many cases far away ; the large hauling-grounds at Southwest Point being about two miles from the nearest rookery. This class of seals is termed "holluschickie" or the "bachelors" by the people : a most fitting and expressive appellation.

The seals of this great subdivision are those with which the natives on the Pribylov group are the most familiar : naturally and especially so, since they are the only ones, with the exception of a few thousand pups, and occasionally an old bull or two, taken late in the fall for food and skins, which are driven up to the killing-grounds at the village for slaughter. The reasons for this exclusive attention to the "bachelors" are most cogent, and will be given hereafter when the "business" is discussed.

Since the "holluschickie" are not permitted by their own kind to land on the rookeries and stop there, they have the choice of two methods of locating, one of which allows them to rest in the rear of the rookeries, and the other on the free beaches. The most notable illustration of the former can be witnessed on Reef Point, where a pathway is left for their ingress and egress through a rookery—a path left by common consent, as it were, between the harems. On these trails of passage they come and go in steady files all day and all night during the season, unmolested by the jealous bulls which guard the seraglios on either side as they travel—all peace and comfort to the young seal if he minds his business and keeps straight on up or down, without stopping to nose about right or left ; all woe and destruction to him, however, if he does not, for in that event he will be literally torn in bloody gripping, from limb to limb, by vigilant "see-catchie."

Since the two and three year old "holluschickie" come up in small squads with the first bulls in the spring, or a few days later, such common highways as those between the rookery ground and the sea are travelled over before the arrival of the cows, and get well defined. A passage for the "bachelors," which I took much pleasure in observing day after day at Polavina, another at Tolstoi, and two on the Reef, in 1872, were entirely closed up by the " sea-catchie " and obliterated when I again searched for them in 1874. Similar passages existed, however, on several of the large rookeries of St. Paul. One of those at Tolstoi exhibits this feature very finely, for here the hauling-ground extends around from English Bay, and lies up back of the Tolstoi rookery, over a flat and rolling summit, from one hundred to one hundred and twenty feet above the sea-level. The young males and yearlings of both sexes come through and between the harems at the height of the breeding season on two of these narrow pathways, and before reaching the ground above, are obliged to climb up an almost abrupt bluff, which they do by following and struggling in the water-runs and washes that are worn into its face. As this is a large hauling-ground, on which, every favorable day during the season, fifteen or twenty thousand commonly rest, a view of skilful seal-climbing can be witnessed here at any time during that period ; and the sight of such climbing as this of Tolstoi is exceedingly novel and interesting. Why, verily, they ascend over and upon places where a lively man might, at first thought, say with great positiveness that it was utterly impossible for him to climb !

The other method of coming ashore, however, is the one most followed and favored. In this case they avoid the rookeries altogether, and repair to unoccupied beaches between them ; and then extend themselves out all the way back from the sea, as far from the water, in some cases, as a quarter and even half of a mile.

ARRIVAL OF THE FUR SEAL MILLIONS

I stood on the Tolstoi sand-dunes one afternoon, toward the middle of July, and had under my eyes, in a straightforward sweep from my feet to Zapadnie, a million and a half of seals spread out on those hauling-grounds. Of these I estimated that fully one-half, at that time, were pups, yearlings, and " holluschickie." The rookeries across the bay, though plainly in sight, were so crowded that they looked exactly as I have seen surfaces appear upon which bees had swarmed in obedience to that din and racket made by the watchful apiarian when he desires to secure a hive of restless honey-makers.

The great majority of yearlings and " holluschickie" are annually hauled out, scattered thickly over the sand-beach and upland hauling-grounds which lie between the rookeries on St. Paul Island. At St. George there is nothing of this extensive display to be seen, for here is only a tithe of the seal-life occupying St. Paul, and no opportunity whatever is afforded for an amphibious parade.

Descend with me from this sand-dune elevation of Tolstoi, and walk into that drove of " holluschickie" below us. We can do it. You do not notice much confusion or dismay as we go in among them. They simply open out before us and close in behind our tracks, stirring, crowding to the right and left as we go, twelve or twenty feet away from us on each side. Look at this small flock of yearlings—some one, others two, and even three years old—which are coughing and spitting around us now, staring up in our faces in amazement as we walk ahead. They struggle a few rods out of our reach, and then come together again behind us, showing no further sign or notice of ourselves. You could not walk into a drove of hogs at Chicago without exciting as much confusion and arousing an infinitely more disagreeable tumult ; and as for sheep on the plains, they would stampede far quicker. Wild animals, indeed ! You can now readily understand how easy it is for two or three men, early in the morning, to come where we are, turn aside from this vast herd in front of and around us two or three thousand of the best examples, and drive them back, up, and over to the village. That is the way they get the seals. There is no " hunting," no " chasing," no " capturing " of fur-seals on these islands.

While the young male seals undoubtedly have the power of going for lengthy intervals without food, they, like the female seals on the breeding grounds, certainly do not maintain any long fasting periods on land. Their coming and going from the shore is frequent and irregular, largely influenced by the exact condition of the weather from day to day. For instance, three or four thick, foggy days seem to call them out from the water by hundreds of thousands upon the different hauling-grounds (which the reader observes recorded on my map). In some cases I have seen them lie there so close together that scarcely a foot of ground, over whole acres, is bare enough to be seen. Then a clear and warmer day follows, and this seal-covered ground, before so thickly packed with animal life, will soon be almost deserted—comparatively so, at least—to be filled up immediately as before, when favorable weather shall again recur. They must frequently eat when here, because the first yearlings and "holluschickie" that appear in the spring are no fatter, sleeker, or livelier than they are at the close of the season. In other words, their condition, physically, seems to be the same from the beginning to the end of their appearance here during the summer and fall. It is quite different, however, with the "see-catch." We know how and where it spends two to three months, because we find it on the ground at all times, day or night, during that period.

A small flock of the young seals, one to three years old generally, will often stray from these hauling-ground margins up and beyond over the fresh mosses and grasses, and there sport and play one with another just as little puppy-dogs do: but, when weary of this gambolling, a general disposition to sleep is suddenly manifested, and they stretch themselves out and curl up in all the positions and all the postures that their flexible spines and ball-and-socket joints will permit. They seem to revel in the unwonted vegetation, and to be delighted with their own efforts in rolling down and crushing the tall stalks of grasses and umbelliferous plants. One will lie upon its back, hold up its hind flippers, and lazily wave them about, while it scratches, or rather rubs, its ribs with the fore-hands alternately, the eyes being tightly closed during the whole performance. The sensation is evidently so luxurious that it does not wish to have any side-issue draw off its blissful self-attention. Another, curled up like a cat on a rug, draws its breath, as indicated by the heaving of its flanks, quickly, but regularly, as though in heavy sleep. Another will lie flat upon its stomach, its hind flippers covered and concealed, while it tightly folds its forefeet back against its sides, just as a fish carries its pectoral fins, and so on to no end of variety, according to the ground and the fancy of the animals.

These "bachelor" seals are, I am sure, without exception, the most restless animals, in the whole brute creation, which can boast of a high organization. They frolic and lope about over the grounds for hours without a moment's cessation, and their sleep after this is exceedingly short, and it is ever accompanied by nervous twitchings and uneasy muscular movements. They seem to be fairly brimful and overrunning with spontaneity, to be surcharged with fervid, electric life.

Another marked feature observed among the multitudes of " holluschickie " which have come under my personal observation and auditory, and one very characteristic of this class, is that nothing like ill-humor appears in all of their playing together. They never growl or bite or show even the slightest angry feeling, but are invariably as happy, one with another, as can be imagined. This is a very singular trait. They lose it, however, with astonishing rapidity when their ambition and strength develops and carries them in due course of time to the rookery.

The pups and yearlings have an especial fondness for sporting on rocks which are just at the water's level and awash, so as to be covered and uncovered as the surf rolls in. On the bare summit of . these wave-worn spots they will struggle and clamber, in groups of a dozen or two at a time, throughout the whole day in endeavoring to push off that one of their number which has just been fortunate enough to secure a landing. The successor has, however, but a brief moment of exultation in victory, for the next roller that comes booming in, together with that pressure by its friends, turns the table, and the game is repeated, with another seal on top. Sometimes, as well as I could see, the same squad of "holluschickie " played for an entire day and night, without a moment's cessation, around such a rock as this off " Nah Speel" rookery; still, in this observation I may be mistaken, because those seals could not be told apart.

That graceful unconcern with which fur-seals sport safely in, among, and under booming breakers, during the prevalence of numerous wild gales at the islands, has afforded me many consecutive hours of spell-bound attention to them, absorbed in watching their adroit evolutions within the foaming surf, that seemingly every moment would, in its fierce convulsions, dash these hardy swimmers, stunned and lifeless, against those iron-bound foundations of the shore which alone checked the furious rush of the waves. Not at all. Through the wildest and most ungovernable mood of a roaring tempest and storm-tossed waters attending its transit I never failed, on creeping out and peering over the bluffs in such weather, to see squads of these perfect watermen, the most expert of all amphibians, gambolling in the seething, creamy wake of mighty rollers which constantly broke in thunder-tones over their alert, dodging heads. The swift succeeding waves seemed every instant to poise those seals at the very verge of death; yet the *Callorhinus*, exulting in his skill and strength, bade defiance to their wrath and continued his diversions.

Fur-seals rising to breathe and look around.
[*Characteristic pelagic attitude of the "holluschickie."*]

The "holluschickie" are the champion swimmers of all the seal tribe; at least, when in the water around the islands, they do nearly every fancy tumble and turn that can be executed. The grave old males and their matronly companions seldom indulge in any extravagant display, as do these youngsters, which jump out of the water like so many dolphins, describing beautiful elliptic curves sheer above its surface, rising three and even four feet from the sea, with the back slightly arched, the fore flippers folded tightly against the sides, and the hinder ones extended and pressed together straight out behind, plumping in head first, to reappear in the same manner, after an interval of a few seconds of submarine swimming, swift as the flight of a bird on its course. Sea-lions and hair-seals never leap in this manner.

All classes will invariably make these dolphin-jumps when they are surprised or are driven into the water, curiously turning their heads while sailing in the air, between the "rises" and "plumps," to take a look at the cause of their disturbance. They all swim rapidly, with the exception of the pups, and may be said to dart under the water with the velocity of a bird on the wing. As they swim they·are invariably submerged, running along horizontally about two or three feet below the surface, guiding their course with the hind flippers, as by an oar, and propelling themselves solely by the fore feet, rising to breathe at intervals which are either very frequent or else so wide apart that it is impossible to see the speeding animal when he rises a second time.*

How long they can remain under water without taking a fresh breath is a problem which I had not the heart to solve, by instituting a series of experiments at the island; but I am inclined to think that, if the truth were known in regard to their ability of going without rising to breathe, it would be considered astounding. On this point, however, I have no data worth discussing, but will say that in all their swimming which I have had a chance to study, as they passed under the water, mirrored to my eyes from the bluff above by the whitish-colored rocks below the rookery waters at

---

* If there is any one faculty better developed than the others in the brain of the intelligent *Callorhinus*, it must be its "bump" of locality. The unerring directness with which it pilots its annual course back through thousands of miles of watery waste to these spots of its birth—small fly-dots of land in the map of Bering Sea and the North Pacific—is a very remarkable exhibition of its skill in navigation. While the Russians were established at Bodega and Ross, Cal., seventy years ago, they frequently shot fur-seals at sea when hunting the sea-otter off the coast between Fuca Straits and the Farallones. Many of these animals, late in May and early in June, were so far advanced in pregnancy that it was deemed certain by their captors that some shore must be close at hand upon which the near-impending birth of the pup took place. Thereupon the Russians searched over every rod of the coast-line of the mainland and the archipelago between California and the peninsula of Alaska, vainly seeking everywhere there for a fur-seal rookery. They were slow to understand how animals so close to the throes of parturition could strike out into the broad ocean to swim fifteen hundred or two thousand miles within a week or ten days ere they landed on the Pribylov group, and, almost immediately after, give birth to their offspring.

Great Eastern rookery, I have not been able to satisfy myself how
they used their long, flexible hindfeet, other than as steering media.
If these posterior members have any perceptible motion, it is so rapid
that my eye is not quick enough to catch it ; but the fore flippers,
however, can be most distinctly seen as they work in feathering forward and sweeping flatly back, opposed to the water, with great rapidity and energy.   They are evidently the sole propulsive power
of the fur-seal in the water, as they are its main fulcrum and lever
combined for progression on land.   I regret that the shy nature of
the hair-seal never allowed me to study its swimming motions, but
it seems to be a general point of agreement among authorities on
the *Phocidæ*, that all motion in water by them arises from that power
which they exert and apply with the hindfeet.   So far as my observations on the hair-seal go, I am inclined to agree with this opinion.

All their movements in water, no matter whether travelling to
some objective point or merely in sport, are quick and joyous, and
nothing is more suggestive of intense satisfaction and pure physical comfort than is that spectacle which we can see every August
a short distance at sea from any rookery, where thousands of
old males and females are idly rolling over in the billows side by
side, rubbing and scratching with their fore and hind flippers,
which are here and there stuck up out of the water by their owners, like so many lateen-sails of Mediterranean feluccas, or, when
their hind flippers are presented, like a "cat-o'-nine tails."   They
sleep in the water a great deal, too, more than is generally supposed,
showing that they do not come ashore to rest—very clearly not.

How fast the fur-seal can swim, when doing its best, I am
naturally unable to state.   I do know that a squad of young "holluschickie" followed the *Reliance*, in which I was sailing, down
from the latitude of the Seal Islands to Akootan Pass with perfect
ease ; playing around the vessel while she was logging, straight
ahead, fourteen knots to the hour.

When the "holluschickie" are up on land they can be readily
separated into their several classes, as to age, by the color of their
coats and size, when noted : thus, as yearlings, two, three, four,
and five years old males.   When the yearlings, or the first class,
haul out, they are dressed just as they were after they shed their
pup-coats and took on the second covering, during the previous
year in September and October ; and now, as they come out in
the spring and summer, one year old, the males and females can-

not be distinguished apart, either by color or size, shape or action ; the yearlings of both sexes have the same steel-gray backs and white stomachs, and are alike in behavior and weight.

Next year those yearling females, which are now trooping out with these youthful males on the hauling-grounds, will repair to the rookeries, but their male companions will be obliged to return alone to this same spot.

About the 15th and 20th of every August they have become perceptibly "stagey," or, in other words, their hair is well under way in shedding. All classes, with the exception of the pups, go through this renewal at this time every year. The process requires about six weeks between the first dropping or falling out of the old over-hair and its full substitution by the new : this change takes place, as a rule, between August 1st and September 28th.

The fur is shed, but it is so shed that the ability of a seal to take to the water and stay there, and not be physically chilled or disturbed during its period of moulting, is never impaired. The whole surface of these extensive breeding-grounds, traversed over by me after the seals had gone, was literally matted with shedded hair and fur. This under-fur or pelage is, however, so fine and delicate, and so much concealed and shaded by the coarser over-hair, that a careless eye or a superficial observer might be pardoned in failing to notice the fact of its dropping and renewal.

The yearling cows retain the colors of the old coat in the new, when they shed for the first time, and so repeat them from that time on, year after year, as they live and grow old. The young three-year-olds and the mature cows look exactly alike, as far as color goes, when they haul up at first and dry out on the rookeries, every June and July.

The yearling males, however, make a radical change when they shed for the first time, since they come out from their "staginess" in a nearly uniform dark gray, and gray and black mixed, and lighter, with dark ochre to whitish on the upper and under parts, respectively. This coat, next year, when they appear as two-year-olds, shedding for the three-year-old coat, is a very much darker gray, and so on to the third, fourth, and fifth seasons ; then after this, with age, they begin to grow more gray and brown, with a rufous-ochre and whitish-tipped "wig" on the shoulders. Some of the very old bulls change in their declining years to a uniform shade, all over, of dull-grayish ochre. The full glory and beauty of the seal's mustache is denied to 'him until he has attained his seventh or eighth year.

The male does not get his full growth and weight until the close of his seventh year, but realizes most of it, osteologically speaking, by the end of the fifth ; and from this it may be perhaps truly inferred that the male seals live to an average age of eighteen or twenty years, if undisturbed in a normal condition, and that the females exist ten or twelve seasons under the same favorable circumstances.   Their respective weights, when fully mature and fat, in the spring, will, in regard to the male, strike an average of from four to five hundred pounds, while the females will show a mean of from seventy to eighty pounds.

The female does not gain a maximum size and weight until the end of her fourth year, so far as I have observed, but she does most of her longitudinal growing in the first two.   After she has passed her fourth and fifth years, she weighs from thirty to fifty pounds more than she did in the days of her youthful maternity.*   In the

* I did not permit myself to fall into error by estimating this matter of weight, because I early found that the apparent huge bulk of a sea-lion bull or fur-seal male, when placed upon the scales, shrank far below my notions: I took a great deal of pains, on several occasions, during the killing-season, to have a platform scale carted out into the field, and as the seals were knocked down, and before they were bled, I had them carefully weighed, constructing the following table from my observations:

Age.	Length.	Girth.	Gross weight of body.	Weight of skin.	Remarks.
	Inches.	Inches.	Pounds.	Pounds.	
One week	12 to 14	10 to 10½	6 to 7½	1¼	A male and female, being the only ones of the class handled, June 20, 1873.
Six months..........	24	25	39	3	A mean of ten examples, males and females, alike in size, November 28, 1872.
One year	38	25	39	4½	A mean of six examples, males and females, alike in size, July 14, 1873.
Two years	45	30	58	5½	A mean of thirty examples, all males, July 24, 1873.
Three years..........	52	36	87	7	A mean of thirty-two examples, all males, July 24, 1873.
Four years..........	58	42	135	12	A mean of ten examples, all males, July 24, 1873.
Five years	65	52	200	16	A mean of five examples, all males, July 24, 1873.
Six years.......... ...	72	64	280	25	A mean of three examples, all males, July 24, 1873.
Eight to twenty years.	75 to 80	70 to 75	400 to 500	45 to 50	An estimate only, calculating on the weight (when fat, and early in the season), of old bulls.

table of these weights given below it will be observed that the adult females correspond with the three years old males ; also, that the younger cows weigh frequently only seventy-five pounds, and many of the older ones go as high as one hundred and twenty, but an average of eighty to eighty-five pounds is the rule. Those specimens just noted which I weighed were examples taken by me for transmission to the Smithsonian Institution, otherwise I should not have been permitted to make this record of their bulk, inasmuch as weighing them means to kill them ; and the law and the habit, or rather the prejudice of the entire community up there, is unanimously in opposition to any such proceeding, for they never touch females, and never go near or disturb the breeding-grounds on such an errand. It will be noticed, also, that I have no statement of the weights of any exceedingly fat and heavy males which appear first on the breeding grounds in the spring ; those which I have referred to, in the table above given, were very much heavier at the time of their first appearance, in May and June, than at the moment when they were in my hands, in July ; but the cows, and the other classes, do not sustain protracted fasting, and therefore their avoirdupois may be considered substantially the same throughout the year.

Thus, from the fact that all the young seals and females do not vary much in weight from the time of their first coming out in the spring, till that of their leaving in the fall and early winter, I feel safe in saying that they feed at irregular but not long intervals, during this period when they are here under our observation, since they are constantly changing from land to water and from water to land, day in and day out. I do not think that the young males fast longer than a week or ten days at a time, as a rule.

By the end of October and November 10th, a great mass of the "holluschickie," the trooping myriads of English Bay, South-west Point, Reef Parade, Lukannon Sands, the table-lands of Polavina, and the mighty hosts of Novastoshnah, at St. Paul, together with the quota of St. George, had taken their departure from these shores, and had gone out to sea, feeding upon the receding schools of fish that were now retiring to the deeper waters of the North Pacific, where, in that vast expanse, over which rolls an unbroken billow, five thousand miles from Japan to Oregon, they spend the winter and the early spring, until they reappear and break up, with their exuberant life, the dreary winter-isolation of the land which gave them birth.

20

A few stragglers remain, however, as late as the snow and ice will permit them to, in and after December ; then they are down by the water's edge, and haul up entirely on the rocky beaches, deserting the sand altogether ; but the first snow that falls in October makes them very uneasy, and a large hauling-ground will be so disturbed by a rainy day and night that its hundreds of thousands of occupants fairly deserted it. The fur-seal cannot bear, and will not endure, the spattering of sand into its eyes, which usually accompanies the driving of a rain-storm ; they take to the water, to reappear, however, when that nuisance shall be abated.

The weather in which the fur-seal delights is cool, moist, foggy, and thick enough to keep the sun always obscured, so as to cast no shadows. Such weather, which is the normal weather of St. Paul and St. George, continued for a few weeks in June and July, brings up from the sea millions of fur-seals. But, as I have before said, a little sunshine, which raises the temperature as high as 50° to 55° Fahr., will send them back from the hauling-grounds almost as quickly as they came. Fortunately, these warm, sunny days on the Pribylov Islands are so rare that the seals certainly can have no ground of complaint, even if we may presume they have any at all. Some curious facts in regard to their selection of certain localities on these islands, and their abandonment of others, are now on record.

I looked everywhere and constantly, when threading my way over acres of ground which were fairly covered with seal-pups and older ones, for specimens that presented some abnormity, i.e., monstrosities, albinos, and the like, such as I have seen in our great herds of stock ; but I was, with one or two exceptions, unable to note anything of the kind. I have never seen any malformations or "monsters" among the pups and other classes of the fur-seals, nor have the natives recorded anything of the kind, so far as I could ascertain from them. I saw only three albino pups among the multitudes on St. Paul, and none on St. George. They did not differ, in any respect, from the normal pups in size and shape. Their hair, for the first coat, was a dull ochre all over ; the fur whitish, changing to a rich brown, the normal hue ; the flippers and muzzle were a pinkish flesh-tone in color, and the iris of the eye sky-blue. After they shed, during the following year, they have a dirty, yellowish-white color, which makes them exceedingly conspicuous when mixed in among a vast majority of black pups, gray yearlings, and "holluschickie" of their kind.

Undoubtedly some abnormal birth-shapes must make their appearance occasionally ; but at no time while I was there, searching keenly for any such manifestation of malformation on the rookeries, did I see a single example. The morphological symmetry of the fur-seal is one of the most salient of its characteristics, viewed as it rallies here in such vast numbers ; but the osteological differentiation and asymmetry of this animal is equally surprising.

It is perfectly plain that a large percentage of this immense number of seals must die every year from natural limitation of life. They do not die on these islands ; that much I am certain of. Not one dying a natural death could I find or hear of on the grounds. They evidently lose their lives at sea, preferring to sink with the *rigor mortis* into that cold, blue depth of the great Pacific, or beneath the green waves of Bering Sea, rather than to encumber and disfigure their summer haunts on the Pribylov Islands.

Prior to the year 1835, no native on the islands seemed to have any direct knowledge, or was even acquainted with a legendary tradition, in relation to the seals, concerning their area and distribution on the land here ; but they all chimed in after that date with great unanimity, saying that the winter preceding this season (1835–36) was one of frightful severity ; that many of their ancestors who had lived on these islands in large barraboras just back of the Black Bluffs, near the present village, and at Polavina, then perished miserably.

They say that the cold continued far into the summer ; that immense masses of clearer and stronger ice-floes than had ever been known to the waters about the islands, or were ever seen since, were brought down and shoved high up on to all the rookery margins, forming an icy wall completely around the island, and loomed twenty to thirty feet above the surf. They further state that this frigid cordon did not melt or in any way disappear until the middle or end of August, 1836.

They affirm that for this reason the fur-seals, when they attempted to land, according to their habit and their necessity, during June and July, were unable to do so in any considerable numbers. The females were compelled to bring forth their young in the water and at the wet, storm-beaten surf-margins, which caused multitudes of mothers and all of the young to perish. In short, the result was a virtual annihilation of the breeding-seals. Hence, at the following season, only a spectral, a shadowy imitation of former multitudes could be observed upon the seal-grounds of St. Paul and St. George.

On the Lagoon rookery, now opposite the village of St. Paul, there were then only two males, with a number of cows. At Nah Speel, close by and right under the village, there were then only some two thousand. This the natives know, because they counted them. On Zapadnie there were about one thousand cows, bulls, and pups; at Southwest Point there were none. Two small rookeries were then on the north shore of St. Paul, near a place called "Maroonitch;" and there were seven small rookeries running round Northeast Point, but on all of these there were only fifteen hundred males, females, and young; and this number includes the "holluschickie," which, in those days, lay in among the breeding-seals, there being so few old males that they were gladly permitted to do so. On Polavina there were then about five hundred cows, bulls, pups, and "holluschickie;" on Lukannon and Keetavie, about three hundred; but on Keetavie there were only ten bulls and so few young males lying in altogether that these old natives, as they told me, took no note of them on the rookeries just cited. On the Reef, and Gorbatch, were about one thousand only. In this number last mentioned some eight hundred "holluschickie" may be included, which laid with the breeding-seals. There were only twenty bulls on Gorbatch, and about ten old males on theReef.

Such, briefly and succinctly, is the sum and the substance of all information which I could gather prior to 1835–36; and while I do not entirely credit these statements, yet the earnest, straightforward agreement of the natives has impressed me so that I narrate it here. It certainly seems as though this enumeration of the old Aleutes was painfully short.

Then, again, with regard to the probable truth of the foregoing statement of the natives, perhaps I should call attention to the fact that the entire sum of seal-life in 1836, as given by them, is just four thousand one hundred, of all classes, distributed as I have indicated above. Now, on turning to Bishop Veniaminov, by whom was published the only statement of any kind in regard to the killing on these islands from 1817 to 1837 (the year when he finished his work), I find that he makes a record of slaughter of seals in the year 1836 of four thousand and fifty-two, which were killed and taken for their skins; but if the natives' statements are right, then only fifty seals were left on the island for 1837, in which year, however, four thousand two hundred and twenty were again killed, according to the bishop's table, and according to which there was also a steady increase in the size of this return from that date along up to 1850, when the Russians governed their catch by the market alone, always having more seals than they knew what to do with.

Again, in this connection, the natives say that until 1847 the practice on these islands was to kill indiscriminately both females and males for skins; but after this year, 1847, that strict respect now paid to the breeding-seals, and exemption of all females, was enforced for the first time, and has continued up to date.

In attempting to form an approximate conception of what the seals were or might have been in those early days, as they spread themselves over the hauling and breeding grounds of these remarkable islands, I have been thrown entirely upon the vague statements given to me by the natives and one or two of the first American pioneers in Alaska. The only Russian record which touches ever so lightly upon the subject* contains a remarkable statement

* Veniaminov: Zapieskie ob Oonalashkenskaho Otdayla, 2 vols., St. Petersburg, 1842. This work of Bishop Innocent Veniaminov is the only one which the Russians can lay claim to as exhibiting anything like a history of Western Alaska, or of giving a sketch of its inhabitants and resources, that has the least merit of truth or the faintest stamp of reliability. Without it we should be simply in the dark as to much of what the Russians were about during the whole period of their occupation and possession of that country. He served, chiefly as a priest and missionary, for nineteen years, from 1823 to 1842, mainly at Oonalashka, having the Seal Islands in his parish, and was made Bishop of all Alaska. He was soon after recalled to Russia, where he became the primate of the national church, ranking second to no man in the Empire save the Czar. He was advanced in life, being more than ninety years of age when he died at Moscow, April 22, 1879. He must have been a man of fine personal presence, judging from the following description of him, noted by Sir George Simpson, who met him at Sitka in 1842, just as he was about to embark for Russia: "His appearance, to which I have already alluded, impresses a stranger with something of awe, while in further intercourse the gentleness which characterizes his every word and deed insensibly moulds reverence into love, and at the same time his talents and attainments are such as to be worthy of his exalted station. With all this, the bishop is sufficiently a man of the world to disdain anything like cant. His conversation, on the contrary, teems with amusement and instruction, and his company is much prized by all who have the honor of his acquaintance." Sir Edward Belcher, who saw him at Kadiak in 1837, said: "He is a formidable-looking man, over six feet three inches in his boots, and athletic. He impresses one profoundly."

which is, in the light of my surveys, simply ridiculous now—that is, that the number of fur-seals on St. George during the first years of Russian occupation was-nearly as great as that on St. Paul. A most superficial examination of the geological character portrayed on the accompanying maps of those two islands will satisfy any unprejudiced mind as to the total error of such a statement. Why, a mere tithe only of the multitudes which repair to St. Paul in perfect comfort over the sixteen or twenty miles of splendid landing ground found thereon could visit St. George, when all of the coast-line fit for their reception on this island is a scant two and a half miles ; but, for that matter, there was at the time of my arrival and in the beginning of my investigation a score of equally wild and incredible legends afloat in regard to the rookeries of St. Paul and St. George. Finding, therefore, that the whole work must be undertaken *de novo*, I went about it without further delay.

Thus it will be seen that there is, frankly stated, nothing that serves as a guide to a fair or even an approximate estimate as to the numbers of the fur-seals on these two islands, prior to the result of my labor.

At the close of my investigation during the first season of my work on the ground in 1872 the fact became evident that the breeding seals obeyed implicitly an imperative and instinctive natural law of distribution—a law recognized by each and every seal upon the rookeries prompted by a fine consciousness of necessity to its own well-being. The breeding-grounds occupied by them were, therefore, invariably covered by the seals in exact ratio, greater or less, as the area upon which they rested was larger or smaller. They always covered the ground evenly, never crowding in at one place here to scatter out there. The seals lie just as thickly together where the rookery is boundless in its eligible area to their rear and unoccupied by them as they do in the little strips which are abruptly cut off and narrowed by rocky walls behind. For instance, on a rod of ground under the face of bluffs which hem it in to the land from the sea there are just as many seals, no more and no less, as will be found on any other rod of rookery ground throughout the whole list, great and small—always exactly so many seals, under any and all circumstances, to a given area of breeding-ground. There are just as many cows, bulls, and pups on a square rod at Nah Speel, near the village, where in 1874, all told, there were only seven or eight thousand, as there are on any square rod at Northeast Point, where a million of them congregate.

This fact being determined, it is evident that just in proportion as the breeding-grounds of the fur-seal on these islands expand or contract in area from their present dimensions, so the seals will increase or diminish in number.

That discovery at the close of the season of 1872 of this law of distribution gave me at once the clue I was searching for, in order to take steps by which I could arrive at a sound conclusion as to the entire number of seals herding on the Pribylov group.

I noticed, and time has confirmed my observation, that the period for taking these boundaries of the rookeries, so as to show this exact margin of expansion at the week of its greatest volume, or when they are as full as they are to be for the season, is between July 10th and 20th of every year—not a day earlier and not many days later. After July 20th the regular system of compact, even organization, breaks up. The seals then scatter out in pods or clusters, the pups leading the way, straying far back: the same number then instantly cover twice and thrice as much ground as they did the day or week before, when they laid in solid masses and were marshalled on the rookery ground proper.

There is no more difficulty in surveying these seal-margins during this week or ten days in July than there is in drawing sights along and around the curbs of a stone fence surrounding a field. The breeding-seals remain perfectly quiet under your eyes all over the rookery and almost within your touch, everywhere on the outside of their territory that you may stand or walk. The margins of massed life, which are indicated on the topographical surveys of these breeding-grounds of St. Paul and St. George, are as clean cut and as well defined against the soil and vegetation as is the shading on my maps. There is not the least difficulty in making such surveys, and in making them correctly.

Without following such a system of enumeration, persons may look over these swarming myriads between Southwest Point and Novastoshnah, guessing vaguely and wildly, at any figure from one million up to ten or twelve millions, as has been done repeatedly. How few people know what a million really is! It is very easy to talk of a million, but it is a tedious task to count it off: this makes a statement as to "millions" decidedly more conservative when the labor has been accomplished. After a thorough sur-

vey of all these great areas of reproduction the following presentation of the actual number of seals massed upon St. Paul is a fair
one :

"Reef rookery" has 4,016 feet of sea-margin, with 150 feet of average depth, making ground for............................... 301,000

"Gorbotch rookery" has 3,660 feet of sea-margin, with 100 feet of
average depth, making ground for......................... 183,000

"Lagoon rookery" has 750 feet of sea-margin, with 100 feet of average depth, making ground for............................... 37,000

"Nah Speel rookery" has 400 feet of sea-margin, with 40 feet of
average depth, making ground for.......................... 8,000

"Lukannon rookery" has 2,270 feet of sea-margin, with 150 feet of
average depth, making ground for.................... 170,000

"Keetavie rookery" has 2,200 feet of sea-margin, with 150 feet of
average depth, making ground for.......................... 165,000

"Tolstoi rookery" has 3,000 feet of sea-margin, with 150 feet of
average depth, making ground for......................... 225,000

"Zapadnie rookery" has 5,880 feet of sea-margin, with 150 feet of
average depth, making ground for......................... 441,000

"Polavina rookery" has 4,000 feet of sea-margin, with 150 feet of
average depth, making ground for......................... 300,000

"Novastoshnah, or Northeast Point" has 15,840 feet of sea-margin,
with 150 feet of average depth, making ground for.......... 1,200,000

A grand total of breeding-seals and young for St. Paul Island
in 1874 of... 3,030,000

The rookeries of St. George are designated and measured as
below :

"Zapadnie rookery" has 600 feet of sea-margin, with 60 feet of
average depth, making ground for......................... 18,000

"Starry Arteel rookery" has 500 feet of sea-margin, with 125 feet of
average depth, making ground for......................... 30,420

"North rookery" has 750 feet of sea-margin, with 150 feet of average depth, and 2,000 feet of sea-margin, with 25 feet of average depth, making ground in all for...................... 77,000

"Little Eastern rookery" has 750 feet of sea-margin, with 40 feet
of average depth, making ground for............... 13,000

"Great Eastern rookery" has 900 feet of sea-margin, with 60 feet
of average depth, making ground for..................... 25,000

A grand total of the seal-life for St. George Island, breeding-
seals and young, of 163,420

Grand sum total for the Pribylov Islands (season of 1873),
breeding-seals and young............................. 3,193,420

The figures thus given show a grand massing of 3,193,420 breeding-scals and their young. This enormous aggregate is entirely exclusive of the great numbers of the non-breeding-seals that, as we have pointed out, are never permitted to come up on those grounds which have been surveyed and epitomized by the table just exhibited. That class of seals, the "holluschickie," in general terms (all males, and those to which the killing is confined), come up on the land and sea-beaches between the rookeries, in immense straggling droves, going to and from the sea at irregular intervals, from the beginning to the closing of an entire season. The method of the "holluschickie" on these hauling-grounds is not systematic—it is not distinct, like the manner and law prescribed and obeyed by the breeding-seals—therefore it is impossible to arrive at a definite enumeration, and my estimate for them is purely a matter of my individual judgment. I think they may be safely rated at 1,500,000 ; thus, we have the wonderful number of 4,700,-000 fur-seals assembled every summer on the rocky rookeries and sandy hauling-grounds of the Pribylov Islands !

No language can express adequately your sensations when you first stroll over the outskirts of any one of those great breeding grounds of the fur-seal on St. Paul's Island. There is no impression on my mind more fixed than is the one stamped thereon during the afternoon of a July day when I walked around the inner margins of that immmense rookery at Northeast Point—indeed, while I pause to think of this subject, I am fairly rendered dumb by the vivid spectacle which rises promptly to my view—I am conscious of my inability to render that magnificent animal-show justice in definition. It is a vast camp of parading squadrons which file and deploy over slopes from the summit of a lofty hill a mile down to where it ends on the south shore—a long mile, smooth and gradual from the sea to that hill-top ; the parade-ground lying between is also nearly three-quarters of a mile in width, sheer and unbroken. Now, upon that area before my eyes, this day and date of which I' have spoken, were the forms of not less than three-fourths of a million seals—pause a moment—think of the number —three-fourths of a million seals, moving in one solid mass from sleep to frolicsome gambols, backward, forward, over, around, changing and interchanging their heavy squadrons, until the whole mind is so confused and charmed by the vastness of mighty hosts that it refuses to analyze any further. Then, too, I remember that the day was one of exceeding beauty for that region—it was a swift alternation overhead of those characteristic rain-fogs, between the succession of which the sun breaks out with transcendent brilliancy through misty halos about it. This parade-field reflected the light like a mirror, and the seals, when they broke apart here and there for a moment, just enough to show its surface, seemed as though they walked upon the water. What a scene to put upon canvas—that amphibian host involved in those alternate rainbow lights and blue-gray shadows of the fog !

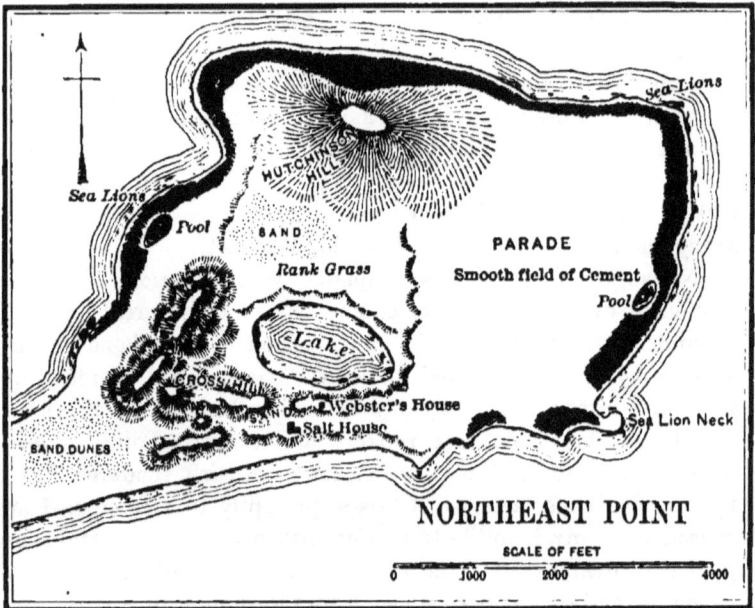

Survey Showing the Immense Breeding Area of Novastoshnah.
[*The shaded belt is that ground wholly covered by Fur-Seal Rookeries.*]

While Novastoshnah is the largest, yet in some respects I consider Tolstoi, with its bluffs and its long sweep which takes in the sands of English Bay, to be the most picturesque, though it be not the most impressive rookery—especially when that parade-ground belonging to it is reached by the climbing seals.

From Tolstoi at this point, circling around three miles to Zapadnie, is the broad sand-beach of English Bay, upon which and back over its gently rising flats are the great hauling-grounds of the "holluschickie," which I have indicated on the general map, and to which I made reference in a previous section of this chapter. Gazing at these myriads of "bachelor-seals" spread out in their restless hundreds and hundreds of thousands upon this ground, one feels the utter impotency of verbal description, and reluctantly shuts his note and sketch books to view it with renewed fascination and perfect helplessness.

Looking from the village across the cove and down upon the lagoon, still another strange contradiction appears—at least it seems a natural contradiction to one's usual ideas. Here we see

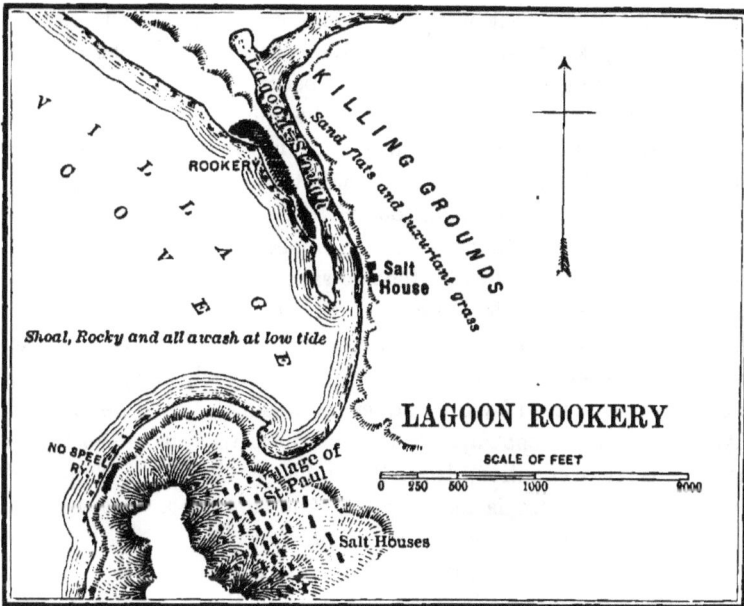

Survey Showing the Close Contact of Village, Slaughter-Field and Breeding Grounds.

the Lagoon rookery, a reach of ground upon which some twenty-five or thirty thousand breeding-seals come out regularly every year during the appointed time, and go through their whole elaborate system of reproduction, without showing the slightest concern for or attention to the scene directly east of them and across that shallow slough not eighty feet in width. There are the great slaughtering fields of St. Paul Island; there are the sand-flats where every seal has been slaughtered for years upon years back, for its skin ; and even as we take this note, forty men are standing there knocking down a drove of two or three thousand "hollu-schickie" for their day's work, and as they labor, the whacking of their clubs and the sounds of their voices must be as plain to those breeding-seals, which are not one hundred feet from them, as it is to us, a quarter of a mile distant! In addition to this enumeration of disturbances, well calculated to amaze, and dismay, and drive off every seal within its influence, are the decaying bodies of the last year's catch—seventy-five thousand or eighty-five thousand un-buried carcasses—that are sloughing away into the sand which, two or three seasons from now, nature will, in its infinite charity, cover with the greenest of all green grasses. The whitened bones and grinning skulls of over three million seals have bleached out on that slaughtering-spot, and are buried below its surface.

Directly under the north face of the village hill, where it falls to the narrow flat between its feet and the cove, the natives have sunk a well. It was excavated in 1857, they say, and subsequently deepened to its present condition in 1868. It is twelve feet deep, and the diggers said that they found bones of the sea-lion and fur-seal thickly distributed every foot down, from top to bottom. How much lower these osteological remains of prehistoric pinnipeds can be found no one knows as yet. The water here, on that ac-count, has never been fit to drink, or even to cook with, but, being soft, was and is used by the natives for washing clothes, etc. Most likely, it records a spot upon which the Russians, during the heyday of their early occupation, drove the unhappy visitors of Nah Speel to slaughter. There is no Golgotha known to man elsewhere in the world as extensive as this one of St. Paul.

Yet, the natives say that this Lagoon rookery is a new feature in the distribution of the seals; that when the people first came here and located a part of the present village, in 1824 up to 1847, there never had been a breeding-seal on that Lagoon rookery of to-day; so they have hauled up here from a small beginning, not very long ago, until they have attained their present numerical ex-pansion, in spite of all these exhibitions of butchery of their kind, executed right under their eyes, and in full knowledge of their nos-trils, while the groans and low moanings of their stricken species, stretched out beneath the clubs of the sealers, must have been and are far plainer in their ears than they are in our own!

Still they come—they multiply, and they increase—knowing so well that they belong to a class which intelligent men never did

molest. To-day at least they know it, or they would not submit to
these manifestations which we have just cited, so close to their
knowledge.

The Lagoon rookery, however, never can be a large one, on account of the very nature of this ground selected by the seals ; it is
a bar simply pushed up beyond the surf-wash of boulders, water-
worn and rounded, which has almost enclosed and cut away the
Lagoon from its parent sea. In my opinion, the time is not far distant when that estuary will be another inland lake of St. Paul,
walled out from salt water and freshened by rain and melting snow,
as are the other pools, lakes, and lakelets on the island.

Zapadnie, in itself, is something like the Reef plateau on its
eastern face, for it slopes up gradually and gently to the parade-
plateau above—a parade-ground not so smooth, however, being
very rough and rocky, but which the seals enjoy. Just around the
point, a low strip of rocky bar and beach connects it with the
ridge-walls of Southwest Point, a very small breeding rookery, so
small that it is not worthy of a survey, is located here. I think,
probably, on account of the nature of the ground, that it will never
hold its own, and is more than likely abandoned by this time.

One of the prehistoric villages, the village of Pribylov's time,
was established here between that point and the cemetery ridge,
on which the northern wing of Zapadnie rests. An old burying-
ground, with its characteristic Russian crosses and faded pictures
of the saints, is plainly marked on the ridge. It was at this little
bight of sandy landing that Pribylov's men first came ashore and
took possession of the island, while others in the same season proceeded to Northeast Point and to the north shore to establish
settlements of their own order. When the indiscriminate sealing of
1868 was in progress, one of the parties lived here, and a salt-house
which was then erected by them still stands. It is in a very fair
state of preservation, although it has never been occupied since,
except by the natives who come over here from the village in the
summer to pick those berries of the *Empetrum* and *Rubus*, which
abound in the greatest profusion around the rough and rocky
flats that environ a little lake adjacent. The young people of St.
Paul are very fond of this berry-festival, so-called among themselves, and they stay there every August, camping out, a week or
ten days at a time, before returning to their homes in the village.

So abundant have been the seals that no driving of animals from the parade-grounds of Zapadnie has ever been made since 1869. It is easily reached, however, if it were desirable to do so.

Polavina has also been an old settlement site, and, for the reason cited at Zapadnie, no "holluschickie" have been driven from this point since 1872, though it is one of the easiest worked. It was in the Russian times a pet sealing-ground with them. The remains of an old village have nearly all been buried in the sand near the lake, and there is really no mark of its early habitation, unless it be the singular effect of a human graveyard being dug out and despoiled by the attrition of seal bodies and flippers. The old cemetery just above and to the right of the barrabkie, near the little lake, was originally established, so the natives told me, far away from the hauling of the "holluschickie." It was, when I saw it in 1876, in a melancholy state of ruin. A thousand young seals (at least) moved off from its surface as I came up, and they had actually trampled out many sandy graves, rolling the bones and skulls of Aleutian ancestry in every direction. Beyond this ancient demesne which the natives established long ago, as a house of refuge during the winter when they were trapping foxes, looking to the west over the lake, is a large expanse of low, flat swale and tundra, which is terminated by the rocky ridge of Kamminista. Every foot of it has been placed there subsequent to the original elevation of the island by direct action of the sea, beyond question. It is covered with a thick growth of the rankest sphagnum, which quakes and trembles like a bog under one's feet, but over which the most beautiful mosses ever and anon crop out, including that characteristic floral display before referred to in speaking of the island. Most of the way from the village up to the Northeast Point, as will be seen by a cursory glance at the map, with the exception of this bluff of Polavina and the terraced table setting back from its face to Polavina Sopka, the whole island is slightly elevated above the level of the sea, and its coast-line is lying just above and beyond the reach of the surf, where great ridges of sand have been piled up by the wind, capped with sheafs and tufts of rank-growing *Elymus*.

Near the village, at that little bight mapped as Zoltoi, is a famous rendezvous for the "holluschickie," and from this place during the season the natives make regular drives, having only to step out from their houses in the morning and walk a few rods to find their fur-bearing quarry.

Passing over Zoltoi on our way down to the point, we quickly come to a basaltic ridge or back-bone over which the sand has been rifted by strong winds, and which supports a rank and luxuriant growth of the *Elymus* and other grasses, with beautiful flowers. A few hundred feet farther along our course brings us in full view, as we look to the south, of one of the most entrancing spectacles that seals afford to man. We glance below upon and survey a full sweep of the Reef rookery along a grand promenade ground, which slopes gently to the eastward and trends southward down to the water from its abrupt walls bordering on the sea to the west ; it is a parade plateau as smooth as the floor of a ball-room, 2,000 feet in length, from 500 to 1,000 feet in width, over which multitudes of "holluschickie" are filing in long strings or deploying in vast platoons, hundreds abreast, in an unceasing march and countermarch. The breath that rises into the cold air from a hundred thousand hot throats hangs like clouds of white steam in the gray fog itself; indeed, it may be said to be a seal-fog peculiar to such a spot, while the din, the roar arising over all, defies adequate description.

We notice to our right and to our left an immense solid mass of the breeding-seals at Gorbotch, and another stretching and trending nearly a mile from our feet, far around to the Reef Point below and opposite that parade-ground, with here and there a neutral passage left open for the "holluschickie" to go down and come up from the waves.

The adaptation of this ground of the Reef rookery to the requirements of the seal is perfect. It so lies that it falls gently from its high Zoltoi Bay margin, on the west, to the sea on the east, and upon its broad expanse not a solitary puddle of mud-spotting is to be seen, though everything is reeking with moisture, and the fog even dissolves into rain as we view the scene. Every trace of vegetation upon this parade has been obliterated. A few tufts of grass, capping the summits of those rocky hillocks, indicated on the eastern and middle slope, are the only signs of botanical life which the seals have suffered to remain.

A small rock, "Seevitchie Kammin," five or six hundred feet right to the southward and out at sea, is also covered with the black and yellow forms of fur-seals and sea-lions. It is environed by shoal-reefs, rough and kelp-grown, which navigators prudently avoid.

At Lukannon and Ketavie there is a joint blending of two large breeding-grounds, their continuity broken by a short reach of sea-wall right under and at the eastern foot of Lukannon Hill. The appearance of these rookeries is, like all the others, peculiar to themselves. There is a rounded, bulging hill, at the foot of Lukannon Bay, which rises perhaps one hundred and sixty or one hundred and seventy feet from the sea, abruptly at the point, but swelling out gently up from the sand-dunes in Lukannon Bay to its summit at the northwest and south. The big rookery rests upon its northern slope. Here is a beautiful adaptation of the finest drainage, with a profusion of those rocky nodules scattered everywhere over it, upon which the female seals so delight in resting.

Standing on the bald summit of Lukannon Hill, we turn to the south, and look over Ketavie* Point, where another large aggregate of rookery life rests under our eye. The hill falls away into

---

* DEFINITIONS FOR RUSSIAN NAMES OF THE ROOKERIES, ETC.—The several titles on my map that indicate the several breeding-grounds, owe their origin and have their meaning as follows:

ZAPADNIE signifies "*westward*," and is so used by the people who live in the village.

ZOLTOI signifies "*golden*," so used to express a metallic shimmering of the sand there.

KETAVIE signifies "*of a whale*," so used to designate that point where a large right whale was stranded in 1849 (?) ; from Russian "*keet*," or "*whale*."

LUKANNON—so named after one Lukannon, a pioneer Russian, that distinguished himself, with one Kaiecov, a countryman, who captured a large number of sea-otters at that point, and on Otter Island, in 1787–88.

TONKIE MEES signifies "*small* (or "*slender*") *cape*" [tonkie, "thin"; mees, "cape"].

POLAVINA literally signifies "*halfway*," so used by the natives because it is practically half way between the salt-houses at Northeast Point and the village. POLAVINA SOPKA, or "*half-way mountain*," gets its name in the same manner.

NOVASTOSHNAH, from the Russian "*novuite*," or "*of recent growth*," so used because this locality in pioneer days was an island to itself; and it has been annexed recently to the mainland of St. Paul.

VESOLIA MISTA, or "*jolly place*," the site of one of the first settlements, and where much carousing was indulged in.

MAROONITCH, the site of a pioneer village, established by one Maroon.

NAHSAYVERNIA, or "*on the north shore*," from Russian, "*sayvernie*."

BOGASLOV, or "*word of God*," indefinite in its application to the place, but is, perhaps, due to the fact that the pious Russians, immediately after landing at Zapadnie, in 1787, ascended the hill and erected a huge cross thereon.

a series of faintly terraced tables, which drop down to a flat that again abruptly descends to the sea at Ketavie Point. Between us and Ketavie rookery is the parade-ground of Lukannon,—a sight almost as grand as is that on the Reef which we have feebly attempted to portray. The sand-dunes to the west and to the north are covered with the most luxuriant grass, abruptly emarginated by sharp abrasions of the hauling-seals: this is shown very clearly on my general map. Ketavie Point is a solid basaltic shelf. Lukannon Hill, the summit of it, is composed of volcanic tufa and cement, with irregular cubes and fragments of pure basalt scattered all over its flipper-worn slopes. This is that place, down along the flat shoals of Lukannon Bay, where the sand-dunes are most characteristic, as they rise in their wind-whirled forms just above surf-wash. Here also is where the natives come from the village during the early mornings of the season for driving, to get any number of "holluschickie" required.

It is a beautiful sight, glancing from the summit of this great rookery hill, up to the north over that low reach of the coast to Tonkie Mees, where the waves seem to roll in with crests which rise in unbroken ridges for a mile in length each, ere they break so grandly and uniformly on the beach. In these rollers the "holluschickie" are playing like sea-birds, seeming to sport the most joyously at the very moment when a heavy billow breaks and falls upon them.

The precipitous shore-line of St. George is enough in itself to explain the small number of seals found there, when contrasted with the swarming myriads of her more favorably adapted sister island. Nevertheless that Muscovitic sailor, Pribylov, not knowing then of the existence of St. Paul, was as well satisfied as if he had possessed the boundless universe when he first found it. As in the case of St. Paul Island, I have been unable to learn much here in regard to the early status of the rookeries, none of the natives having any real information. The drift of their sentiment goes to show that there never was a great assemblage of fur-seals on St. George

---

EINAHNUHTO, an Aleutian word, signifying the "*three mammæ.*"

TOLSTOI, a Russian name, signifying "*thick*"; it is given to at least a hundred different capes and headlands throughout Alaska, being applied as indiscriminately as we do the term "Bear Creek" to little streams in our Western States and Territories.

21

—in fact, never as many as there are to-day, insignificant as the exhibit is, compared with that of St. Paul. They say that at first the
sea-lions owned this island, and that the Russians, becoming cogni-
zant of the fact, made a regular business of driving off the "sec-
vitchie," in order that fur-seals might be encouraged to land.
Touching this statement, with my experience on St. Paul, where
there is no conflict at all between the fifteen or twenty thousand
sea-lions which breed around on the outer edge of the seal rookeries
there and at Southwest Point, I cannot agree to the St. George
legend. I am inclined to believe, however—indeed, it is more than
probable—that there were a great many more sea-lions on and
about St. George before it was occupied by men—a hundred-fold
greater, perhaps, than now, because a sea-lion is an exceedingly
timid, cowardly creature when it is in the proximity of man, and
will always desert any resting-place where it is constantly brought
into contact with him.*

The rookeries on this island, being so much less in volume, are
not especially noted—still, one of them, "Starry Arteel," is unique
indeed, lying as it does in a bold sweep from the sea up a very
steep slope to a point where the bluffs bordering it seaward are
over four hundred feet in vertical declination. The seals crowd
just as closely to the edge of this precipice along its entire face as
they do at the tide-level. It is a very strange sight for that visitor
who may sail under these bluffs with a boat in fair weather for landing, and, as you walk the beach, above which the cliff-wall frowns
a sheer five hundred feet, there, directly over your head, the cran-
ing necks and twisting forms of restless seals, appear as if ready to
launch out and fall below, ever and anon, as you glance upward,
so closely and boldly do they press to the very edge of the preci-

---

* One of the natives, "stareek," Zachar Oostigov, told me that the "Russians, when they first landed, came ashore in a thick fog" at Tolstoi Mees,
near the present sea-lion rookery site. As the water is deep and "bold" there,
Pribylov's sloop, the *St. George*, must have jammed her bowsprit against those
lofty cliffs ere the patient crew had intimation of their position. The old
Aleut then showed me that steep gully there, up which the ardent discoverers
climbed to a plateau above : and, to demonstrate that he was not chilled or
weakened by age, he nimbly scrambled down to the surf below, some three
hundred and fifty vertical feet, and I followed, half stepping and half sliding
over Pribylov's path of glad discovery and proud possession, trodden one June
day by him nearly a hundred years ago.

pice. I have been repeatedly astonished at an amazing power possessed by the fur-seal of resistance to shocks which would certainly kill any other animal. To explain clearly, the reader will observe by reference to the maps that there are a great many cliffy places between the rookeries on the shore-lines of the islands. Some of these bluffs are more than one hundred feet in abrupt elevation above the surf and rocks awash below. Frequently "holluschickie," in ones, or twos, or threes, will stray far away back from the great masses of their kind and fall asleep in the thick grass and herbage which covers these mural reaches. Sometimes they will repose and rest very close to the edge, and then as you come tramping along you discover and startle them and yourself alike. They, blinded by their first transports of alarm, leap promptly over the brink, snorting, coughing, and spitting as they go. Curiously peering after them and looking down upon the rocks, fifty to one hundred feet below, instead of seeing their stunned and motionless bodies, you will invariably catch sight of them rapidly scrambling into the water, and, when in it, swimming off like arrows from the bow. Three "holluschickie" were thus inadvertently surprised by me on the edge of the west face to Otter Island. They plunged over from an elevation there not less than two hundred feet in sheer descent, and I distinctly saw them fall, in scrambling, whirling evolutions, down, thumping upon the rocky shingle beneath, from which they bounded as they struck, like so many rubber balls. Two of them never moved after the rebound ceased ; but the third one reached the water and swam away swift as a bird on the wing.

While they seem to escape without bodily injury incident to such hard falls as ensue from dropping fifty or sixty feet upon pebbly beaches and rough boulders below, and even greater elevations, yet I am inclined to think that some internal injuries are necessarily sustained in almost every case, which soon develop and cause death. The excitement and the vitality of the seal at the moment of the terrific shock are able to sustain and conceal a real injury for the time being.

Driving the "holluschickie" on St. George, owing to the relative scantiness of hauling area for those animals there, and consequent small numbers found upon these grounds at any one time, is a very arduous series of daily exercises on the part of the natives who attend to it. Glancing at the map, the marked considerable distance over an exceedingly rough road will be noticed between

Zapadnie aud the village, yet in 1872 eleven different drives across the island of four hundred to five hundred seals each were made in the short four weeks of that season.

The peculiarly rough character to this trail is given by large, loose, sharp-edged basaltic boulders which are strewn thickly over all those lower levels that bridge the island between the high bluffs at Starry Arteel and the slopes of Ahluckeyak Hill. The summits of the two broader, higher plateaux, east and west respectively, are comparatively smooth and easy to travel over, and so is the sea-level flat at Zapadnie itself. On the map of St. George a number of very small ponds will be noticed. They are the fresh-water reservoirs of the island. The two largest of these are near the summit of this rough divide. The seal-trail from Zapadnie to the village runs just west of them and comes out on the north shore, a little to the eastward of the hauling-grounds of Starry Arteel, where it forks and unites with that path. A direct line between the village and Zapadnie, though nearly a mile shorter on the chart, is equal to five miles more of distance by reason of its superlative rocky inequalities.

One question is always sure to be asked in this connection. The query is: "At the present rate of killing seals it will not be long ere they are exterminated—how much longer will they last?" My answer is now as it was then : "Provided matters are conducted on the Seal Islands in the future as they are to-day, 100,000 male seals under the age of five years and over one may be safely taken every year from the Pribylov Islands without the slightest injury to the regular birth-rate or natural increase thereon ; provided also that the fur-seals are not visited by plagues or by pests, or any such abnormal cause for their destruction, which might be beyond the control of men, and to which, like any other great body of animal life, they must ever be subjected to the danger of."* From my calculations

---

* The thought of what a deadly epidemic would effect among these vast congregations of *Pinnipedia* was one that was constant in my mind when on the ground and among them. I have found in the "British Annals" (Fleming's), on page 17, an extract from the notes of Dr. Trail: "In 1833 I inquired for my old acquaintances, the seals of the Hole of Papa Westray, and was informed that about four years before they had totally deserted the island, and had only within the last few months begun to reappear. . . . About fifty years ago multitudes of their carcasses were cast ashore in every bay in the north of Scotland, Orkney, and Shetland, and numbers were found at sea in a sickly state." This note of Trail is the only record which I can find of a

given above it will be seen that 1,000,000 pups or young seals, in round numbers, are born upon these islands of the Pribylov group every year. Of this million, one-half are males. These 500,000 young males, before they leave the islands for sea during October and November, and when they are between five or six months old, fat, and hardy, have suffered but a trifling loss in numbers—say one per cent.—while on and about the islands of their birth, surrounding which and upon which they have no enemies whatever to speak of; but after they get well down to the Pacific, spread out over an immense area of watery highways in quest of piscatorial food, they form the most helpless of their kind to resist or elude the murderous teeth and carnivorous attacks of basking sharks * and killer-whales. †

fatal epidemic among seals. It is not reasonable to suppose that the Pribylov rookeries have never suffered from distempers in the past, or are not to in the future, simply because no occasion seems to have arisen during the comparatively brief period of their human domination.

* *Somniosus microcephalus.* Some of these sharks are of very large size, and when caught by the Indians of the northwest coast, basking or asleep on the surface of the sea, they will, if transfixed by the natives' harpoons, take a whole fleet of canoes in tow and run swiftly with them several hours before exhaustion enables the savages to finally despatch them. A Hudson Bay trader, William Manson (at Fort Alexander in 1865), told me that his father had killed one in the smooth waters of Millbank Sound which measured twenty-four feet in length, and its liver alone yielded thirty-six gallons of oil. The *Somniosus* lies motionless for long intervals in calm waters of the North Pacific, just under and at the surface, with its dorsal fin clearly exposed above. What havoc such a carnivorous fish would be likely to effect in a "pod" of young fur-seals can be better imagined than described.

† *Orca gladiator.* While revolving this particular line of inquiry in my mind when on the ground and among the seals, I involuntarily looked constantly for some sign of disturbance in the sea which would indicate the presence of an enemy, and, save seeing a few examples of the *Orca*, I never detected anything. If the killer-whale was common here, it would be patent to the most casual eye, because it is the habit of this ferocious cetacean to swim so closely at the surface as to show its peculiar sharp dorsal fin high above the water. Possibly a very superficial observer could and would confound that long trenchant fluke of the *Orca* with the stubby node upon the spine of a humpback whale, which that animal exhibits only when it is about to dive. Humpbacks feed around the islands, but not commonly ; they are the exception. They do not, however, molest the seals in any manner whatever, and little squads of these pinnipeds seem to delight themselves by swimming in endless circles around and under the huge bodies of those whales, frequently leaping out and entirely over the cetacean's back, as witnessed on one occasion by myself and the crew of the *Reliance* off the coast of Kadiak, June, 1874.

By these agencies, during their absence from the islands until their reappearance in the following year, and in July, they are so perceptibly diminished in number that I do not think, fairly considered, more than one-half of the legion which left the ground of their birth last October come up the next July to these favorite landing-places—that is, only 250,000 of them return out of the 500,000 born last year. The same statement in every respect applies to the going and the coming of the 500,000 female pups, which are identical in size, shape, and behavior.

As yearlings, however, these 250,000 survivors of last year's birth have become strong, lithe, and active swimmers, and when they again leave the hauling grounds, as before, in the fall, they are fully as able as are the older class to take care of themselves, and when they reappear next year, at least 225,000 of them safely return in the second season after birth. From this on I believe that they live out their natural lives of fifteen to twenty years each, the death-rate now caused by the visitation of marine enemies affecting them in the aggregate but slightly. And, again, the same will hold good touching the females, the average natural life of which, however, I take to be only nine or ten years each.

Out of these two hundred and twenty-five thousand young males we are required to save only one-fifteenth of their number to pass over to the breeding-grounds, and meet there the two hundred and twenty-five thousand young females; in other words, the polygamous habit of this animal is such that, by its own volition, I do not think that more than one male annually out of fifteen born is needed on the breeding-grounds in the future ; but in my calculations, to be within the margin and to make sure that I save two-year-old males enough every season, I will more than double this proportion, and set aside every fifth one of the young males in question. That will leave one hundred and eighty thousand seals, in good condition, that can be safely killed every year, without the slightest injury to the perpetuation of the stock itself forever in all of its original integrity.

In the above showing I have put a very extreme estimate upon that loss sustained at sea by the pup-seals—too large, I am morally certain ; but, in attempting to draw this line safely, I wish to place the matter in the very worst light in which it can be put, and to give the seals the full benefit of every doubt. Surely I have clearly presented the case, and certainly no one will question the premises after they have studied the habit and disposition of the rookeries ; hence, it is a positive and tenable statement, that no danger of the slightest appreciable degree of injury to the interests of the Government on the Seal Islands of Alaska exists, as long as the present law protecting it, and the management executing it, continues.

These fur-seals of the Pribylov group, after leaving the islands in the autumn and early winter, do not visit land again until the time of their return, in the following spring and early summer, to these same rookery and hauling-grounds, unless they touch, as they are navigating their lengthened journey back, at the Russian Islands, Copper and Bering, seven hundred miles to the westward of the Pribylov group. They leave our islands by independent squads, each one looking out for itself. Apparently all turn by common consent to the south, disappearing toward the horizon, and are soon lost in the vast expanse below, where they spread themselves over the entire Pacific as far south as the 48th and even the 47th parallels of north latitude : within this immense area between Japan and Oregon, doubtless, many extensive submarine fishing-shoals and banks are known to them ; at least, it is definitely understood that Bering Sea does not contain them long when they depart from the breeding rookeries and the hauling-grounds therein. While it is carried in mind that they sleep and rest in the water with soundness and with the greatest comfort on its surface, and that even when around the land, during the summer, they frequently put off from the beaches to take a bath and a quiet snooze just beyond the surf, we can readily agree that it is no inconvenience whatever, when the reproductive functions have been discharged, and their coats renewed, for them to stay the balance of the time in their most congenial element—the briny deep.

That these animals are preyed upon extensively by killer-whales (*Orca gladiator*), in especial, and by sharks, and probably other submarine foes now unknown, is at once evident ; for, were they not held in check by some such cause, they would, as they exist to-day on St. Paul, quickly multiply, by arithmetical progression, to so great an extent that the island, nay, Bering Sea itself, could not contain them. The present annual killing of one hundred thousand out of a yearly total of over a million males does not, in an appreciable degree, diminish the seal-life, or interfere in the slightest with its regular, sure perpetuation on the breeding-grounds every year. We may, therefore, properly look upon this aggregate of four and five millions of fur-seals, as we see them every season on these Pribylov Islands, as that maximum limit of increase assigned to them by natural law. The great equilibrium which nature holds in life upon this earth must be sustained at St. Paul as well as elsewhere.

Think of the enormous food-consumption of these rookeries and hauling-grounds; what an immense quantity of finny prey must pass down their voracious throats as every year rolls by! A creature so full of life, strung with nerves, muscles like bands of steel, cannot live on air, or absorb it from the sea. Their food is fish, to the practical exclusion of all other diet. I have never seen them touch, or disturb with the intention of touching it, one solitary example in the flocks of water-fowl which rest upon the surface of the water all about the islands. I was especially careful in noting this, because it seemed to me that the canine armature of their mouths must suggest flesh for food at times as well as fish; but fish we know they eat. Whole windrows of the heads of cod and wolf-fishes, bitten off by these animals at the nape, were washed up on the south shore of St. George during a gale in the summer of 1873. This pelagic decapitation evidently marked the progress and the appetite of a band of fur-seals to the windward of the island, as they passsed into and through a stray school of these fishes.

How many pounds per diem is required by an adult seal, and taken by it when feeding, is not certain in my mind. Judging from the appetite, however, of kindred animals, such as sea-lions fed in confinement at Woodward's Gardens, San Francisco, I can safely say that forty pounds for a full-grown fur-seal is a fair allowance, with at least ten or twelve pounds per diem to every adult female, and not much less, if any, to the rapidly growing pups and young "holluschickie." Therefore, this great body of four and five millions of hearty, active animals which we know on the Seal Islands, must consume an enormous amount of such food every year. They cannot average less than ten pounds of fish per diem, which gives the consumption, as exhibited by their appetite, of over six million tons of fish every year! What wonder, then, that nature should do something to hold these active fishermen in check.*

---

* I feel confident that I have placed this average of fish eaten per diem by each seal at a starvation allowance, or, in other words, it is a certain minimum

During the winter solstice—between the lapse of the autumnal and the verging of the vernal equinoxes—in order to get this enormous food-supply, the fur-seals are necessarily obliged to disperse over a very large area of fishing-ground, ranging throughout the North Pacific, five thousand miles across between Japan and the Straits of Fuca. In feeding, they are brought to the southward all this time ; and, as they go, they come more and more in contact with those natural enemies peculiar to the sea of these southern

of the whole consumption. If the seals can get double the quantity which I credit them with above, startling as it seems, still I firmly believe that they eat it every year. An adequate realization by icthyologists and fishermen as to what havoc the fur-seal hosts are annually making among the cod, herring, and salmon of the northwest coast and Alaska, would disconcert and astonish them. Happily for the peace of political economists who may turn their attention to the settlement and growth of the Pacific coast of America, it bids fair to never be known with anything like precision. The fishing of man, both aboriginal and civilized, in the past, present, and prospective, has never been, is not, nor will it be, more than a drop in the bucket contrasted with those piscatorial labors of these icthyophagi in the waters adjacent to their birth. What catholic knowledge of fish and fishing-banks any one of those old "seecatchie" must possess, which we observe hauled out on the Pribylov rookeries each summer ! It has, undoubtedly, during the eighteen or twenty years of its life, explored every fish-eddy, bank, or shoal throughout the whole of that vast immensity of the North Pacific and Bering Sea. It has had more piscine sport in a single twelvemonth than Izaak Walton had in his whole life.

An old sea-captain, Dampier, cruising around the world just about two hundred years ago, wrote diligently thereof (or, rather, one Funnel is said to have written for him), and wrote well. He had frequent reference to meeting hair-seals and sea-lions, fur-seals, etc., and fell into repeating this maxim, evidently of his own making: "For wherever there be plenty of fysh, there be seals." I am sure that, unless a vast abundance of good fishing-ground was near by, no such congregation of seal-life as is that under discussion on the Seal Islands could exist. The whole eastern half of Bering Sea, in its entirety, is a single fish-spawning bank, nowhere deeper than fifty to seventy-five fathoms, averaging, perhaps, forty ; also, there are great reaches of fishing-shoals up and down the northwest coast, from and above the Straits of Fuca, bordering the entire southern, or Pacific coast, of the Aleutian Islands. The aggregate of cod, herring, and salmon which the seals find upon these vast icthyological areas of reproduction, must be simply enormous, and fully equal to a most extravagant demand of the voracious appetites of *Callorhini*.

When, however, the fish retire from spawning here, there, and everywhere over these shallows of Alaska and the northwest coast, along by the end of September to the 1st of November, every year, I believe that the young fur-

latitudes, which are almost strangers and are really unknown to the waters of Bering Sea; for I did not observe, with the exception of ten or twelve perhaps, certainly no more, killer-whales, a single marine disturbance, or molestation, during the three seasons which I passed upon the islands, that could be regarded in the slightest degree inimical to the peace and life of the *Pinnipedia;* and thus, from my observation, I am led to believe that it is not until they descend well to the south of the Aleutian Islands, and in the North Pacific, that they meet with sharks to any extent, and are diminished by the butchery of the killer whales.

But I did observe a very striking exhibition, however, of this character one afternoon while looking over Lukannon Bay. I saw a "killer" chasing the alert "holluschickie" out beyond the breakers, when suddenly, in an instant, the cruel cetacean was turned toward the beach in hot pursuit, and in less time than this is read the ugly brute was high and dry upon the sands. The natives were called, and a great feast was in prospect when I left the carcass.

But this was the only instance of the orca in pursuit of seals that came directly under my observation; hence, though it does undoubtedly capture a few here every year, yet it is an insignificant cause of destruction, on account of its rarity.

The young fur-seals going out to sea for the first time, and following in the wake of their elders, are the clumsy members of the family. When they go to sleep on the surface of the water, they rest much sounder than the others; and their alert and wary nature, which is handsomely developed ere they are two seasons old, is in its infancy. Hence, I believe that vast numbers of them are easily captured by marine foes, as they are stupidly sleeping, or awkwardly fishing.

I must not be understood as saying that fish alone constitute the diet of the Pribylov pinnipeds; I know that they feed, to a limited

---

seal, in following them into the depths of the great Pacific, must have a really arduous struggle for existence—unless it knows of fishing-banks unknown to us. The yearlings, however, and all above that age, are endowed with sufficient muscular energy to dive rapidly in deep soundings, and to fish with undoubted success. The pup, however, when it goes to sea, five or six months old, is not lithe and sinewy like the yearling; it is podgy and fat, a comparatively clumsy swimmer, and does not develop, I believe, into a good fisherman until it has become pretty well starved after leaving the Pribylovs.

extent, upon crustaceans and upon the squid (*Loligo*), also eating tender algoid sprouts; I believe that the pup-seals live for the first five or six months at sea largely, if not wholly, upon crustaceans and squids; they are not agile enough, in my opinion, to fish successfully, in any great degree, when they first depart from the rookeries.

In this connection I wish to record an impression very strongly made upon my mind, in regard to their diverse behavior when out at sea away from the islands, and when congregated thereon. As I have plainly exhibited in the foregoing, they are practically without fear of man when he visits them on the land of their birth and recreation; but the same seal that noticed you with quiet indifference at St. Paul, in June and July, and the rest of the season while he was there, or gambolled around your boat when you rowed from the ship to shore, as a dog will play about your horses when you drive from the gate to the house, that same seal, when you meet him in one of the passes of the Aleutian chain, one hundred or two hundred miles away from here, as the case may be, or to the southward of that archipelago, is the shyest and wariest creature your ingenuity can define. Happy are you in getting but a single glimpse of him, first; you will never see him after, until he hauls out, and winks and blinks across Lukannon sands.

But the companionship and the exceeding number of the seals, when assembled together annually, makes them bold; largely due, perhaps, to their fine instinctive understanding, dating, probably, back many years, seeming to know that man, after all, is not wantonly destroying them; and what he takes, he only takes from the ravenous maw of the killer-whale or the saw-tipped teeth of the Japan shark. As they sleep in the water, off the Straits of Fuca, and the northwest coast as far as Dixon's Sound, the Indians belonging to that region surprise them with spears and rifle, capturing quite a number every year, chiefly pups and yearlings.

When fur-seals were noticed, by myself, far away from these islands, at sea, I observed that then they were as shy and as wary as the most timorous animal would be, in dreading man's proximity—sinking instantly on apprehending the approach or presence of the ship, seldom to reappear to my gaze. But, when gathered in such immense numbers at the Pribylov Islands, they are suddenly metamorphosed into creatures wholly indifferent to my person: It must cause a very curious sentiment in the mind of

him who comes for the first time, during a summer season, to the
Island of St. Paul—where, when the landing boat or lighter carries
him ashore from the vessel, this whole short marine journey is enlivened by the gambols and aquatic evolutions of fur-seal convoys
to the "bidarrah," which sport joyously and fearlessly round and
round his craft, as she is rowed lustily ahead by the natives; the
fur-seals then, of all classes, "holluschickie" principally, pop their
dark heads up out of the sea, rising neck and shoulders erect above
the surface, to peer and ogle at him and at his boat, diving quickly
to reappear just ahead or right behind, hardly beyond striking distance from the oars. These gymnastics of *Callorhinus* are not wholly
performed thus in silence, for it usually snorts and chuckles with
hearty reiteration.

The sea-lion up here also manifest much the same marine interest, and gives the voyager an exhibition quite similar to the one
which I have just spoken of, when a small boat is rowed in the
neighborhood of its shore rookery; it is not, however, so bold, confident, and social as the fur-seal under the circumstances, and utters
only a short, stifled growl of surprise, perhaps; its mobility, however, of vocalization is sadly deficient when compared with the scope
and compass of its valuable relative's polyglottis.

The hair-seals (*Phoca vitulina*) around these islands never approached our boats in this manner, and I seldom caught more than
a furtive glimpse of their short, bull-dog heads when traversing the
coast by water.

The walrus (*Rosmarus obesus*) also, like *Phoca vitulina*, gave undoubted evidence of sore alarm over the presence of my boat and
crew anywhere near its proximity in similar situations, only showing itself once or twice, perhaps, at a safe distance, by elevating
nothing but the extreme tip of its muzzle and its bleared, popping
eyes above the water; it uttered no sound except a dull, muffled
grunt, or else a choking, gurgling bellow.

What can be done to promote the increase of fur seals? We
cannot cause a greater number of females to be born every year
than are born now; we do not touch or disturb these females as
they grow up and live; and we never will, if the law and present
management is continued. We save double—we save more than
enough males to serve; nothing more can be done by human agency;
it is beyond our power to protect them from their deadly marine
enemies as they wander into the boundless ocean searching for food.

Zapadnie.　　S. W. Point.　　Upper Zapadnie.　　Cox.'s Hill.

NATIVES GATHERING A "DRIVE"

In view, therefore, of all these facts, I have no hesitation in saying, quite confidently, that under the present rules and regulations governing the sealing interests on these islands, the increase or diminution of the seal-life thereon will amount to nothing in the future ; that the seals will exist, as they do exist, in all time to come at about the same number and condition recorded by this presentation of the author.

By reference to the habit of the fur-seal, which I have discussed at length, it is now plain and beyond doubt that two-thirds of all the males which are born, and they are equal in numbers to the females born, are never permitted by the remaining third, strongest by natural selection, to land upon the same breeding-ground with the females, which always herd thereupon *en masse.* Hence this great band of "bachelor" seals, or "holluschickie," so fitly termed, when it visits the island, is obliged to live apart wholly— sometimes and in some places, miles away from the rookeries ; and, by this admirable method of nature are those seals which can be killed without injury to the rookeries selected and held aside of their own volition, so that the natives can visit and take them without disturbing, to the least degree, that entire quiet of those breeding-grounds where the stock is perpetuated.

The manner in which the natives capture and drive up "holluschickie " from the hauling-grounds to the slaughter-fields near the two villages of St. Paul and St. George, and elsewhere on the islands, cannot be improved upon. It is in this way : At the beginning of every sealing-season, that is, during May and June, large bodies of the young " bachelor" seals do not haul up on land very far from the water—a few rods at the most—and, when these first arrivals are sought after, the natives, to capture them, are obliged to approach slyly and run quickly between the dozing seals and the surf, before they can take alarm and bolt into the sea ; in this manner a dozen Aleutes, running down the sand beach of English Bay, in the early morning of some June day, will turn back from the water thousands of seals, just as the mould-board of a plough lays over and back a furrow of earth. When the sleeping seals are first startled, they arise, and, seeing men between them and the water, immediately turn, lope and scramble rapidly back up and over the land ; the natives then leisurely walk on the flanks and in the rear of this drove thus secured, directing and driving it over to the killing-grounds, close by the village. The task of getting up early of a morning, and going out to the several hauling-grounds, closely adjacent, is really all there is of that labor expended in securing the number of seals required for a day's work on the killing-grounds. The two, three, or four natives upon whom, in rotation, this duty is devolved by the order of their chief, rise at first glimpse of dawn, between one and two o'clock, and hasten over to Lukannon, Tolstoi, or Zoltoi, as the case may be, "walk out" their "holluschickie," and have them duly on the slaughtering field before six or seven o'clock, as a rule, in the morning. In favorable weather the "drive" from Tolstoi consumes from two and a half to three hours' time ; from Lukannon, about two hours, and is often done in an hour and a half ; while Zoltoi is so near by that the time is merely nominal.

A drove of seals on hard or firm grassy ground, in cool and moist weather, may be driven with safety at the rate of half a mile an hour ; they can be urged along, with the expenditure of a great many lives, however, at the speed of a mile or a mile and a quarter per hour ; but that is seldom done. An old bull-seal, fat and unwieldy, cannot travel with the younger ones, though it can lope or gallop as it starts across the ground as fast as an ordinary man can run, over one hundred yards—then it fails utterly, falls to the earth supine, entirely exhausted, hot, and gasping for breath.

The "holluschickie" are urged along over paths leading to the killing-ground with very little trouble, and require only three or four men to guide and secure as many thousand at a time. They are permitted frequently to halt and cool off, as heating them injures their fur. These seal-halts on the road always impressed me with a species of sentimentalism and regard for the creatures themselves. When the men drop back for a few moments, that awkward shambling and scuffling of the march at once ceases, and the seals stop in their tracks to fan themselves with their hind flippers, while their heaving flanks give rise to subdued panting sounds. As soon as they apparently cease to gasp for want of breath, and are cooled off comparatively, the natives step up once more, clatter a few bones, with a shout along the line, and this seal-shamble begins again—their march to death and the markets of the world is taken up anew.[*]

---

[*] I heard a great deal of talk among the white residents of St. Paul, when I first landed and the sealing-season opened, about the necessity of "resting" the hauling-grounds ; in other words, they said if the seals were driven in repeated daily rotation from any one of the hauling-grounds, that this would so

I was also impressed by the singular docility and amiability of these animals when driven along the road. They never show fight any more than a flock of sheep would do ; if, however, a few old seals get mixed in, they usually grow so weary that they prefer to come to a stand-still and fight rather than move ; otherwise no sign whatever of resistance is made by the drove from the moment it is intercepted, and turned up from the hauling-grounds, to the time of its destruction at the hands of the sealing-gang.

This disposition of the old seals to fight rather than endure the panting torture of travel, is of great advantage to all parties concerned, for they are worthless commercially, and the natives are only too glad to let them drop behind, where they remain unmolested, eventually returning to the sea. The fur on them is of little or no value ; their under-wool being very much shorter, coarser, and more scant than in the younger ; especially so on the posterior parts along the median line of the back.

This change for the worse or deterioration of the pelage of the

---

disturb these animals as to prevent their coming to any extent again thereon, during the rest of the season. This theory seemed rational enough to me at the beginning of my investigations, and I was not disposed to question its accuracy ; but subseqent observation directed to this point particularly satisfied me, and the sealers themselves with whom I was associated, that the driving of the seals had no effect whatever upon the hauling which took place soon or immediately after the field, for the hour, had been swept clean of seals by the drivers. If the weather was favorable for landing, i. e., cool, moist, and foggy, the fresh hauling of the "holluschickie" would cover the bare grounds again in a very short space of time : sometimes in a few hours after the driving of every seal from Zoltoi sands over to the killing-fields adjacent, those dúnes and the beach in question would be swarming anew with fresh arrivals. If, however, the weather is abnormally warm and sunny, during its prevalence, even if for several consecutive days, no seals to speak of will haul out on the emptied space ; indeed, if these "holluschickie" had not been taken away by man from Zoltoi or any other hauling-ground on the islands when "tayopli" weather prevailed, most of those seals would have vacated their terrestrial loafing-places for the cooler embraces of the sea.

The importance of clearly understanding this fact as to the readiness of the "holluschickie" to haul promptly out on steadily "swept" ground, provided the weather is inviting, is very great ; because, when not understood, it was deemed necessary, even as late as the season of 1872, to "rest" the hauling-grounds near the village (from which all the driving has been made since), and make trips to far-away Polavina and distant Zapadnie—an unnecessary expenditure of human time, and a causeless infliction of physical misery upon phocine backs and flippers.

fur-seal takes place, as a rule, in the fifth year of their age—it is thickest and finest in texture during the third and fourth year of life ; hence, in driving the seals on St. Paul and St. George up from the hauling-grounds the natives make, as far as practicable, a selection only from males of that age. It is quite impossible, however, to get them all of one age without an extraordinary amount of stir and bustle, which the Aleutes do not like to precipitate ; hence the drive will be found to consist usually of a bare majority of three and four-year-olds, the rest being two-year-olds principally, and a very few, at wide intervals, five-year-olds, the yearlings seldom ever getting mixed up in it.

As this drove progresses along that path to those slaughtering-grounds, the seals all move in about the same way ; they go ahead with a kind of walking step and a sliding, shambling gallop. The progression of the whole caravan is a succession of starts, spasmodic and irregular, made every few minutes, the seals pausing to catch their breath, making, as it were, a plaintive survey and mute protest. Every now and then a seal will get weak in the lumbar region, then drag its posteriors along for a short distance, finally drop breathless and exhausted, quivering and panting, not to revive for hours—days, perhaps—and often never. During the driest driving-days, or those days when the temperature does not combine with wet fog to keep the earth moist and cool, quite a large number of the weakest animals in the drove will be thus laid out and left on the track. If one of these prostrate seals is not too much heated at the time, the native driver usually taps the beast over the head and removes its skin.

This prostration from exertion will always happen, no matter how carefully they are driven ; and in the longer drives, such as two and a half and five miles from Zapadnie on the west, or Polavina on the north, to the village at St. Paul, as much as three or four per cent. of the whole drive will be thus dropped on the road ; hence I feel satisfied, from my observation and close attention to this feature, that a considerable number of those that are thus rejected from the drove, and are able to rally and return to the water, die subsequently from internal injuries sustained on the trip, superinduced by this over-exertion. I therefore think it highly improper and impolitic to extend drives of the "holluschickie" over any distance on St. Paul Island exceeding a mile, or a mile and a half—it is better for all parties concerned, and the business

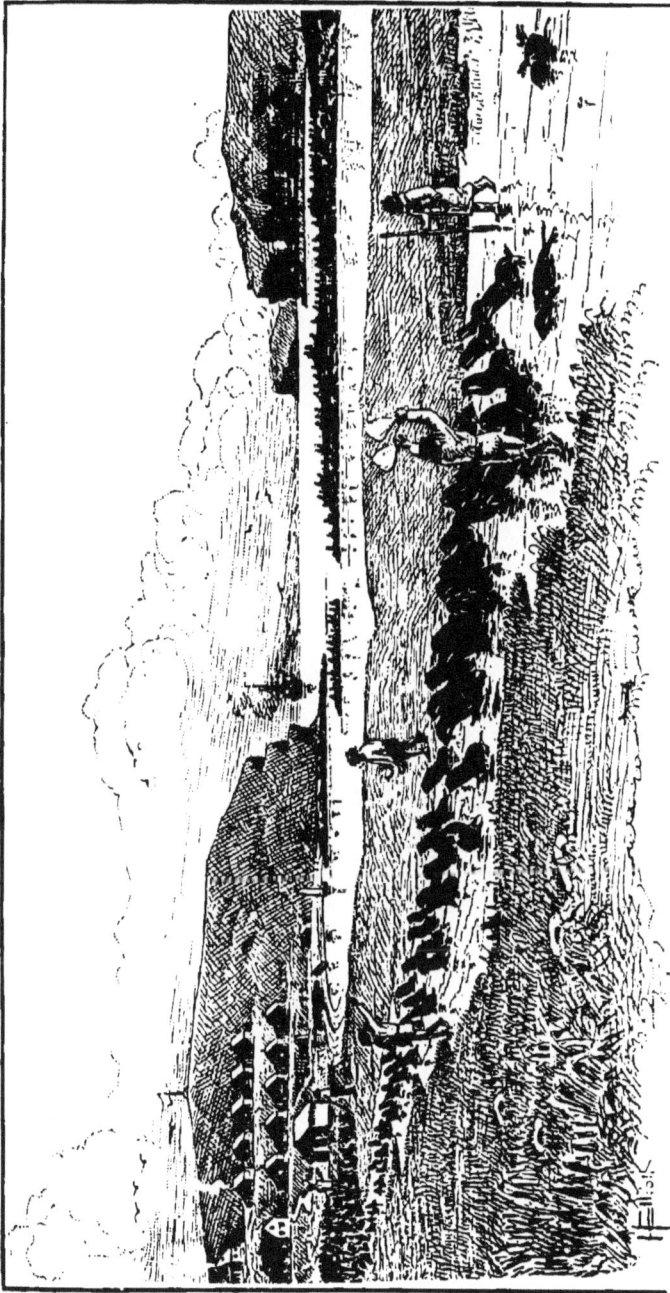

NATIVES DRIVING "HOLLUSCHICKIE"

The Drove passing over the Lagoon Flats to the Killing Grounds under the Village of St. Paul. Looking S. W. over the Village Cove and the Lagoon Rookery

too, that salt-houses be erected, and killing-grounds established contiguous to all of the great hauling-grounds, two miles distant from the village on St. Paul Island, should the business ever be developed above the present limit, or should the exigencies of the future require a quota from all these places in order to make up the hundred thousand which may be lawfully taken.

As matters are to-day, one hundred thousand seals alone on St. Paul can be taken and skinned in less than forty working days, within a radius of one mile and a half from the village, and from the salt-house at Northeast Point; hence the driving, with the exception of two experimental droves which I witnessed in 1872, has never been made from longer distances than Tolstoi to the eastward, Lukannon to the northward, and Zoltoi to the southward of the killing-grounds at St. Paul village. Should, however, an abnormal season recur, in which the larger portion of days during the right period for taking the skins be warmish and dry, it might be necessary, in order to get even seventy-five thousand seals within the twenty-eight or thirty days of their prime condition, for drives to be made from the other great hauling-grounds to the westward and northward, which are now, and have been for the last ten years, entirely unnoticed by our sealers.*

The seals, when finally driven up on those flats between the east landing and the village, and almost under the windows of the dwellings, are herded there until cool and rested. Such drives are usually made very early in the morning, at the first breaking of day, which is half-past one to two o'clock of June and July in these latitudes. They arrive, and cool off on the slaughtering-grounds,

---

* The fur-seal, like all of the pinnipeds, has no sweat-glands; hence, when it is heated, it cools off by the same process of panting which is so characteristic of the dog, accompanied by the fanning that I have hitherto fully described; the heavy breathing and low grunting of a tired drove of seals, on a warmer day than usual, can be heard several hundred yards away. It is surprising how quickly the hair and fur will come out of the skin of a blood-heated seal—literally rubs bodily off at a touch of the finger. A fine specimen of a three-year-old "holluschak" fell in its tracks at the head of the lagoon while being driven to the village killing-grounds. I asked that it be skinned with special reference to mounting; accordingly a native was sent for, who was on the spot, knife in hand, within less than thirty minutes from the moment that this seal fell in the road, yet soon after he had got fairly to work patches of the fur and hair came off here and there wherever he chanced to clutch the skin.

22

so that by six or seven o'clock, after breakfast, the able-bodied male population turn out from the village and go down to engage in the work of killing them. These men are dressed in their ordinary laboring-garb of thick flannel shirts, stout cassimere or canvas pants, over which the "tarbossar" boots are drawn. If it rains they wear

Peter Peeshenkov: Pribylov Sealer.
[*Attired in the costume of the killing gang, when at work in wet weather.*]

their "kamlaykas," made of the intestines and throats of the sea-lion and fur-seal. Thus dressed, they are each armed with a club, a stout oaken or hickory bludgeon, which has been made particularly for the purpose at New London, Conn., and imported here for this especial service. Those sealing-clubs are about five or six feet in length, three inches in diameter at their heads, and the thickness of a man's forearm where they are grasped by the

THE KILLING GANG AT WORK

hands. Each native also has his stabbing-knife, his skinning-knife and his whetstone : these are laid upon the grass convenient, when the work of braining or knocking the seals down is in progress : this is all the apparatus which they employ for killing and skinning.

When the men gather for work they are under the control of their chosen foremen or chiefs ; usually, on St. Paul, divided into two working parties at the village, and a sub-party at North-east Point, where another salt-house and slaughtering-field is established. At the signal of the chief the labor of the day begins by the men stepping into that drove corralled on the flats and driv-ing out from it one hundred or one hundred and fifty seals at a time, making what they call a "pod," which they surround in a circle, huddling the seals one on another as they narrow it down, until they are directly within reach and under their clubs. Then the chief, after he has cast his experienced eye over the struggling, writhing "kautickie" in the centre, passes the word that such and such a seal is bitten, that such and such a seal is too young, that such and such a seal is too old ; the attention of his men being called to these points, he gives the word "Strike!" and instantly the heavy clubs come down all around, and every animal eligible is stretched out stunned and motionless, in less time, really, than I take to tell it. Those seals spared by order of the chief now struggle from under and over the bodies of their insensible companions and pass, hustled off by the natives, back to the sea.

The clubs are dropped, the men seize the prostrate seals by the hind flippers and drag them out so they are spread on the ground without touching each other, then every sealer takes his knife and drives it into the heart at a point between the fore flippers of each stunned form ; its blood gushes forth, and the quivering of the animal presently ceases. A single stroke of a heavy oak bludgeon, well and fairly delivered, will crush in at once the slight, thin bones of a fur-seal's skull, and lay the creature out almost lifeless. These blows are, however, usually repeated two or three times with each animal, but they are very quickly done. The bleeding, which is immediately effected, is so speedily undertaken in order that the strange reaction, which the sealers call "heating," shall be delayed for half an hour or so, or until the seals can all be drawn out and laid in some disposition for skinning.

I have noticed that within less than thirty minutes from the time a perfectly sound seal was knocked down, it had so "heated," owing to the day being warmer and drier than usual, that, when touching it with my foot, great patches of hair and fur scaled off. This is rather exceptionally rapid metamorphosis—it will, however, take place in every instance, within an hour, or an hour and a half on these warm days, after the first blow is struck, and the seal is quiet in death ; hence no time is lost by a prudent toyone in directing the removal of the skins as rapidly as the seals are knocked down and dragged out. If it is a cool day, after bleeding the first "pod" which has been prostrated in the manner described, and after carefully drawing the slain from the heap in which they have fallen, so that the bodies will spread over the ground just free from touching one another, they turn to and strike down another "pod ;" and so on, until a whole thousand or two are laid out, or the drove, as corralled, is finished. The day, however, must be raw and cold for this wholesale method. Then, after killing, they turn to work and skin ; but if it is a warm day every pod is skinned as soon as it is knocked down.

The labor of skinning is exceedingly severe, and is trying even to an expert, demanding long practice ere the muscles of the back and thighs are so developed as to permit a man to bend down to, and finish well, a fair day's work. The knives used by the natives for skinning are ordinary kitchen or case-handle butcher-knives. They are sharpened to cutting edges as keen as razors, but something about the skins of the seal, perhaps fine comminuted sand along the abdomen, so dulls these knives, as the natives work, that they are obliged to whet them constantly.

The body of the seal, preparatory to skinning, is rolled over and balanced squarely on its back ; then the native makes a single swift cut through the skin down along the neck, chest, and belly, from the lower jaw to the root of the tail : he uses for this purpose his long stabbing-knife.* The fore and hind flippers are then successively lifted, as the man straddles a seal and stoops down to

---

* When turning the stunned and senseless carcasses, the only physical danger of which the sealers run the slightest risk, during the whole circuit of their work, occurs thus: at this moment the prone and quivering body of the "holluschak" is not wholly inert, perhaps, though it is nine times out of ten; and as the native takes hold of a fore flipper to jerk the carcass over on to its back, the half-brained seal rouses, snaps suddenly and viciously, often biting the hands or legs of unwary skinners: they then come leisurely and unconcernedly up into the surgeon's office at the village, for bandages, etc. A few

his work over it, and a sweeping circular incision is made through the skin on them just at the point where the body-fur ends ; then, seizing a flap of the hide on either one side or the other of the abdomen, the man proceeds with his smaller, shorter butcher-knife,

---

men are bitten every day or two during the season on the islands, in this manner, but I have never learned of any serious result following any case.

The sealers, as might be expected, become exceedingly expert in keeping their knives sharp, putting edges on them as keen as razors, and in an instant detect any dulness by passing the balls of their thumbs over the suspected edges to such blades.

The white sealers of the Antarctic always used an orthodox butcher's " steel " in sharpening their knives, but these natives never have, and probably never will abandon those little whetstones above referred to.

During the Russian management, and throughout the strife in killing by our own people in 1868, a very large number of the skins were cut through, here and there, by the slipping of the natives' knives, when they were taking them from the carcasses, and "flensing " them from the superabundance, in spots, of blubber. These knife-cuts through the skin, no matter how slight, give great annoyance to the dresser, hence they are always marked down in price. The prompt scrutiny of each skin on the islands by an agent of the Alaska Commercial Company, who rejects every one of them thus injured, has caused the natives to exercise greater care, and the number now so damaged, every season, is absolutely trifling.

Another source of small loss is due to a habit which the "holluschickie" have of occasionally biting each other when they are being urged along in the drives, and thus crowded once in a while one upon the other. Usually these examples of "zoobäden" are detected by the natives prior to the "knocking down," and spared ; yet those which have been nipped on the chest or abdomen cannot be thus noticed, and, until the skin is lifted, the damage is not apprehended.

The aim and force with which the native directs his blow determines the death of a fur-seal. If struck direct and violently, a single stroke is enough. The seals' heads are stricken so hard sometimes that those crystalline lenses to their eyes fly out from the orbital sockets like hail-stones, or little pebbles, and frequently struck me sharply in the face, or elsewhere, while I stood near by watching a killing-gang at work.

A singular lurid green light suddenly suffuses the eye of a fur-seal at intervals when it is very much excited ; as the "podding" for the clubbers is in progress and at the moment when last raising its head it sees the uplifted bludgeons on every hand above, fear seems then for the first time to possess it and to instantly gild its eye in this strange manner. When the seal is brained in this state of optical coloration I have noticed that the opalescent tinting remained well defined for many hours or a whole day after death. These remarkable flashes are very characteristic to the eyes of the old males during their hurly-burly on the rookeries, but never appear in the younger classes unless as just described, as far as I could observe.

rapidly to cut the skin, clean and free from the body and blubber, which he rolls over and out from the hide by hauling up on it as he advances with his work, standing all this time stooped over the carcass so that his hands are but slightly above it, or the ground. This operation of skinning a fair-sized "holluschak" takes the best men only one minute and a half, but the average time made by the gang on the ground is about four minutes to the seal.

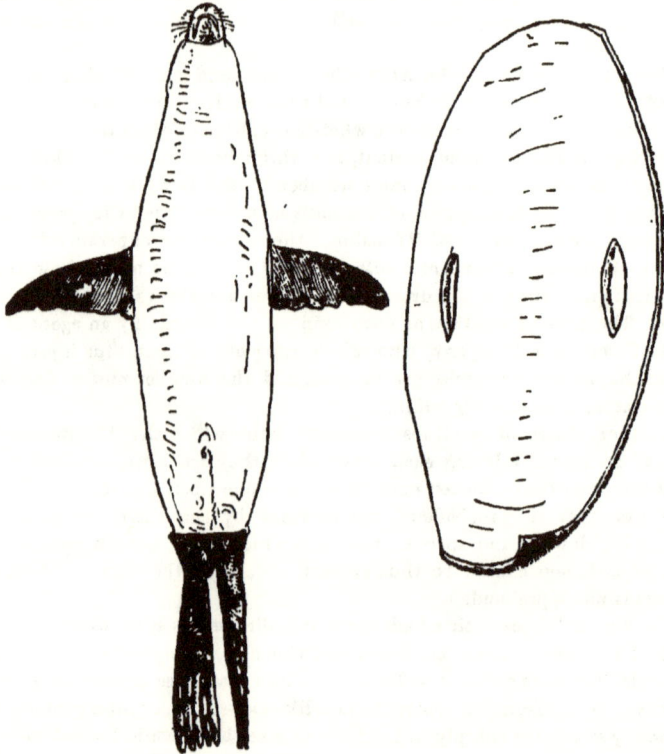

The Carcass after Skinning—The Skin as taken therefrom.

Nothing is left of the skin upon the carcass, save a small patch of each upper lip on which the coarse mustache grows, the skin on the tip of the lower jaw, and its insignificant tail. After removal of the skin from the body of a fur-seal, the entire surface of the carcass is covered with a more or less dense layer, or envelope, of soft, oily blubber, which in turn completely conceals the muscles or flesh of the trunk and neck. This fatty substance, which we now see, resembles that met with in such seals everywhere, only possessing that strange peculiarity not shared by any other of its kind, of being positively overbearing and offensive in odor to an unaccustomed human nostril. The rotting, sloughing carcasses around about did not, when stirred up, affect me more unpleasantly than did this strong, sickening smell of the fur-seal blubber. It has a character and appearance intermediate between those belonging to the adipose tissue found on the flesh of cetacea and some carnivora.

This continuous envelope of blubber to the bodies of the "holluschickie" is thickest in deposit at those points upon the breast between the fore flippers, reaching entirely around and over the shoulders, where it is from one inch to a little over in depth. Upon the outer side of the chest it is not half an inch in thickness, frequently not more than a quarter, and it thins out considerably as it reaches the median line of the back. The neck and head are clad by an unbroken continuation of the same material, which varies from one-half to one-quarter of an inch in depth. Toward the middle line of the abdominal region there is a layer of relative greater thickness. This is coextensive with the sterno-pectoral mass ; but it does not begin to retain its volume as it extends backward, where this fatty investment of the carcass upon the loins, buttocks, and hinder limbs fades out finer than on the pectoroabdominal parts, and assumes a thickness corresponding to its depth on the cervical and dorsal regions. As it descends on the limbs this blubber thins out very preceptibly ; and, when reaching the flippers, it almost entirely disappears, giving way to a glistening aureolar tissue, while the flipper skin finally descends in turn to adhere closely and firmly to the tendinous ligamentary structures beneath, which constitute the tips of the swimming-palms.

The flesh and the muscles are not lined between or within by fat of any kind : this blubber envelope contains it all, with one exception—that which is found in the folds of the small intestine and about the kidneys, where there is an abundant secretion of a harder, whiter, though still offensive-smelling fat.

It is quite natural for our people when they first eat a meal on the Pribylov Islands to ask questions in regard to what seal-meat looks and tastes like. Some of the white residents will answer, saying that they are very fond of it cooked so and so ; others will reply that in no shape or manner can they stomach the dish. An inquirer must himself try the effect on his own palate. I frankly confess that I had a slight prejudice against seal-meat at first, having preconceived ideas that it would be fishy in flavor; but I soon satisfied myself to the contrary, and found that the flesh of young seals not over three years old was as appetizing and toothsome as some of the beef, mutton, and pork I was accustomed to at home. The following precautions must be rigidly observed, however, by the cook who prepares fur-seal steaks and sausage-balls for our delectation and subsistence. He will fail if he does not:

1. The meat must be perfectly cleaned of every vestige of blubber or fat, no matter how slight.

2. Cut the flesh then into very thin steaks or slices and soak them from six to twelve hours in salt and water, a tablespoon of fine salt to a quart of fresh water. This whitens the meat and removes the residuum of dark venous blood that will otherwise give a slightly disagreeable taste, hardly definable, though existing.

3. Fry these steaks, or stew them à la mode, with a few thin slices of sweet "breakfast" bacon, seasoning with pepper and salt. A rich brown gravy follows the cooking of the meat. Serve hot, and it is, strictly judged, a very excellent meat for the daintiest feeder, and I hereby recommend it confidently as a safe venture for any newcomer to make.

The flesh of young sea-lions is still better than that of the fur-seal, while the natives say that the meat of the hair-seal (*Phoca vitulina*) is superior to both, being more juicy. Fur-seal meat is exceedingly dry; hence the necessity of putting bacon into the frying-pan or stew-pot with it. Sea-lion flesh is an improvement in this respect, and also that its fat, strange to say, is wholly clear, white, and inodorous, while the blubber of the "holluschickie" is sickening to the smell, and will, nine times out of ten, cause any civilized stomach to throw it up as quickly as it is swallowed. The natives, however, eat a great deal of it, simply because they are too lazy to clean their fur-seal cuts and not because they really relish it.

In this connection it may be well to add that the liver of both *Callorhinus* and *Eumetopias* is sweet and wholesome; or, in other words, it is as good as liver usually is in Fulton Market. The tongues are small, white, and fat. They are regularly cut out to some extent and salted in ordinary water-buckets for exportation to curious friends. They have but slight claim to gastronomic favor. The natives are, however, very partial to the liver ; but though they like the tongues, yet they are too lazy to prepare them. A few of them, in obedience to pressing and prayerful appeals from relatives at Oonalashka, do exert themselves enough every season to undergo the extra labor of putting up several barrels of fresh salted seal-meat, which, being carried down to Illoolook by the company's vessels, affords a delightful variation to the steady and monotonous codfish diet of those Aleutian Islanders.

The final acts of curing and shipping pelts of fur-seals from the warehouses of the villages, rapidly follow work upon the killing-grounds. The skins are taken from the field to the salt-house,

Interior of Salt House, Village of St. Paul.

[*Showing the method of receiving, selecting, kenching and salting "green" fur-seal skins.*]

where they are laid out, after being again carefully examined, one upon another, "hair to fat," like so many sheets of paper, with salt profusely spread upon the fleshy sides as they are piled up in the "kenches," or bins. The salt-house is a large barn-like frame structure, so built as to afford one-third of its width in the centre, from end to end, clear and open as a passage-way : while on each side are rows of stanchions, with sliding planks, which are taken down and put up in the form of deep bins or boxes—"kenches," the sealers call them. As the pile of skins is laid up from the bottom of an empty "kench" and salt thrown in on the outer edges, these planks are also put in place, so that the salt may be kept intact until that bin is filled as high up as a man can toss the skins. After lying two or three weeks in this style they become "pickled," and they are suited then at any time to be taken up and rolled into bundles of two skins to the package, with the hairy side out, tightly corded, ready for shipment from the islands.

The average weight of a two-year-old skin is five and one-half pounds; of a three-year-old skin, seven pounds, and of a four-year-old skin, twelve pounds, so that, as the major portion of the catch is two or three year olds, these bundles of two skins each have a general weight of from twelve to fifteen pounds. In this form they go into the hold of the company's steamer at St. Paul, and are counted out from it in San Francisco. Then they are either at once shipped to London by the Isthmus of Panama in the same shape, only packed up in large hogsheads of from twenty to forty bundles to the package, or expressed by railroad, via New York, to a similar destination.

The work of bundling the skins is not usually commenced by the natives until the close of the last week's sealing; or, in other words, those skins which they first took, three weeks ago, are now so pickled by the salt in which they have been lying ever since, as to render them eligible for this operation and immediate shipment. The moisture of the air dissolves and destroys a very large quantity of that saline preservative which the company brings up annually · in the form of rock-salt, principally obtained at Carmen Island, Lower California.

The Alaska Commercial Company, by the provisions of law under which they enjoy their franchise, are permitted to take one hundred thousand male seals annually, and no more, from the Pribylov Islands. This they do in June and July of every year. After that season the skins rapidly grow worthless, as the animals enter into shedding, and, if taken, would not pay for transportation and the tax.

The bundled skins are carried from the salt-houses to the beach, when an order for shipment is given, pitched into a bidarrah, one by one, and rapidly stowed; seven hundred to twelve hundred bundles make an average single load; then, when alongside the steamer, they are again tossed up from the lighter and onto her deck, whence they are stowed in the hold.*

* The shallow depths of Bering Sea give rise to a very bad surf, and though none of the natives can swim, as far as I could learn, yet they are quite creditable surfmen, and work the heavy "baidar" in and out from the landing adroitly and circumspectly. They put a sentinel upon the bluffs over Nah

The method of air-drying which the old settlers employed is well portrayed by the practice of the natives now, who treat a few hundred sea-lion skins to that process every fall, preparing them thus for shipment to Oonalashka, where they are used by brother Aleutes in covering their bidarkies or kayaks.

The natives, in speaking to me of this matter, said that whenever the weather was rough and the wind blowing hard, these air-dried seal-skins, as they were tossed from the bidarrah to the ship's deck, numbers of them, would frequently turn in the wind and fly clean over the vessel into the water beyond, where they were lost.

Under the old order of affairs, prior to the present management, the skins were packed up and carried on the backs of the boys and girls, women and old men, to the salt-houses, or drying-frames. When I first arrived, season of 1872, a slight variation was made in this respect by breaking a small Siberian bull into harness and hitching it to a cart, in which the pelts were hauled. Before the cart was adjusted, however, and the "buik" taught to pull, it was led out to the killing-grounds by a ring in its nose, and literally covered with the green seal-hides, which where thus packed to the kenches. The natives were delighted with even this partial assistance ; but now they have no further concern about it at all, for several mules and carts render prompt and ample service.

The common or popular notion in regard to seal-skins is, that they are worn by those animals just as they appear when offered for sale ; that the fur-seal swims about, exposing the same soft coat with which our ladies of fashion so delight to cover their tender

---

Speel, and go and come between the rollers as he signals. They are not graceful oarsmen under any circumstances, but can pull heartily and coolly together when in a pinch. The apparent ease and unconcern with which they handled their bidarrah here in the "baroon" during the fall of 1869 so emboldened three or four sailors of the United States Revenue Marine cutter *Lincoln* that they lost their lives in such surf through sheer carelessness. The "gig" in which they were coming ashore "broached to" in the breakers just outside the cove, and their lifeless forms were soon after thrown up by merciless waves on the Lagoon rookery. Three graves of these men are plainly marked on a western slope of the Black Bluffs.

There is a false air of listlessness and gentleness about an open sea, or roadstead roller, that is very apt to deceive even watermen of good understanding. The crushing, overwhelming power with which an ordinary breaker will hurl a large ship's boat on rocks awash, must be personally experienced ere it is half appreciated.

forms during inclement winter. This is a very great mistake; few skins are less attractive than a seal-skin is when it is taken from the creature. The fur is not visible; it is concealed entirely by a coat of stiff over-hair, dull, gray-brown, and grizzled. It takes three of them to make a lady's sack and boa; and in order that a reason for their costliness may be apparent, I take great pleasure in submitting a description of the tedious and skilful labor necessary to their dressing by the furriers ere they are fit for use: a leading manufacturer, writing to me, says:—

"When the skins are received by us in the salt, we wash off the salt, placing them upon a beam somewhat like a tanner's beam, removing the fat from the flesh side with a beaming-knife, care being required that no cuts or uneven places are made in the pelt. The skins are next washed in water and placed upon the beam with the fur up, and the grease and water removed by the knife. The skins are then dried by moderate heat, being tacked out on frames to keep them smooth. After being fully dried, they are soaked in water and thoroughly cleansed with soap and water. In some cases they can be unhaired without this drying process, and cleansed before drying. After the cleansing process they pass to the picker, who dries the fur by stove-heat, the pelt being kept moist. When the fur is dry he places the skin on a beam, and while it is warm he removes the main coat of hair with a dull shoe-knife, grasping the hair with his thumb and knife, the thumb being protected by a rubber cob. The hair must be pulled out, not broken. After a portion is removed the skin must be again warmed at the stove, the pelt being kept moist. When the outer hairs have been mostly removed, he uses a beaming-knife to work out the fine hairs (which are shorter), and the remaining coarser hairs. It will be seen that great care must be used, as the skin is in that soft state that too much pressure of the knife would take the fur also; indeed, bare spots are made. Carelessly cured skins are sometimes worthless on this account. The skins are next dried, afterward dampened on the pelt side, and shaved to a fine, even surface. They are then stretched, worked, and dried, afterward softened in a fulling-mill, or by treading them with the bare feet in a hogshead, one head being removed and the cask placed nearly upright, into which the workman gets with a few skins and some fine, hardwood sawdust, to absorb the grease while he dances upon them to break them into leather. If the skins have been shaved thin, as required when finished, any defective spots or holes must now be mended, the skin smoothed and pasted with paper on the pelt side, or two pasted together to protect the pelt in dyeing. The usual process in the United States is to leave the pelt sufficiently thick to protect them without pasting.

"In dyeing, the liquid dye is put on with a brush, carefully covering the points of the standing fur. After lying folded, with the points touching each other, for some time, the skins are hung up and dried. The dry dye is then removed, another coat applied, dried, and removed, and so on, until the required shade is obtained. One or two of these coats of dye are put on much heavier and pressed down to the roots of the fur, making what is called the ground. From eight to twelve coats are required to produce a good color. The skins are then washed clean, the fur dried, the pelt moist. They are shaved down to the required thickness, dried, working them some while drying, then softened in a hogshead, and sometimes run in a revolving cylinder with fine sawdust to clean them. The English process does not have the washing after dyeing."

On account of the fact that all labor in this country, especially skilled labor, commands so much more per diem in the return of wages than it does in London or Belgium, it is not practicable for the Alaska Commercial Company, or any other company here, to attempt to dress and put upon the market its catch of Bering Sea, which is in fact the entire catch of the whole world. Our people understand the theory of dressing these skins perfectly; but they cannot compete with the cheaper labor of the Old World. Therefore, nine-tenths, nearly, of the fur-seal skins taken every year are annually purchased and dressed in London, and from thence distributed all over the civilized world where furs are worn and prized.

The great variation in the value of seal-skin sacks, ranging from seventy-five dollars up to three hundred and fifty dollars, and even five hundred dollars, is not often due to a variance in the quality of the fur originally; but it is due to that quality of the work whereby the fur was treated and prepared for wear. For instance, cheap sacks are so defectively dyed that a little moisture causes them to soil the collars and cuffs of their owners, and a little exposure makes them speedily fade and look ragged. A properly dyed skin, one that has been conscientiously and laboriously finished (for it is a labor requiring great patience and great skill), will

not rub off or "crock" the whitest linen when moistened ; and it
will wear the weather, as I have myself seen it on the form of a sea-
captain's wife, for six and seven successive seasons, without showing the least bit of dimness or raggedness. I speak of dyeing
alone ; I might say the earlier steps of unhairing, in which the over-
hair is deftly combed out and off from the skin, heated to such a
point that the roots of the fur are not loosened, while those to the
coarser hirsute growth are. If this is not done with perfect uniformity, the fur will never lie smooth, no matter how skilfully dyed;
it will always have a rumpled, ruffled look.   Therefore the hastily-
dyed sacks are cheap ; and are enhanced in order of value just as
the labor of dyeing is expended upon them.

Another singular and striking characteristic of the Island of St.
Paul, is the fact that this immense slaughtering-field, upon which
seventy-five thousand to ninety thousand fresh carcasses lie every
season, sloughing away into the sand beneath, does not cause any
sickness among the people who live right over them, so to speak.
A cool, raw temperature, and strong winds, peculiar to the place,
seem to prevent any unhealthy effect from that fermentation of decay. An *Elymus* and other grasses once more take heart and
grow with magical vigor over the unsightly spot, to which the scal-
ing-gang again return, repeating their work upon this place which
we have marked before, three years ago. In that way this strip of
ground, seen on my map between the village, the east landing, and
the lagoon, contains the bones and the oil-drippings and other fragments thereof, of more than three million seals slain since 1786
thereon, while the slaughter-fields at Novashtoshnah record the end
of a million more! .

I remember well those unmitigated sensations of disgust which
possessed me when I first landed, April 28, 1872, on the Pribylov
Islands, and passed up from the beach, at Lukannon, to the village
over the killing-grounds ; though there was a heavy coat of snow on
the fields, yet each and every one of seventy-five thousand decaying
carcasses was there, and bare, having burned, as it were, their way
out to the open air, polluting the same to a sad degree. I was
laughed at by the residents who noticed my facial contortions, and
assured that this state of smell was nothing to what I should soon
experience when the frost and snow had fairly melted. They were
correct ; the odor along by the end of May was terrific punishment
to my olfactories, and continued so for several weeks until my sense

of smell became blunted and callous to such stench by long familiarity. Like the other old residents I then became quite unconscious
of the prevalence of this rich "funk," and ceased to notice it.

Those who land here, as I did, for the first time, nervously and
invariably declare that such an atmosphere must breed a plague or
a fever of some kind in the village, and hardly credit the assurance
of those who have resided in it for the whole period of their lives,
that such a thing was never known to St. Paul, and that the island
is remarkably healthy. It is entirely true, however, and, after a
few weeks' contact, or a couple of months' experience, at the longest, the most sensitive nose becomes used to that aroma, wafted as
it is hourly, day in and out, from decaying seal-flesh, viscera, and
blubber ; and, also, it ceases to be an object of attention. The
cool, sunless climate during the warmer months has undoubtedly
much to do with checking too rapid decomposition and consequent
trouble therefrom, which would otherwise arise from those killing-
grounds.

The freshly-skinned seal bodies of this season do not seem to rot
substantially until the following year ; then they rapidly slough
away into the sand upon which they rest ; the envelope of blubber
left upon each body seems to act as an air-tight receiver, holding
most of the putrid gases that evolved from the decaying viscera
until their volatile tension causes it to give way ; fortunately the
line of least resistance to that merciful retort is usually right where
it is adjacent to the soil, so both putrescent fluids and much of the
stench within is deodorized and absorbed before it can contami-
nate the atmosphere to any great extent. The truth of my observation will be promptly verified, if the sceptic chooses to tear open
any one of the thousands of gas-distended carcasses in the fall, that
were skinned in the killing-season ; if he does so, he will be smitten
by the worst smell that human sense can measure ; and should he
chance to be accompanied by a native, that callous individual, even,
will pinch his grimy nose and exclaim, it is a "keeshla pahknoot !"

At the close of the third season after skinning, a seal's body
will have so rotted and sloughed down, as to be marked only by
the bones and a few of the tendinous ligaments ; in other words,
it requires from thirty to thirty-six months' time for such a carcass
to rot entirely away, so that nothing but whitened bones remain
above ground. The natives govern their driving of the seals and
laying out of the fresh bodies according to this fact—they can,

and do, spread this year a whole season's killing out over the same
spot of the field previously covered with such fresh carcasses three
summers ago ; by alternating with the seasons thus, the natives are
enabled to annually slaughter all of the "holluschickie" on a relatively small area, close by the salt-houses and the village, as I have
indicated on my map of St. Paul.

The St. Paul village site is located wholly on the northern slope
of the village hill, where it drops from its greatest elevation, at the
flagstaff of one hundred and twenty-five feet, gently down to those
sandy killing-flats below and between it and the main body of the
island.   The houses are all placed facing north at regular intervals along the terraced streets, which run southeast and northwest.
There are seventy-four or eighty native houses, ten large and
smaller buildings of the company, a Treasury agent's residence,
a church, cemetery crosses, and a school building which are all
standing here in coats of pure white paint.   No offal or decaying
refuse of any kind is allowed to rest around the dwellings or lie
in the streets.   It required much determined effort on the part of
the whites to effect this sanitary reform ; but now most of the natives take equal pride in keeping their surroundings clean and un-
polluted.   The killing-ground of St. Paul is a bottomless sand-flat
only a few feet above high water, and which unites the village hill
and the reef with the island itself.   It is not a stone's throw from
the heart of the settlement ; in fact, it is right in town, not even
suburban.

The site of the St. George settlement is more exposed and bleak
than is the one we have just referred to on St. Paul.   It is planted
directly on a rounded summit of one of the first low hills that
rise from the sea on the north shore.   Indeed, it is the only hill
that does slope directly and gently to the salt water on the island.
Here are twenty-four to thirty native cottages, laid with their doors
facing the opposite sides of a short street between, running also
east and west, as at St. Paul.   There, however, each house looks
down upon the rear of its neighbor in front and below—here the
houses face each other on the top of the hill.   The Treasury agent's
quarters, the company's six or seven buildings, the school-house, and
the church, are all neatly painted : therefore this settlement, by its
prominent position, shows from the sea to a much better advantage
than the larger one of St. Paul does.   The same municipal sanitary
regulations are enforced here.   Those who may visit the St. George

and St. Paul of to-day will find their streets dry and hard as floors for they have been covered with a thick layer of volcanic cinders on both islands.

On St. George the "holluschickie" are regularly driven to that northeast slope of the village hill which drops down gently to the sea, where they are slaughtered, close by and under the houses, as at St. Paul. Those droves which are brought in from the North rookery to the west, and also Starry Arteel, are frequently driven right through the village itself. This killing-field of St. George is hard tufa and rocky, but it slopes away to the ocean rapidly enough to drain itself well; hence the constant rain and humid fogs of summer carry off that which would soon clog and deprive the natives from using the ground year after year in rotation, as they do. Several seasons have occurred, however, when this natural cleansing of the place, above mentioned, has not been as thorough as must be, so as to be used again immediately: then the seals were skinned back of the village hill and in that ravine to the westward on the same slope from its summit.

This village site of St. George to-day, and the killing-grounds adjoining, used to be, during early Russian occupation, in Pribylov's time, a large sea-lion rookery, the finest one known to either island, St. Paul or St. George. Natives are living now, who told me that their fathers had been employed in shooting and driving these sea-lions, so as to deliberately break up a breeding-ground, and thus rid the island of what they considered a superabundant supply of the *Eumetopias*, and thereby to aid and encourage a fresh and increased accession of fur-seals from the vast majority peculiar to St. Paul, which could not ensue while big sea-lions held the land.

23

CHAPTER XI.

THE ALASKAN SEA-LION.

A Pelagic Monarch.—Marked Difference between the Sea-lion and the Fur-seal.—The Imposing Presence and Sonorous Voice of the "Sea-king."—Terrible Combats between old Sea-lion Bulls.—Cowardly in the Presence of Man, however.—Sea-lions Sporting in the Fury of Ocean Surf.—It has no Fur on its Huge Hide.—Valuable only to the Natives, who Cover their "Bidarrah" with its Skin.—Its Sweet Flesh and Inodorous Fat.—Not such Extensive Travellers as the Fur-seals.—The Difficulty of Capturing Sea-lions.—How the Natives Corral them.—The Sea-lion "Pen" at Northeast Point.—The Drive of Sea-lions.—Curious Behavior of the Animals.—Arrival of the Drove at the Village.—A Thirteen-mile Jaunt with the Clumsy Drove.—Shooting the old Males.—The Bloody "Death-whirl."—The Extensive Economic Use made of the Carcass by the Natives.—Chinese Opium Pipes Picked with Sea-lion Mustache bristles.

THE sea-lion is also a characteristic pinniped of the Pribylov Islands, but ranks much below the fur-seal in perfected physical organization and intelligence. It can, as well as its more sagacious and valuable relative, the *Callorhinus*, be seen, perhaps, to better advantage on these islands than elsewhere in the whole world that I know of. The marked difference between a sea-lion and the fur-seal up here is striking, the former being twice the size of its cousin.

The size and strength of a northern sea-lion, *Eumetopias stelleri*, its perfect adaptation to its physical surroundings, unite with a singular climatic elasticity of organization. It seems to be equally well satisfied with the ice-floes of the Kamchatka Sea to the northward, or with the polished boulders and the hot sands of the coast of California. It is an animal as it appears upon its accustomed breeding grounds at Northeast Point, where I first saw it, that commanded my involuntary admiration by its imposing presence and sonorous voice, as it reared itself before me, with head, neck, and chest upon its powerful forearms, over six feet in height, while its heavy bass voice drowned the booming of the surf that thundered on the rocks beneath its flanks.

A GROUP OF SEA-LIONS

Young   Adult Male and Female   Old Bull Roaring

[*Eumetopias stelleri*: a Life Study made at St. Paul's Island. July 16, 1872]

The bulk and power of the adult sea-lion male will be better appreciated when I say that it has an average length of ten and eleven feet osteologically, with an enormous girth of eight to nine feet around the chest and shoulders; but while the anterior parts of its frame are as perfect and powerful on land as in sea, those posterior are ridiculously impotent when the huge beast leaves its favorite element. Still, when hauled up beyond the reach of the brawling surf, as it rears itself, shaking the spray from its tawny chest and short grizzly mane, it has a leonine appearance and bearing, greatly enhanced as the season advances by a rich golden-rufous color of its coat; the savage gleam of its expression is due probably, to the sinister muzzle, and cast of its eye. This optical organ is not round and full, soft and limpid, like the fur-seal's, but it is an eye like that of a bull-dog : it is small and clearly shows under its heavy lids the white or sclerotic coat, with a light-brown iris. Its teeth gleam and glisten in pearly whiteness against a dark tongue and the shadowy recesses of its wide, deep mouth. The long, sharp, broad-based canines, when bared by the wrathful snarling of its gristled lips, glittered more wickedly, to my eye, than the keenest sword ever did in the hand of man.

With these teeth alone, backed by the enormous muscular power of a mighty neck and broad shoulders, the sea-lion confines its battles to its kind, spurred by terrible energy and heedless and persistent brute courage. No animals that I have ever seen in combat presented a more savage or more cruelly fascinating sight than did a brace of old sea-lion bulls which met under my eyes near the Garden Cove at St. George.

Here was a sea-lion rookery the outskirts of which I had trodden upon for the first time. Two aged males, surrounded by their meek, polygamous families, were impelled towards each other by those latent fires of hate and jealousy which seemed to burst forth and fairly consume the angry rivals. Opening with a long, round, vocal prelude, they gradually came together, as the fur-seal bulls do, with averted heads, as though the sight of each other was sickening—but fight they must. One would play against the other for an unguarded moment in which to assume the initiative, until it had struck its fangs into the thick skin of its opponent's jowl; then, clinching its jaws, was not shaken off until the struggles of its tortured victim literally tore them out, leaving an ugly, gaping wound—for the sharp eye-teeth cut a deeper gutter in the skin and

flesh than would have held my hand ; fired into almost supernatural
rage, the injured lion retaliated, quick as a flash, in kind ; the hair
flew from both of them into the air, the blood streamed down in
frothy torrents, while high above the boom of the breaking waves,
and shrill deafening screams of water-fowl over head, rose the
ferocious, hoarse, and desperate roar of these combatants.

Though provided with flippers, to all external view, as the fur-
seal is, the sea-lion cannot, however, make use of them at all in the
same free manner.   The fur-seal may be driven five or six miles in
twenty-four hours under the most favorable conditions of cool, moist
weather ; the "seevitchie," however, can only go two miles, the
weather and roadway being the same.   When driven, a sea-lion bal-
ances and swings its long and heavy neck, as a lever, to and fro,
with every hitching up behind of its posterior limbs, which it seldom raises from the ground, drawing them up after the fore-feet
with a sliding drag over the grass or sand and rocks, as the case
may be, ever and anon pausing to take a sullen and savage survey
of the field and the natives who are urging it.

The sea-lion is polygamous, but it does not maintain any regular system and method in preparing for and attending to its harem,
like that so finely illustrated on the breeding-grounds of the fur-
seal ; and it is not so numerous, comparatively speaking.   There
are not, according to my best judgment, over ten or twelve thousand of these animals altogether on the breeding-grounds of the
Pribylov Islands.   It does not haul more than a few rods anywhere
or under any circumstances back from the sea.   It cannot be visited
and inspected by men as the fur-seals are, for it is so shy and suspicious that on the slightest warning of such an approach, a stampede into the water is sure to result.

That noteworthy, intelligent courage of a fur-seal, though it
does not possess half the size nor one-quarter of the muscular
strength of a sea-lion, is entirely wanting in the huge bulk and
brain of the *Eumetopias*.   A boy with a rattle or a pop-gun could
stampede ten thousand sea-lion bulls in the height of a breeding-
season to the water, and keep them there for the rest of the time.*

---

* That the sea-lion bull should be so cowardly in the presence of man,
yet so ferocious and brave toward one another and other amphibious animals,
struck me as a line of singular contrast with the undaunted bearing of a fur-
seal "seecatch," which, though being not half the size or possessing muscular
power to anything like its development in the "seevitchie," nevertheless

Old males come out and locate themselves over the narrow belts of rookery-grounds (sometimes, as at St. Paul, on the immediate sea-margin of fur-seal breeding-places), two or three weeks in advance of the females, which arrive later, i.e., between the 1st to the 6th of June ; and these females are never subjected to that intense, jealous supervision so characteristic of the fur-seal harem. Big sea-lion bulls, however, fight savagely* among themselves, and turn off from the breeding-ground all younger and weaker males.

A cow sea-lion is not quite half the size of an adult male ; she will measure from eight to nine feet in length osteologically, with a weight of four or five hundred pounds ; she has the same general cast of countenance and build of the bull ; but, as she does not sustain any fasting period of over a week or ten days consecutively, she never comes out so grossly fat as he does. With reference to the weight of the latter, I was particularly unfortunate in not being able to get one of those big bulls on the scales before it had been bled, and in bleeding I know that a flood of blood poured out which should have been recorded in the weight. Therefore I can only estimate this aggregate avoirdupois of one of the finest-conditioned adult male sea-lions at fourteen to fifteen hundred pounds ; an average weight, however, might safely be recorded as touching twelve hundred pounds.†

will unflinchingly face on its station at the rookery any man to the death. The sea-lion bulls certainly fight as savagely and as desperately one with another, as the fur-seal males do. There is no question about that, and their superior strength and size only makes the result more effective in the exhibition of gaping wounds and attendant bloodshed. I have repeatedly seen examples of these old warriors of the sea which were literally scarred from their muzzles to their posteriors so badly and so uniformly as to have fairly lost all the color or general appearance even of hair anywhere on their bodies.

* I recall in this connection the sight of an aged male sea-lion which had been defeated by a younger and more lusty rival, perhaps. It was hauled upon a lava shelf at Southwest Point, solitary and alone ; the rock around it being literally covered with pools of pus, that was oozing out and trickling down from a score of festering wounds ; the victim stood planted squarely on its torn fore flippers, with head erect and thrown back upon its shoulders; its eyes were closed, and it gently swayed its sore neck and shoulders in a sort of troubled, painful day-dreaming or dozing. 'Like the fur seal, the sea-lion never notices its wounds to nurse and lick them, as dogs do, or other carniv ora ; it never pays the slightest attention to them, no matter how grievously it may be injured.

† Often when the fur-seal and sea-lion bulls haul up in the beginning of the season examples among them which are inordinately fat will be seen ;

You will notice that if you disturb and drive off any portion of the rookery, by walking up in plain sight, those nearest to you will take to the water instantly, swim out to a distance of fifty yards or so, leaving their pups behind, helplessly sprawled around and about the rocks at your feet. Huddled up all together in the surf in two or three packs or squads, the startled parents hold their heads and necks high out of the sea, and peer keenly at you : then, all roaring in an incessant concert, they make an orchestra to which those deep sonorous tones of the organ in that great Mormon tabernacle, at Salt Lake City, constitute the fittest and most adequate resemblance.

You will witness an endless tide of these animals travelling to the water, and a steady stream of their kind coming out, if you but keep in retirement and do not disturb them. When they first issue from the surf they are a dark chocolate brown-and-black, and glisten ; but, as their coats dry off, the color becomes an iron-gray, passing into a bright golden rufous, which covers the entire body alike—shades of darker brown on the pectoral patches and sterno-pectoral region. After getting entirely dry, they seem to grow exceedingly uneasy, and act as though oppressed by heat, until they plunge back into the sea, never staying out, as the fur-seal does, day after day, and week after week. The females and the young males frolic in and out of the water, over rocks awash, incessantly, one with another, just as puppies play upon a green sward ; and, when weary, stretch themselves out in any attitude that will fit the character of that rock, or the lava-shingle upon which they may happen to be resting. The movements of their supple spines, and ball-and-socket joint attachments, permit of the most extraordinary contortions of a trunk and limbs, all of which, no matter how distressing to your eyes, they seem actually to relish. But the old battle-scarred bulls of the harem stand or lie at their positions day and night without leaving them, except to take a short bath when the coast is clear, until the end of the season.

When swimming, the sea-lion lifts its head only above the surface long enough to take a deep breath, then drops down a few

---

their extra avoirdupois renders them very conspicuous, even among large gatherings of their kind ; they seem to exhibit a sense of self-oppression then, quite as marked as is that subsequent air of depression worn when, later, they have starved out this load of surplus blubber, and are shambling back to the sea, for recuperation and rest.

feet below, and propels itself, for about ten or fifteen minutes, like a cigar-steamer, at the rate of six or seven knots, if undisturbed; but, if chased or alarmed, it seems fairly to fly under water, and can easily maintain for a long time a speed of fourteen or fifteen miles per hour. Like the fur-seal, its propulsion through water is the work entirely of its powerful fore-flippers, which are simultaneously struck out, both together, and back against the water, feathering forward again to repeat, while the hind flippers are simply used as a rudder oar in deflecting an ever-varying swift and abrupt course of the animal. On land its hind flippers are employed just as a dog uses its feet in scratching fleas—the long peculiar toe-nails thereof seeming to reach and comb those spots affected by vermin, which annoys it, as the fur-seal is, to a great extent, and causes them both to enjoy a protracted scratching.

Again, both genera, *Callorhinus* and *Eumetopias*, are happiest when the surf is strongest and wildest. Just in proportion to the fury of a gale, so much the greater joy and animation of these animals. They delight in riding on the crests of each dissolving breaker up to a moment when it fairly foams over iron-bound rocks. At that instant they disappear like phantoms beneath the creamy surge, to reappear on the crown of the next mighty billow.

When landing, they always ride on the surf, so to speak, to an objective point: and, it is marvellous to see with what remarkable agility they will worm themselves up steep, rocky landings, having an inclination greater than forty-five degrees, to flat bluff-tops above, which have an almost perpendicular drop to water.

As the sea-lion is without fur, its skin has little or no commercial value.* The hair is short, an inch to an inch and a half in

---

* The sea-lion and hair-seals of Bering Sea, having no commercial value in the eyes of civilized men, have not been subjects of interest enough to the pioneers of those waters for mention in particular; such record, for instance, as that given of the walrus, the sea-otter, and the fur-seal. Steller was the first to draw the line clearly between them and seals in general, especially defining their separation from the fur-seal; still his description is far from being definite or satisfactory in the light of our present knowledge of the animal.

In the South Pacific and Atlantic the sea-lion has been curiously confounded by many of our earliest writers with the sea-elephant, *Macrorhinus leoninus*, and its reference is inextricably entangled with the fur-seal at the Falklands, Kerguelen's Land, and the Crozettes. The proboscidean seal, however, seems to be the only pinniped which visits the Antarctic continent: but

length, being longest over the nape of the neck; straight, and somewhat coarse, varying in color as the season comes and goes. For instance, when the *Eumetopias* makes its first appearance in the spring and dries out after landing, it has then a light-brownish rufous tint, with darker shades back and under the fore flippers and on the abdomen. By the expiration of a month or six weeks, about June 15th, generally, this coat will then be weathered into a glossy rufous, or ochre yellow; this tinting remains until shed along by the middle of August, or a little earlier. After a new coat has fairly grown, and just before an animal leaves the island rookery in November, it is a light sepia or Vandyke brown, with deeper shades, almost black, upon its abdomen. The cows after shedding never color up so darkly as the bulls; but when they come back to the land next year they return identically the same in tinting; so that the eye, in glancing over a sea-lion rookery during June and July, cannot discern any dissimilarity in color, at all note-worthy, existing between the coats of the bulls and the cows; also, the young males and yearlings appear in that same golden-brown and ochre, with here and there an animal which is noted as being spotted somewhat like a leopard—a yellow rufous ground predominating, with patches of dark-brown, blotched and mottled, irregularly interspersed over the anterior regions down to those posterior. I have never seen any of the old bulls or cows thus mottled, and this is likely due to some irregularity of shedding in the younger animals; for I have not noticed it early in the season, and it seems to fairly fade away so as not to be discerned on the same animal at the close of its summer solstice. Many of the old bulls have a grizzled or "salt and pepper" look during the shedding period, which is from August 10th up to November 10th or 20th. The pups, when born, are a rich dark-chestnut brown. This coat they shed in October, and take one much lighter in its stead, still darker, however, than their parents.

The time of arrival at, stay on, and departure from the islands, is about the same as that which I have recorded as characteristic of the fur-seal; but, if a winter is an open, mild one, some of the sea-lions will frequently be seen about the shores during the whole

that is a mere inference of mine, because so little is known of those ice-bound coasts, and Wilkes, who gives the only record made of the subject, saw no other animal there save that one.

year; and then the natives occasionally shoot them, long after the fur-seals have entirely disappeared. Again, it does not confine its landing to the Pribylov Islands alone, as the fur-seal unquestionably does, with reference to such terrestrial location in our own country. On the contrary, it is a frequent visitor to almost all of the Aleutian Islands: it ranges, as I have said before, over the mainland coast of Alaska, south of Bristol Bay, and about the Siberian shores to the westward, throughout the Kuriles and the Japanese northern waters.*

---

* The winter of 1872–73, which I passed on the Pribylov Islands, was so rigorous that those shores were ice-bound, and the sea covered with floes from January until May 28th; hence I did not have an opportunity of seeing, for myself, whether the sea-lion remains about its breeding-grounds there throughout that period. The natives say that a few of them, when the sea is open, are always to be found, at any day during the winter and early spring, hauled out at Northeast Point, on Otter Island, and around St. George. They are, in my opinion, correct; and, being in such small numbers, the "seevitchie" undoubtedly find enough subsistence in local crustacea, pisces, and other food. The natives, also, further stated that none of the sea-lions which we observe on the islands during the breeding-season leave the waters of Bering Sea from the date of their birth to the time of their death. I am also inclined to agree with this proposition, as a general rule, though it would be strange if Pribylov sea-lions did not occasionally slip into the North Pacific, through and below the Aleutian chain, a short distance, even to travelling as far to the eastward as Cook's Inlet. *Eumetopias stelleri* is well known to breed at many places between Attoo and Kadiak Islands. I did not see it at St. Matthew, however, and I do not think it has ever bred there, although this island is only two hundred miles away to the northward of the Seal Islands—too many polar bears. Whalers speak of having shot it in the ice-packs in a much higher latitude, nevertheless, than that of St. Matthew. I can find no record of its breeding anywhere on the islands or mainland coast of Alaska north of the fifty-seventh parallel or south of the fifty-third parallel of north latitude. It is common on the coast of Kamchatka, the Kurile Islands, and the Commander group, in Russian waters.

There are vague and ill-digested rumors of finding *Eumetopias* on the shores of Prince of Wales and Queen Charlotte Islands in breeding rookeries; I doubt it. If it were so, it would be authoritatively known by this time. We do find it in small numbers on the Farallones Rocks, off the entrance to the harbor of San Francisco, where it breeds in company with, though sexually apart from, an overwhelming majority of *Zalophus;* and it is credibly reported as breeding again to the southward, on the Santa Barbara, Guadaloupe, and other islands of Southern and Lower California, consorting there, as on the Farallones, with an infinitely larger number of the lesser-bodied *Zalophus.*

There is no record made which shows that the fur-seals, even, have any

When I first returned, in 1873, from the Seal Islands, those authors, whose conclusions were accepted prior to my studies there, had agreed in declaring that the sea-lion, so common off the port of San Francisco, was the same animal also common in Alaska, and the Pribylov Islands in especial; but my drawings from life, and studies, quickly pointed out the error, for it was seen that the creature most familiar to the Californians was an entirely different animal from my subject of study on the Seal Islands. In other words, while scattered examples of the *Eumetopias* were, and are, unquestionably about and off the harbor of San Francisco, yet nine tenths of the sea-lions there observed were a different animal—they were the *Zalophus californianus*. This *Zalophus* is not much more than half the size of *Eumetopias*, relatively; it has the large, round, soft eye of the fur-seal, and the more attenuated Newfoundland-dog-like muzzle; and it never roars, but breaks out incessantly with a *honk, honk, honking* bark, or howl.

No example of *Zalophus* has ever been observed in the waters of Bering Sea, nor do I believe that it goes northward of Cape Flattery, or really much above Mendocino, Cal.

According to the natives of St. George, some sixty or seventy years ago the *Eumetopias* held almost exclusive possession of that island being there in great numbers, some two or three hundred

---

regular or direct course of travel up or down the northwest coast. They are principally seen in the open sea, eight or ten miles from land, outside the heads of the Straits of Fuca, and from there as far north as Dixon Sound. During May and June they are aggregated in greatest numbers here, though examples are reported the whole year around. The only fur-seal which I saw, or was noted by the crew of the *Reliance*, in her cruise, June 1st to 9th, from Port Townsend to Sitka, was a solitary "holluschak" that we disturbed at sea well out from the lower end of Queen Charlotte's Islands; then, from Sitka to Kadiak, we saw nothing of the fur-seal until we hauled off from Point Greville, and coming down to Ookamok Islet, a squad of agile "holluschickie" suddenly appeared among a school of humpback whales, sporting in the most extravagant manner around, under, and even leaping over the wholly indifferent cetacea. From this eastern extremity of Kadiak Island clear up to the Pribylov group we daily saw them here and there in small bands, or also as lonely voyageurs, all headed for one goal. We were badly outsailed by them; indeed, the chorus of a favorite "South Sea pirate's" song, as incessantly sung on the cutter's "'tween decks," seemed to have special adaptation to them:

"For they bore down from the wind'wiard,
A sailin' seven knots to our four'n."

thousand strong ; and they aver, also, that the fur-seals then were barely permitted to land by these animals, and in no great number ; therefore, they assert they were directed by the Russians (*i.e.*, their own ancestry) to hunt and worry the sea-lions off from the island : the result was that, as the sea-lions left, the fur-seals came, so to-day *Callorhinus* occupies nearly the same ground which *Eumetopias* alone covered sixty years ago. I call attention to this statement of the people because it is, or seems to be, corroborated in the notes of a French naturalist and traveller, who, in his description of the Island of St. George, which he visited sixty years ago, makes substantially the same representation.*

That great intrinsic value to the domestic service of the Aleutes rendered by the flesh, fat, and sinews of this animal, together with its skin, arouses the natives of St. Paul and St. George, who annually make drives of "seevitchie," by which they capture two or three hundred, as the case may be. On St. George driving is positively difficult, owing to the character of the land itself : hence, a few only are secured there ; but at St. Paul unexceptional advantages are found on Northeast Point for the capture of these shy and timid brutes. The natives of St. Paul, therefore, are depended upon to secure the necessary number of skins required by both islands for their boats and other purposes. This capture of the sea-lion is the only serious business which the people have on St. Paul. It is a labor of great care, industry, and some physical risk for the Aleutian hunters.†

---

* Choris : Voyage Pittoresque autour du Monde.

† A curious, though doubtless authentic, story was told me in this connection illustrative of the strength and energy of the sea-lion bull when at bay. Many years ago (1847), on St. Paul Island, a drive of September sea-lions was brought down to the village in the usual style; but when the natives assembled to kill them, on account of a great scarcity at that time of powder on the island, it was voted best to lance the old males also, as well as the females, rather than shoot them in the customary style. The people had hardly set to work at the task when one of their number, a small, elderly, though tough, able-bodied Aleut, while thrusting his lance into the "life" of a large bull, was suddenly seen to fall on his back directly under that huge brute's head. Instantly the powerful jaws of the "seevitchie" closed upon the waistband, apparently, of the native, and, lifting the yelling man aloft as a cat would a kitten, the sea-lion shook and threw him high into the air, away over the heads of his associates, who rushed up to the rescue and quickly destroyed the animal by a dozen furious spear-thrusts; yet death did not loosen its clinched jaws, in which were the tattered fragments of Ivan's clothing.

By reference to my sketch-map of Northeast Point rookery the reader will notice a peculiar neck or boot-shaped point, which I have designated as Sea-lion Neck. That area is a spot upon which a large number of sea-lions are always to be found during the season. As they are so shy and sure to take to water upon the appearance or presence of man near by, the natives adopt this plan : Along by the middle or end of September, as late sometimes as November, and after the fur-seal rookeries have broken up for the year, fifteen or twenty of the very best men in the village are selected by one of their chiefs for a sea-lion rendezvous at Northeast Point. They go up there with their provisions, tea, and sugar, and blankets, and make themselves at home in the *barrabora* and house which I have located on the sketch-map of Novastoshnah, prepared to stay, if necessary, a month, or until they shall get the whole drove together of two or three hundred sea-lions.

The "seevitchie," as the natives call those animals, cannot be approached successfully by daylight, so these hunters lie by in this house of Webster's until a favorable night comes along, one in which the moon is partially obscured by drifting clouds and the wind blows over them from the rookery where the sea-lions lie. Such an opportunity being afforded, they step down to the beach at low water and proceed to creep on all fours across surf-beaten sand and boulders up between the dozing herd, and the high-water mark where it rests. In this way a small body of natives, crawling along in Indian file, may pass unnoticed by sea-lion sentries, which doubtless in that uncertain light see, but confound the forms of their human enemies with those of seals. When the creeping Aleutes have all reached the strip of beach that is left bare by ebbtide, and is between the surf and those unsuspecting animals, at a given signal from their crawling leader they at once leap to their feet, shout, yell, brandishing their arms and firing off pistols, while the astonished and terrified lions roar, and flounder in every direction.

If at the moment of surprise seevitchie are sleeping with their heads pointed toward the water, as they rise up in fright they charge straight on in that direction, right over the men themselves ; but those which have been resting at this instant, when startled, pointed landward, up they rise and follow that course just as desperately, and nothing will turn them either one way or the other. These sea-lions which charged for the water are lost, of

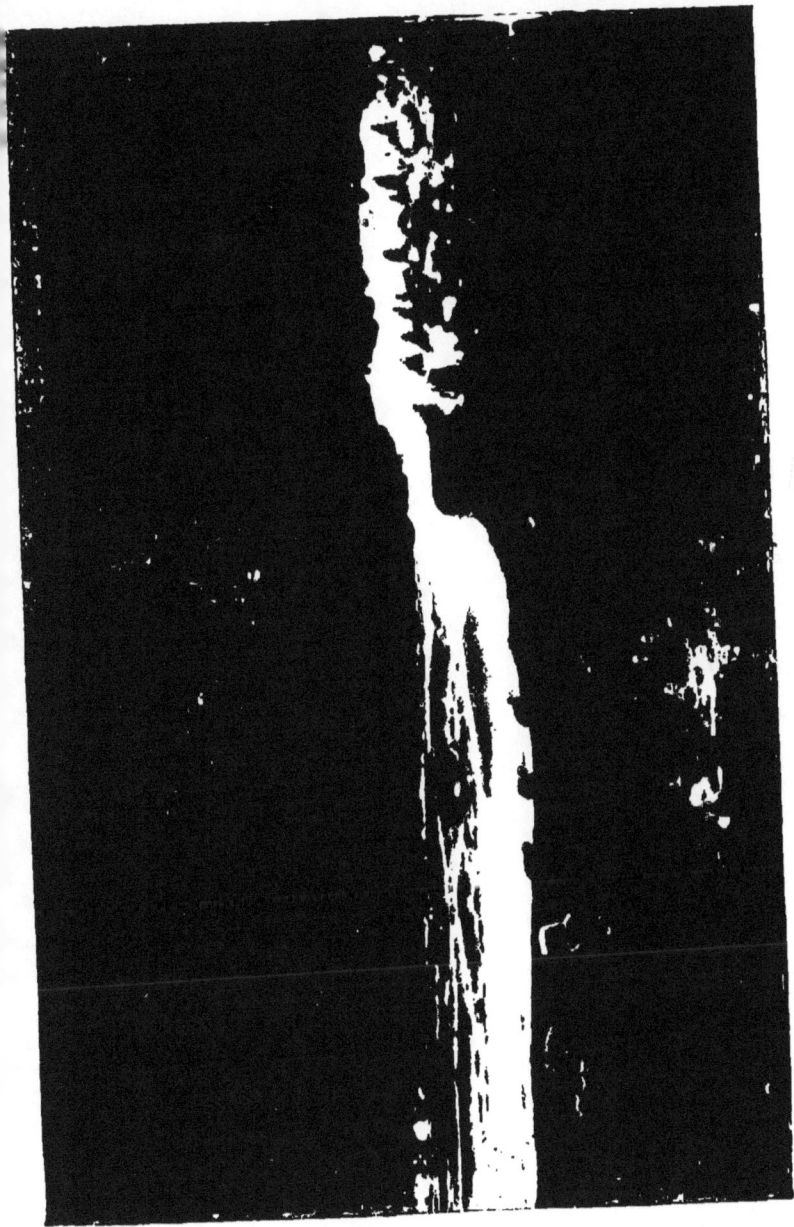

ALEUTES CAPTURING SEA-LIONS

Natives creeping up between a herd of dozing Sea-lions and the water, at low tide during a moonlight night, at Garden Cove, St. George's Island; getting into position for "springing the alarm"

THE SEA-LION PEN

Method of corralling Sea-lions at Novastoshnah, St. Paul's Island, while the Natives are getting a Drove together for driving to the Village

course ; * but the natives promptly follow up the land-turned animal with a rare combination of horrible noises and demoniacal gesticulations until the first frenzied spurt and exertions of the terrified creatures so completely exhaust them that they fall panting, gasping, prone upon the earth, extended in spite of their bulk and powerful muscles, helpless, and at the mercy of their cunning captors, who, however, instead of slaying them as they lie, rudely rouse them up again and urge the herd along to the house in which they have been keeping watch during the several days past.

Here at this point is a curious stage in such proceeding. The natives drive up to that " Webster's " house those twenty-five or thirty or forty sea-lions, as the case may be, which they have just captured —they seldom get more at any one time—and keep them in a corral or pen close by the barrabora, on the flattened surface of a sand-ridge, in the following comical manner : When they have huddled up the "pod," they thrust stakes down around it at intervals of ten to twenty feet, to which strips of cotton cloth are fluttering as flags, and a line or two of sinew-rope or thong of hide is strung from pole to pole around the group, making a circular cage, as it were. Within this flimsy circuit the stupid sea-lions are securely imprisoned, and, though they are incessantly watched by two or three men, the whole period of caging and penning which I observed, extending over nine or ten days and nights, passed without a single

* The natives appreciate this peculiarity of the sea-lion very keenly, for good and sufficient cause, though none of them have ever been badly injured in driving or " springing the alarm." I camped with them for six successive nights of September, 1872, in order to witness the whole procedure. During the several drives made while I was with them I saw but one exciting incident. Everything went off in an orthodox manner, as described in the text above. The exceptional incident occurred during the first drive of the first night and rendered those natives so cautious that it was not repeated. When the alarm was sprung, old Luka Mandrigan was leading the van, and at that moment down upon him, despite his wildly gesticulating arms and vociferous yelling, came a squad of bull "seevitchie." The native saw instantly that they were pointed for the water, and, in his sound sense, turned to run from under. His tarbosars slipped upon a slimy rock awash ; he fell flat as a flounder just as a dozen or more big sea-lions plunged over and on to his prostrate form in the shallow water. In less time than this can be written the heavy pinnipeds had disappeared, while the bullet-like head of old Luka was quickly raised, and he trotted back to us with an alternation of mirth and chagrin in his voice. He was not hurt in the least.

effort being made by the "seevitchie" to break out of their flimsy bonds, and it was passed by these animals, not in stupid quiescence, but in alert watchfulness, roaring, writhing, twisting, turning one upon and over the other.

By this method of procedure, after the lapse usually of two or three weeks, a succession of favorable nights will have occurred : then the natives secure their full quota, which, as I have said before, is expressed by a herd of two or three hundred of these animals.

The Sea-lion Caravan.

[*Natives driving a drove over the plain of Polavina, en route from Northeast Point to St. Paul.*]

When that complement is filled, the natives prepare to drive their herd back to the village over the grassy and mossy uplands and intervening stretches of sand-dune tracts, fully eleven miles : preferring thus to take the trouble of prodding such clumsy brutes, wayward and obstinate as they are, rather than to pack their heavy hides in and out of boats, making in this way each sea-lion carry its own skin and blubber down to the doors of their houses in the village. If the weather is normally wet and cold, this drive or caravan of sea-lions can be driven to its point of destination in five or six days ; but should it be dry and warmer than usual, three weeks, and even longer, will elapse before the circuit is traversed.

When the drive is started, the natives gather around the herd on all sides, save an opening which they leave pointing to that

SPRINGING THE ALARM

Natives surprising a Herd of Sea-lions at Tolstoi, St. George's Island.   August 3. 1873

direction they desire the animals to travel ; in this manner they escort and urge the "seevitchie" along to their final resting and slaughter near the village. The young lions and the females, being much lighter than old males, less laden with fat or blubber, take the lead, for they travel twice and thrice as easy and as fast as the latter ; these, by reason of their immense avoirdupois, are incapable of moving ahead more than a few rods at a time, then they are completely checked by sheer loss of breath, though the vanguard of the females allures them on ; but when an old sea-lion feels his wind coming short, he is sure to stop, sullenly and surlily turning upon the drivers, not to move again until his lungs are clear.

In this method and manner of direction the natives stretch a herd out in extended file, or as a caravan, over the line of march, and as the old bulls pause to savagely survey the field and catch their breath, showing their wicked teeth, the drivers have to exercise every art and all their ingenuity in arousing them to fresh efforts. This they do by clapping boards and bones together, firing fusees, and waving flags ; and of late, and best of all, the blue gingham umbrella repeatedly opened and closed in the face of an old bull has been a more effective starter than all the other known artifices or savage expedients of the natives. *

---

* The curious behavior of sea-lions in the Big Lake when they are *en route* and driven from Novastoshnah to the village deserves mention. After the drove gets over the sand dunes and beach between Webster's house and the extreme northeastern head of the lake, a halt is called and the drove " penned " on the bank there. Then, when the sea-lions are well rested, they are started up and pell-mell into the water. Two natives in a bidarka keep them from turning out from shore into the broad bosom of Meesulkmahnee, while another bidarka paddles in their rear and follows their swift passage right down the eastern shore. In this method of procedure the drive carries itself nearly two miles by water in less than twenty minutes from the time the sea-lions are first turned in at the north end to that moment when they are driven out at the southeastern elbow of the Big Pond. The shallowness of the water here accounts probably for the strange failure of these sea-lions to regain their liberty, and it so retards their swimming as to enable the bidarka, with two men, to keep abreast of their leaders easily, as they plunge ahead ; and, "as one goes, so all go sheep," it is not necessary to pay attention to those which straggle behind in the wake. They are stirred up by a second bidarka, and none make the least attempt to diverge from that track which the swifter mark out in advance. If they did, they could escape " scot-free " in any one of the twenty minutes of this aquatic passage.

By consulting the map of St. Paul it will be observed that in a direct line

The procession of sea-lions, managed in this strange manner day and night—for the natives never let up—is finally brought to rest within a stone's throw of the village, which has pleasurably anticipated for days and for weeks its arrival, and rejoices in its appearance. The men get out their old rifles and large sea-lion lances, and sharpen their knives, while the women look well to their oil-pouches, and repair to the field of slaughter with meat-baskets on their heads.

No attempt is made, even by the boldest Aleut, to destroy an adult bull sea-lion by spearing the enraged, powerful beast, which, now familiar with man and conscious, as it were, of his puny strength, would seize the lance between its jaws and shake it from the hands of the stoutest one in a moment. Recourse is had to a rifle. The herd is started up those sloping flanks of the Black Bluff hillside; the females speedily take the front, while the old males hang behind. Then the marksmen, walking up to within a few paces of each animal, deliberately draw gun-sight upon their heads and shoot them just between the eye and the ear. The old males thus destroyed, the cows and females are in turn surrounded by the natives, who, dropping their rifles, thrust big heavy iron lances into their trembling bodies at a point behind the fore flip-

between the village and Northeast Point there are quite a number of small lakes, including this large one of Meesulkmahnee. Into all of these ponds the sea-lion drove is successively driven. This interposition of fresh water at such frequent intervals serves to shorten the time of that journey fully ten days in warmish weather, and at least four or five under the best of climatic conditions.

This track between Webster's house and the village killing-grounds is strewn with the bones of *Eumetopias*. They will drop in their tracks now and then, even when carefully driven, from cerebral or spinal congestion principally, and when they are hurried the mortality *en route* is very great. The natives when driving them keep them going day and night alike, but give them frequent resting-spells after every spurt ahead. The old bulls flounder along for a hundred yards or so, then sullenly halt to regain breath, five or ten minutes being allowed them; then they are stirred up again, and so on, hour after hour, until the tedious transit is completed.

The younger sea-lions and the cows which are in the drove carry themselves easily far ahead of the bulls, and, being thus always in the van, serve unconsciously to stimulate and coax the heavy males to travel. Otherwise I do not believe that a band of old bulls exclusively could be driven down over this long road successfully.

pers, touching the heart with a single lunge. It is an unparalleled spectacle, dreadfully cruel and bloody.*

This surrounding of the cows is, perhaps, the strangest procedure on the islands. To fully appreciate this subject the reader must first call to his mind's eye the fact that these female sea-lions, though small beside the males, are yet large animals ; seven and eight feet long and weighing each as much as any four or five average men. But, in spite of their strength and agility, fifteen or twenty Aleutes, with rough, iron-tipped lances in their hands, will surround a drove of fifty or one hundred and fifty of them by forming a noisy, gesticulating circle, gradually closing up, man to man, until the sea-lions are literally piled in a writhing, squirming, struggling mass, one above the other, three or four deep, heads, flippers, bellies, backs, all so woven and interwoven in this panic-stricken heap of terrified creatures that it defies adequate description. The natives spear those cows on top, which, as they sink in

---

* When slowly sketching, by measurements, the outlines of a fine adult bull sea-lion which the ball from Booterin's rifle had just destroyed, an old "starooka" came up abruptly ; not seeming to see me, she deliberately threw down a large, greasy, skin meat-bag, and whipping out a knife, went to work on my specimen. Curiosity prompted me to keep still, in spite of the first sensations of annoyance, so that I might watch her choice and use of the animal's carcass. She first removed the skin, being actively aided in this operation by an uncouth boy ; she then cut off the palms to both fore flippers; the boy at the same time pulled out its mustache-bristles ; she then cut out its gullet, from the glottis to its junction with the stomach, carefully divested it of all fleshy attachments and fat ; she then cut out the stomach itself, and turned it inside out, carelessly scraping its gastric walls free of copious biliary secretions, the inevitable bunch of *ascaris ;* she then told the boy to take hold of the duodenum end of the small intestine, and, as he walked away with it, she rapidly cleared it of its attachments, so that it was thus uncoiled to its full length of at least sixty feet ; then she severed it and then it was re-coiled by the "melchiska," and laid up with the other members just removed, except the skin, which she had nothing more to do with. She then cut out the liver and ate several large pieces of that workhouse of the blood before dropping it into her meat-pouch. She then raked up several handfuls of the " leaf-lard," or hard, white fat that is found in moderate quantity around the viscera of all these pinnipeds, which she also dumped into the flesh-bag ; she then drew her knife through the large heart, but did not touch it otherwise, looking at it intently, however, as it still quivered in unison with the warm flesh of the whole carcass She and the boy then poked their fingers into the tumid lobes of the immense lungs, cutting out portions of them only, which

24

death, are mounted in turn by the live animals underneath ; these meet the deadly lance, in order, and so on until the whole herd is quiet and stilled in the fatal ebbing of their hearts' blood.

Although the sea-lion has little or no commercial value for us, yet to the service of the natives themselves, who live all along the Bering Sea coast of Alaska, Kamchatka, and the Kuriles, it is invaluable ; they set great store by it. It supplies them with its hide, mustaches, flesh, fat, sinews, and intestines, which they make up into as many necessary garments, food-dishes, etc. They have abundant reason to treasure its skin highly, since it is the covering to their neat *bidarkas* and *bidarrahs*, the former being the small *kayak* of Bering Sea, while the latter is a boat of all work, exploration and transportation. These skins are unhaired by sweating in a pile, then they are deftly sewed and carefully stretched over a light keel and frame of wood, making a perfectly water-tight boat that will stand, uninjured, the softening influence of water for a day or two

were also put into the grimy pouch aforesaid ; then she secured the gall-bladder and slipped it into a small yeast-powder tin, which was produced by the urchin ; then she finished her economical dissection by cutting the sinews out of its back in unbroken bulk from the cervical vertebra to the sacrum ; all these were stuffed into that skin bag, which she threw on her back and supported it by a band over her head ; she then trudged back to the barrabkie from whence she sallied a short hour ago, like an old vulture to the slaughter. She made the following disposition of its contents: The palms were used to sole a pair of tarbosars, or native boots, of which the uppers and knee-tops were made of the gullets—one sea-lion gullet to each boot-top; the stomach was carefully blown up and left to dry on the barrabkie roof, eventually to be filled with oil rendered from sea-lion or fur-seal blubber. The small intestine was carefully injected with water and cleansed, then distended with air, and pegged out between two stakes, sixty feet apart, with little cross-slats here and there between to keep it clear of the ground. When it is thoroughly dry it is ripped up in a straight line with its length and pressed out into a broad band of parchment gut, which she cuts up and uses in making a water-proof "kamlayka," sewing it with those sinews taken from the back. The liver, leaf-lard, and lobes of the lungs were eaten without further cooking, and the little gall-bag was for some use in poulticing a scrofulous sore. The mustache-bristles were a venture of the boy, who gathers all that he can, then sends them to San Francisco, where they find a ready sale to the Chinese, who pay about one cent apiece for them. When the natives cut up a sea-lion carcass, or one of a fur-seal, on the killing-grounds for meat, they take only the hams and the loins. Later in the season they eat the entire carcass, which they hang up by its hind flippers on a "laabas" by their houses.

at a time, if properly air-dried and oiled. After being used during the day these skin boats are always drawn out on the beach, turned bottom-side up and air-dried during the night—in this way made ready for employment again on the morrow.

A peculiar value is attached to the intestines of the sea-lion, which, after skinning, are distended with air and allowed to dry in that shape; then they are cut into ribbons and sewed strongly together into that most characteristic rain-proof garment of the world, known as the "kamlayka," which, while being fully as water-repellant as india-rubber, has far greater strength, and is never affected by grease and oil. It is also transparent in its fitting over dark clothes. The sea-lions' throats are treated in a similar manner, and when cured, are made into boot tops, which

The " Bidarrah."

[*Characteristic Alaskan boat, made by fitting sea-lion skins over a wooden frame and keel.*]

are in turn soled by very tough skin that composes the palms of this animal's fore flippers.

The Aleutian name for this garment is unpronounceable in our language, and equally so in the more flexible Russian; hence the Alaskan "kamlayka," derived from the Siberian "kamliia." That is made of tanned reindeer skin, unhaired, and smoked by larch bark until it is colored a saffron yellow; and is worn over a reindeer-skin undershirt, which has the hair next to its owner's skin, and the obverse side stained red by a decoction of alder-bark. The kamliia is closed behind and before, and a hood, fastened to the back of the neck, is drawn over the head, when leaving shelter; so is the Aleutian kamlayka; only the one of Kolyma is used to keep out piercing dry cold, while the garment of the Bering Sea is a perfect water-tight affair.

Around the natives' houses, on St. Paul and St. George, constantly appear curious objects which, to an unaccustomed eye, resemble overgrown gourds or enormous calabashes with attenuated necks ; examination proves them to be the dried, distended stomach-walls of a sea-lion, filled with its oil—which (unlike the offensive blubber of the fur-seal) boils out clear and inodorous from its fat. The flesh of an old sea-lion, while not very palatable, is tasteless and dry ; but the meat of a yearling is very much like veal, and when properly cooked I think it is just as good ; but the superiority of sea-lion meat over that of the fur-seal is decidedly marked. It requires some skill in the *cuisine* ere sausage and steaks of the *Callorhinus* are accepted on the table ; while it does not, however, require much art, experience, or patience for good cooks to serve up the juicy ribs of a young sea-lion so that the most fastidious palate will not fail to relish it.

The carcass of a sea-lion, after it is stripped of its hide, and disembowelled, is hung up in cool weather by its hind flippers, over a rude wooden frame or "labaas," as the natives call such a structure, where, together with many more bodies of fur-seals treated in the same manner, it serves from November until the following season of May, as the meat-house for an Aleut on St. Paul and St. George. Exposed in this manner to open weather, the natives keep their seal-meat almost any length of time, in winter, for use ; and, like our old duck and bird-hunters, they say they prefer to have this flesh tainted rather than fresh, declaring that it is most tender and toothsome when decidedly "loud."

The tough, elastic mustache-bristles of a sea-lion are objects of great commercial activity by the Chinese, who prize them highly as pickers for their opium pipes, and several ceremonies peculiar to their joss-houses. Such lip-bristles of the fur-seal are usually too small and too elastic for this service. The natives, however, always carefully pluck them out of the *Eumetopias*, and get their full value in exchange.

The sea-lion also, as in the case of the fur-seal, is a fish-eater, pure and simple, though he, like the latter, occasionally varies his diet by consuming a limited amount of juicy sea-weed fronds, and tender marine crustaceans ; but he hunts no animal whatever for food, nor does he ever molest, up here, the sea-fowl that incessantly hover over his head, or sit in flocks without any fear on the surface of the waters around him. He, like *Callorhinus*, is, without question, a mighty fisherman, familiar with every submarine haunt of his piscatorial prey ; and, like his cousin, rejects the heads of all those fish which have hard horny mouths or are filled with teeth or bony plates.

Many authorities who are quoted in regard to the habits of hair-seals and southern sea-lions speak with much fine detail of having witnessed the capture of water-birds by *Phocidæ* and *Otariidæ*. To this point of inquiry on the Pribylov Islands I gave continued close attention ; because, off and around all of the rookeries, large flocks of auks, arries, gulls, shags, and choochkies were swimming upon the water, and shifting thereupon incessantly, day and night, throughout the late spring, summer, and early fall. During the four seasons of my observation I never saw the slightest motion made by a fur-seal or sea-lion, a hair-seal or a walrus, toward intentionally disturbing a single bird, much less of capturing and eating it. Had these seals any appetite for sea-fowl, this craving could have been abundantly satisfied at the expense of absolutely no effort on their part. That none of these animals have any taste for water-birds I am thoroughly assured.

In concluding this recitation of that wonderful seal life belonging to those islets of Pribylov, it is well to emphasize the fact that, with an exception of the Russian and American seal islands of Bering Sea, there are none elsewhere in the world of the slightest importance to-day ; the vast breeding-grounds of fur-seals bordering on the Antarctic have been, by the united efforts of all nationalities—misguided, short-sighted, and greedy of gain—entirely depopulated ; only a few thousand unhappy stragglers are now to be seen throughout all that southern area, where millions once were found, and a small rookery, protected and fostered by the government of a South American State, north and south of the mouth of the Rio de la Plata. When, therefore, we note the eagerness with which our civilization calls for seal-skin fur, the fact that in spite of fashion and its caprices this fur is and always will be an article of intrinsic value and in demand, the thought at once occurs that the Government is exceedingly fortunate in having this great amphibious stock-yard, far up and away in the quiet seclusion of Bering Sea, from which it shall draw an everlasting revenue, and on which its wise regulation and its firm hand can continue the seals forever.

CHAPTER XII.

INNUIT LIFE AND LAND.

"Nooshagak " is not a very euphonious name, yet it is employed in
Alaska to express the whole of an immense area that backs the
borders of Bristol Bay ; but, when strictly applied, it is the designation of a small trading-post at the head of a large, brackish estuary of the sea, into which the Nooshagak River pours its heavy
flood. A cruise of three hundred and eighty miles to the northeast
from Oonalashka in a trim little trading-schooner, which alone can
make the landing, takes you to this old and well-known Russian
outpost ; but the mariner who pilots that vessel must be well acquainted with those perilous shoals and tide-rips of Bristol Bay, or
you will never disembark at the foot of that staircase which leads
up to the doors of Alexandrovsk. The river here is a broad arm of
the sea, full of shifting sand-bars and mud-flats which try the tem-

per of the most patient and skilful navigator. It runs over these shallows at certain turns of the tide, like the ebb and flow in the Bay of Fundy, with a big, booming tidal wave, or " bore." The current of this river may be discerned for a long distance out into Bristol Bay, easily traced at the season of high water by its turbidity. Above the settlement of Nooshagak that river rapidly narrows into a width of half a mile between banks for a long distance up its winding course. It is very deep, with a succession of ripples, or bars, that prevent navigation. When the northern bend is reached, then it changes to a brawling, swift, and shoal current, with higher rocky banks up to its source in the big lake which bears its guttural name. It is clear and pure here, and is not muddy until it reaches the shelving, alluvial banks of its lower course, which precipitate, by their caving and washing out, large quantities of soil and timber into the stream. Its shores are, and all the country back is, thickly wooded by spruce forests, and parked with grassy slopes which reach out here and there, planted sparsely with thickets and clumps of graceful birch- and poplar-trees. These nod and wave their tremulous foliage as the summer gusts sweep now and then over them. Countless pools, ponds, and lakes nestle in the moors and in the forest hollows, upon which flocks of geese, ducks, and all other kinds of hardy water-fowl breed and moult their plumage during the short, hot summer. The traders say that this river is the only one in Alaska, of the least magnitude, which has banks on both sides of firm soil throughout its entire course.

This site of Nooshagak village was an initial point of Russian influence and trade among the great Innuit people of Alaska, who live extended in their numerous settlements from the head of Bristol Bay clear to the Arctic Ocean. Kolmakov established the post in 1834, and named it Alexandrovsk. A simple cylindrical wooden shaft, twenty feet high, surmounted with a globe, stands erected to his memory on a small hillock overlooking the post below. The village itself is located on the abrupt slopes of a steep, grassy hillside which rises from the river's edge. The trading-stores and the residence of the priest, the church, log-huts of the natives and their barraboras are planted on a succession of three earthen terraces, one rising immediately behind the other. All communication from flat to flat is by slippery staircases, which are fraught with great danger to a thoughtless pedestrian, especially when fogs moisten the steps and darkness obscures his vision.

The red-roofed, yellow-painted walls of the old Russian buildings, the smarter, sprucer dwellings of our traders, with lazy, curling wreaths of bluish smoke, are brought into very picturesque relief by the verdant slopes of Nooshagak's hillside, caught up and reflected deeply by the swiftly flowing current of the river below. The natives have festooned their long drying-frames with the crimson-tinted flesh of salmon ; bleached drift-logs are scattered in profusion upon a bare sandy high-water bench that stretches like a buff-tinted ribbon just beneath them, and above, the dark, turbid whirl of flood and eddy so characteristic of a booming, rising river. A gleam of light falls upon a broad expanse of the estuary beyond that point under which the schooner lies at anchor, and brings out the thickly wooded banks of an opposite shore, causing us to note the fact that, for some reason or other, no timber seems ever to have spread down so far toward the sea on this side of the stream, or where the settlement stands, since nothing but scattered copses of alder- and willow-bushes grow on its suburbs or anywhere else as far as an eye can range up the valley.

We notice a decided difference in bearing and expression among the natives here—nothing like what we have studied at Oonalashka, Kadiak, or Sitka. They are Innuits, or representatives of the most populous savage family indigenous to Alaska, and are as nomadic as Bedouins. They are the least changed or altered by contact with our race. They are Eskimo, strictly speaking, and the natives of Kadiak are almost strictly related to them. In portraying the physique, physiognomy, and disposition of these people, we find in an average Innuit a man who stands about five feet six or seven inches in his heelless boots ; his skin is fair, slightly Mongolian in its complexion and facial expression ; a broad face, prominent cheek-bones, a large mouth with full lips, small black eyes, but prominently set in their sockets—not under a lowering brow, as in the case of true Indian faces. The nose is very insignificant and much depressed, having between the eyes scarcely any bridge at all. He has an abundance of coarse black hair ; never any of a reddish hue, as frequently noted among the Aleutes when first discovered and described by the Russians. Up to the age of thirty years an Innuit usually keeps his hair cut pretty close to his scalp ; some of them shave the occiput, so that it shines like a billiard-ball. After this period in life he lets it grow as it will, wearing it in ragged, unkempt locks. He sometimes will sport a well-developed mus-

tache and chin-whisker, of which he is as proud as though a
Caucasian. He has shapely hands and feet; his limbs are well
made, formed, and muscled. An Innuit woman is proportionately
smaller than the man, and, when young, sometimes she is not unpleasant to look at. The skin of her cheeks then will be faintly
suffused with blushes of natural color, her lips pouting and red,
with small, tapering hands and high-instepped feet. She rarely
pierces her lips or disfigures her nose; she lavishes upon her child
or children a wealth of affectionate attention—endows them with
all her ornaments. She allows her hair to grow to its full length,
gathers it up behind into thick braids, or else it is bound up in
ropes lashed by copper wire or sinews. She seldom tattooes her

An Innuit Woman.

skin in any place; a faint drawing of transverse blue lines upon the
chin and cheeks is usually made by her best friend when she is
married.

We are not reminded of the clothing stores of San Francisco
when we meet Innuits everywhere between Point Barrow and Nooshagak; they are clad in the primitive garments of their remote
ancestry, as a rule—a few exceptions to this generalization being
those individuals who are living constantly about the widely scattered trading-posts, and the chapels, or missions, located in their
territory, where they act as servants or interpreters. The conventional coat of these people is the "parka," made of marmot and
muskrat-skins, or of tanned reindeer-hides, with enormous winter
hoods, or collars, of dog-hair or fox-fur. This parka has sleeves,
and compasses the body of the wearer, without an opening either

before or behind, from his neck to his feet.   His head is thrust
through an aperture left for it, with a puckering string which draws
it up snugly around the neck.   In winter the heavy hood-collar, or
cowl, is fitted so as to be drawn over his entire head and pulled
down to the eyes.   This parka is worn with singular ease and
abandon ; frequently the arms are withdrawn from the big, baggy
sleeves and stowed under the waist-slack of the garment, leaving
these empty appendages to dangle.   Natives, as they sit down,
draw the parka out and over the knees, still keeping their arms underneath ; or, when on the trail, and the wet grass and bushes make
it imperative, the parka is gathered up and bound by a leather
thong-strap or girdle of sinews, so as to keep its bottom border dry
and as high as the knees of a tramping native ; the baggy folds
of it then give its wearer a grotesque and clumsy figure as they
bulge out over his hips and abdomen.   The most favored and valuable parka is that one made out of alder-bark tanned reindeer-skin,
for winter use ; the hair is worn inside, next to the skin.   For summer styles those fashioned out of the breasts of water-fowl, of mar-
mot- and mink-skins, are most common.   The hood is never attached
to the parka in the warmer months of the year.   It is a very capa-
cious pouch which, when not in service, is resting in thick folds
back of the head and upon the shoulders.   It is ornamented in a
variety of ways, but usually a thick fringe of long-haired dog- or
fox-fur forms its border, and when drawn into position encircles
the wearer's face and gives it a wild and unkempt air.

The only underwear which a Mahlemoot affects is limited to
that garment which we call a shirt, made of light skins or of cheap
cotton drillings ; if it is of skin, it is worn from father to son, and
becomes a real heirloom highly polished and redolent.   Their trousers are, for both sexes, a pair of thin skin or cotton drawers, puckered at the ankles and bound about with the uppers of their
moccasons, or else enclosed by the tops to their reindeer-boots,
which are the prevalent covering for their feet.   Such are the characteristics of a costume worn by much more than half the entire
aboriginal population of Alaska ; but when we come to inspect
their dwellings we find a greater variety of housing than indexed in
dressing.

A very great majority of the Innuits live in a house that outwardly resembles a circular mound of earth, seven or eight feet high,
and thirty or forty feet in circumference.   It is overgrown with

"CHAMI"

An Innuit Girl, about 14 years old

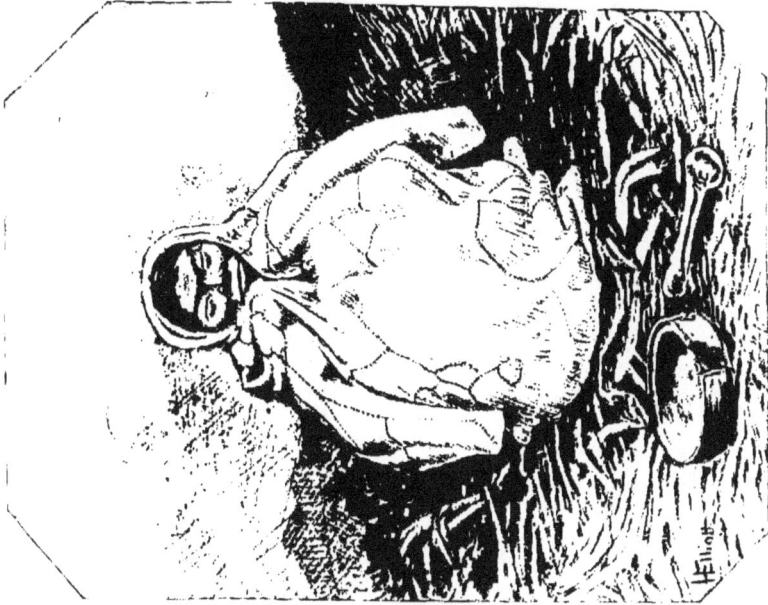

AFTER DINNER—GOOD DIGESTION

Favorite position of Innuits

rank grasses, littered with all sorts of utensils, weapons, sleds, and other Eskimo furniture. A small spiral coil of smoke rises from a hole in its apex, a dog or two are crouching upon it, and children climb up and roll down its sides, scattering bones and fragments of fish and meat as they eat in the irregular fashion of these people. A rude pole scaffolding stands close by, upon which, high above the reach of dogs, is a wooden cache, containing all winter stores of dried provision, "ukali," and the like. This hut is usually right down upon the sea-beach, just above high tide, or high-water mark, on the river banks, for these savages draw their sustenance largely, even wholly in many instances, from the piscine life of those northern streams.

An Innuit Home on the Kuskokvim.

All these tribes have summer dwellings distinct from those used during the winter. For the winter houses a square excavation of about ten feet or more is made, in the corners of which posts of drift-wood or whale-ribs from eight to ten feet in height are set up; the walls are formed by laying posts of drift-wood one above the other against the corner-posts; outside of this another wall is built, sometimes of stone, sometimes of logs, the intervals being filled with earth or rubble; the whole of the structure, including the roof, is covered with sods, leaving a small opening on top, that can be closed by a frame over which a thin, transparent seal-skin is tightly drawn. The entrance to one of these houses consists of a narrow, low, underground passage from ten to twelve feet in length, through which an entrance can only be accomplished on hands and knees. The interior arrangement of such a winter house is simple, and is nearly the same with all these tribes. A piece of bear- or reindeer-skin is hung before an inner opening of the doorway; in the centre of the enclosure is a fireplace, which is a square excavation directly under that smoke-hole in the roof; the floor is rarely planked, and frequently two low platforms, about four feet in width, extend along the sides of the house from the entrance to the back, and covered with mats and skins which serve as beds at night. In the larger dwellings, occupied by more than one family, the sleeping-places of each are separated from each other by suspended mats, or simply by a piece of wood. All the bladders containing oil, the wooden vessels, kettles, and other domestic utensils, are kept in the front part of the dwelling, and before each sleeping- place there is generally a block of wood upon which is placed the oil-lamp used for heating and cooking.

The only ingress or egress is afforded by a small, low, irregularly shaped aperture (it cannot rightfully be called a door), through which the natives stoop and enter, passing down a foot or two through a short, depressed passage that is created by the thickness of the walls to the hut; the floor is hard-tramped earth, and the ground-plan of it a rude circle, or square, twelve, fifteen, or twenty feet in diameter, as the case may be, and in which the only light of day comes feebly in from a small smoke-opening at the apex of the roof, the ceiling of which rises tent-like from the floor. A faint, smouldering fire is always made directly in the centre, and the atmosphere of the apartment is invariably thick and surcharged with its combustion.

Hard and rude are the beds of the Innuit—a clumsy shelf of poles is slightly elevated above the earth, and placed close against the walls; upon this staging the skins of bears and reindeer, seals, and even walrus-hides, together with mats of plaited sedge and bark, are laid; sometimes these bedsteads are mere platforms of sod and peat. If the hut stands in a situation where it is exposed to the full force of boisterous storms, then the architect builds a rough hallway of earth and sods, with a bulging expansion, whereby room is given in which to shelter his dogs and keep many utensils and traps under cover. He also, in warm weather, lives outside of this winter hut, to a great degree, when at home; and, for that purpose, he builds a summer cook-house, or kitchen, which resembles the igloo itself, only it is not more than five or six feet square, and no higher than a stooping posture within warrants. This is also a great resort for his dogs, which renders the place very offensive to us.

The summer houses are erected above ground, and are generally slight pole frames, roofed with skins and open in front; fire is rarely made in them, and therefore they have no opening in the roof, all cooking being done in the open air during fine weather. They seldom have flooring, but otherwise the interior arrangements resemble those of the winter houses. The store-houses of all our Eskimo tribes are set on posts at a height of from eight to ten feet above the ground, to protect them from foxes, wolves, and dogs. They have generally a small square opening in front that can be closed with a sliding board, and which is reached by means of a notched stick of wood. These boxes are seldom more than eight feet square by three or four feet in height.

The routine of life which these natives of the Nooshagak and Kuskokvim valleys and streams follow is one of much activity—they are on the tramp or are paddling up and down the rivers pretty much all of the time. A year is divided up by them about as follows : In February they prepare to go to the mountains, and go then most of them do, though some will be as late as April in getting away on account of their children, or of sheer laziness. They move with the entire family outfit, bag and baggage, dogs, sleds, and boats. They settle down along by the small mountain streams, trap martens, shoot deer, and dig out beaver. February and March are the best months for marten, April and May for the beaver, bear, and land-otter.

By June 10th they return to their winter villages and visit the trading-posts. They then begin their preparations for salmon-fishing, getting their traps into shape so as to be used effectively when those fish begin to run. They air-dry salmon on frames, and put the heads in holes and allow them to rot slightly before eating ; also the spawn, which, however, is preserved in oil, and used as a great delicacy during their own festivals in the midwinter season. The salmon-fishing is all over about July 20th. By August 10th these nomads return to the mountains, leaving the old women and youngest children with their mothers in charge of the caches at the villages. This time they go for reindeer, which have just shed their hair and are in the full beauty of new, fine, sleek coats. They hunt these animals from that time until the middle of September, when the fur of the beaver is again in prime condition; then *Castor canadensis* receives their undivided attention. They catch these giant rodents in wooden "dead-falls," and also by breaking open the dams, which causes the water to suddenly leave the beavers fully exposed to the spears of their savage human enemies.

When the first snow flies in October they rig up rude deer-skin boats, like the "bull-boats" on the Missouri, and float all their traps and rude equipage down the river back from whence they started. They all return for the winter by the middle of October; then, without going far from the vicinity of their settlements, they renew and set up fresh dead-fall traps for marten—they never go any distance from home for this little animal, and when ice forms on the rivers, about the end of October or early in November, they put their white-fish traps under it. The marten-trapping is abandoned in December, because the intense, stormy, and cold weather then drives these pine-weasels into winter holes, where they remain semi-dormant until the end or middle of February. During this period of severe wintry weather the Innuit gives himself up to unrestrained loafing and vigorous dancing festivals, which last until the year is again renewed by going out to the mountains in February.

These natives of the Nooshagak and Kuskokvim regions have a large and varied natural food-supply. They have reindeer-meat, the flesh of moose, of bears, and of all the smaller fur-bearing animals found in this territory—the list is a full one, comprising land-otters, cross, red, and black foxes, the mink, the marten, the marmot, and the ground-squirrel, or "yeavrashka," which last is the most abundant. The bears are all brown in this country—no black ones. They also secure large gray and white wolves, while those who live right on the coast of Bering Sea get walrus, the big "mahklok" seal, and a little harbor *phoca*, or "nearpah."

They have a great abundance of water-fowl, such as geese, ducks, and the small waders, and they occasionally kill a beluga, or white grampus, and at still more rare intervals they find a stranded whale, which is set upon and eaten. They save carefully all the oil which comes from marine mammals; they treasure it up in seal-skin bags that are placed high up above the reach of dogs and foxes on a frame scaffold which adjoins every hut. Fish-oil is also secured in the same manner; it answers a threefold purpose—it serves for food, for fuel, and for light, and it is a luxurious skin and hair dressing for them all, old and young.

Fish they capture in the greatest abundance, and the variety is quite fair. Salmon is the staff, and is found in all of the thousand and one lakes and sluggish or rapid streams that run from them into the greater rivers, where a mighty rush of the same fish is annually made up in June and July from Bering Sea. In all of the deeper lakes, and the big rivers, a variety of large white-fish and trout are found, especially prized and searched for by these people in midwinter, when they are trapped there in wicker-work baskets and pole weirs under ice.

The Big Mahklok.

In round numbers these Eskimo, or Innuits, of Alaska, number nearly eighteen thousand souls; they inhabit the entire coast-line of Bering Sea and the Arctic Ocean, with an exception of the Aleutian chain and that portion of the peninsula west of Oogashik. The numerous subdivisions of this great family are based wholly upon dialectic differentiation, and as its elaboration would entail a dreary and uninteresting chapter upon any reader save a studious ethnologist, it will not be itemized here. These Eskimo are all hunters and fishermen; those land animals to which we have made allusion are pursued by them at the proper seasons of the year. They do not have much, in the aggregate, of value to a trader; it is chiefly oil and walrus-ivory. Their proximity to a relatively warm coast renders the furs which they get of small value comparatively, since these pelts are paler and lighter-haired than those brought in from the distant interior, where the winters are vastly colder and longer. But an Innuit does not require a great deal from the trader—he is very much more independent than is his semi-civilized Aleutian brother ; his wants are only a small supply of lead and powder, of sugar and of tobacco, a little red cloth and a small sack of flour which suffice for a large Innuit family during the year. The flour he makes up into pancakes and fries them in rancid oil ; but, as a rule, all cooking is a mere boiling or stewing of fish and meat in sheet-iron or copper kettles. In those huts where they can afford to use tea, a small number of earthenware cups and saucers will be found carefully treasured in a little cupboard ; but they never set a table or think of such a thing, except those highly favored individuals who live as servants about the trading-posts and missions, where they do boil a "samovar" (tea-urn) and spread a cloth over the top of a box or rude table upon which to place their teacups.

Down here at Nooshagak these natives have earned a distinction of being the most skilful sculptors of the whole northern range. Their carvings in walrus-ivory are exceedingly curious, and beautifully wrought in many examples. The patience and fidelity with which they cut from walrus-tusks delicate patterns furnished them by the traders are equal in many respects to that remarkable display made in the same line by the Chinese, and so much admired. Time to them, at Nooshagak, is never reckoned, and it does not raise a ripple of concern in the Innuit's mind when, as he carves upon a tusk of white ivory, he pauses to think whether he shall be six hours or six months engaged upon the task. Shut up as he is from December until the end of February in his dark and smoky hut, he welcomes the task as one which enables him to "kill time" most agreeably, and bring in a trifle, at least, to him from the trader in the way of credit or of direct revenue.

All of these people, when they go hunting, use fire-arms of modern patterns and many old flint-lock muskets ; for fish and bird-capture they never waste any precious ammunition ; they employ spears and arrows of most artful construction and effective service. But a large number of those very primitive Eskimo, the Togiaks, just west and north of Nooshagak, use nothing at all in the chase other than the same antique bows and spears of a remote ancestry.

The disposition of these people is one of greater *bonhomie* than that evidenced by the Aleutes or the Koloshians, who are rather taciturn. The Innuit is very independent in his bearing, without being at all vindictive or ugly. He is light-hearted, enjoys conversation with his fellows, tells jokes with great gusto, sings rude songs with much animation, in excellent time but with no music, and dances with exceeding exhilaration during the progress of those savage festivals which he calls in to enliven a long dreary winter solstice.

Such a man is naturally quite sociable. Hence we find in every Innuit settlement, big or little, a town hall, or "kashga." This is a building put up after the pattern of all winter houses in the village, but of very much larger dimensions ; some of the more populous hamlets boast of a kashga which will measure as much as sixty

The Kashga.

feet square, and be from twenty to thirty feet high under its smoky rafters. A raised platform from the earth, of rough-hewn planks, runs all around the walls of the interior, and in the largest council-houses a series of three tiers of such staging is observed. The fireplace in the centre is large, often three or four feet deep and eight feet square ; on ordinary days in the spring, and during the summer and early fall, when no fire is wanted, it is covered with planks. An underground tunnel-entrance to the kashga is made just as it is into some of the family huts, only here it is divided at the end ; one branch leads to a fireplace below the flooring, and the other rises to the main apartment. The natives are obliged to crawl on all fours when they enter that underground passage or leave the kashga through its dark opening.

This is the great and sole rendezvous of the men and older boys of most settlements. The bachelors and widowers sleep here and

25

prepare their simple meals; the village guest and visitors of the
male sex are all quartered here; the discussion of all the town affairs is conducted here; the tanning of skins, the plaiting and
weaving of wicker-work fish-traps, and the manufacture of sleds
and dog-harness, spear- and arrow-heads, and carving of wood and
ivory—in fact, everything done by these people under shelter, of
that kind, is executed on the platforms of a kashga. It is the
theatre for the absurd and vigorous masked dances and mummery
of their festivals, and above all, it is the spot chosen for that vile
ammoniacal bath of the Eskimo, the most popular of all their rec-
reations.

The daily routine of living as practised by an Innuit family is
exceedingly simple. The head of the household usually sleeps over

Section showing Subterranean Entrance and Interior of a Kashga.

night in the kashga, as do all of his peers. His wife in the early
morning rolls out of her rude deer-skins, retucks her parka about
her hips, and starts up the smouldering fire which she banked with
ashes before going to sleep. A little meat or fish is soon half-
boiled, and a small kantag of oil is decanted, a handful of dried
berries thrown into it, and perhaps she has a modicum of rotten
fish-roe to add. This she takes out to her husband in the kashga,
rousing him, if he is not awake, with a gentle but firm admonition.
A large bowl of fresh water is also brought by her, and then everything is before the husband for his breakfast. She returns to her
hut after he has finished, and feeds her children and herself. If
she or her husband has a male visitor, he is served in the same way.
When the evening meal is ready, sometimes the men go home and

dine with their families; but the women and children invariably eat at home, and when they wait upon the males in the town hall they always turn their backs to them while the men are dining, it being considered a gross breach of good manners for a woman to look at a man when he is eating.

After breakfast the male Innuits start out, if the weather permits, to hunt or fish, as the case may be. If a driving storm prevents them, then in-door work is resumed or recourse to sleep again assumed. At some time in the afternoon the fire is usually drawn from the hot stoves on the hearth, the water and a kantag of chamber-lye poured over them, which, arising in dense clouds of vapor, gives notice (by its presence and its horrible ammoniacal odor) to the delighted inmates that the bath is on. The kashga is heated to suffocation, it is full of smoke, and the outside men run in from their huts, with wisps of dry grass for towels, and bunches of alder-twigs to flog their naked bodies. They throw off their garments; they shout and dance and whip themselves into profuse perspiration as they caper in the hot vapor. More of their disgusting substitute for soap is rubbed on, and produces a lather which they rinse off with cold water; and, to cap the full enjoyment of this satanic bath, these naked actors rush out and roll in a snow-bank or plunge into the icy flood of some lake or river adjoining, as the season warrants. This is the most enjoyable occasion of an Innuit's existence, so he solemnly affirms. Nothing else affords him a tithe of the infinite pleasure which this orgie gives him. To us, however, there is nothing so offensive about him as that stench which such a performance arouses.

When a bath is over, the smoke-hole is reopened (it was closed during the process!), and fresh air descends upon those men who sit around upon the platforms stupefied by that smoke and weak from their profuse perspiration. Slowly these terrible odors leave the kashga, and only the minor ones remain, rendering it quite habitable once more. Night comes on: the huge stone lamps are filled with seal-oil and lighted; the men soon lop down for sleep in their reindeer-skins or parkas, removing their trousers only, which they roll up and use as pillows, tucking the parka snugly over and around their bended knees, which are drawn up tightly to the abdomen. In the morning whoever happens to awake first relights the lamp, if any of the fluid remains over; if not, he goes to his own cache and gets a supply. If he is a bachelor, he attends then to making a fresh fire in the hearth below and prepares his coarse breakfast.

The women assist their husbands in harnessing and unharness-ing the dogs ; they go out and gather the firewood, and employ themselves in sewing, patching, and making thread from deer-ten-dons. They plait grass mats and weave grass stockings, because nearly all of the coast Innuits wear socks very skilfully made of dried grass. The boys and girls scatter about the vicinity looking after their snares and traps, or engage, in hilarious groups, playing at ball and leap-frog games, tag, and jumping matches. They har-ness up the young dogs and the pups, and sport for hours at a time with them.

" Tatlah ; " an Innuit Dog.

These people are savages, and not at all affected by the earnest and persistent attempts of the Russian priests to Christianize them. They are even less influenced by the teachings of missionaries than the Siwashes of the Sitkan archipelago, and that is saying a great deal for their hardness of heart. They are a brave race, and have displayed the utmost physical courage in fighting their way up the great rivers, Yukon, Kuskokvim, and Nooshagak, whereby they dis-placed and destroyed the Indians who once lived there. The Kolt-chanes, or Ingaleeks of the interior, who disputed that privilege with them, bear cheerful witness to this fact. But all such strife between the two great families is only known to us by legends which they recite of ancient time. No trace of recent war can be found among them.

They have no ear for music ; they are not fond of it like the Aleutes, yet they keep perfect time to cultivated tunes and melodies of our own order. The song of an Innuit is essentially like that of his Sitkan relative : it is usually a weird dirge, monotonous, and long-drawn out, accompanied by a regular and rhythmic beating of a rude drum, or a dry stick, or resonant bag. Some of the native Innuit chantings, when rendered intelligible to us, have a plaintive pathos running through them which is attractive and are simple in composition ; but such ballads are very, very rare. The majority are tedious and boastful recitations of a singer's achievements on land or water when engaged in hunting or fishing. Their mythology is the rudest and the least ornate of all savage races, unless it be that perfect vacuum of the Australasians and Terra del Fuegians.

These savages respect the dead, but they fear the sick. When death invades an Innuit family, taking the husband, or the wife, or a child, the survivors eat nothing, after the decease of the relative, but sour or last year's food, and refrain from going out or from work of any sort for a period of twenty days. They seat themselves in one corner of the hut, or "kahsime," with their backs toward the door. Every five days they wash themselves, otherwise death would promptly come to them again. The body of the dead native is composed in a sitting position, with its knees drawn up to the stomach and its arms clasped around them. It is placed in one corner, with its head against the wall. The inhabitants of that village where the dead man has lived voluntarily bring to the hut dresses of reindeer-skin, in one of which the corpse is shrouded. A coffin, or box, is prepared at some selected spot outside of the village, set up a few feet from the ground, on four stoutly driven posts, and in it the body is deposited. Near by is planted a square board or smoothly hewn plank, upon which rude figures are painted of the animals that the deceased was most fond of hunting, such as a beaver, a deer, a fish, or seal. A few of his most cherished belongings are laid in the coffin with him, but the balance of his property is divided among his family.*

---

* The Indians, or Koltchanes, of the Alaskan interior burn their dead. If anyone dies in the winter, the relatives carry that corpse everywhere with them, use it at night in the place of a pillow, and only burn it at the commencement of warm weather.

A festival in honor of the spirits of land and sea, and in memory
of deceased kinsmen, is celebrated annually in the month of October or November. Lieutenant Zagoskin,* who spent five years
among these people exploring the Yukon and Kuskokvim Rivers,
has given us full details of that strange mummery and capers
which characterize Innuit festivals and dances. What he saw between 1842 and 1845, and so graphically narrated, is to be seen substantially the same now everywhere among these people, who are
almost wholly unchanged from their primeval habits as they live
to-day.

Of the tribal organization of these people but little is known :
yet, there seems to be no recognized chieftainship—each isolated
settlement generally contains one man who makes himself prominent by superintending all intercourse and traffic with visitors.
The profits accruing to him from this position give him some slight
influence among his people ; but the *oomailik* (*oomuialik* of Zagoskin), as these middlemen or spokesmen are called, possess no authority over the people of their village, who pay far more attention
to the advice or threats of sorcerers, shamans, or "medicine men."
In the festivals, consisting of feasting, singing, and dancing, with
which these hyperboreans while away the long winter nights, the
shamans also play a prominent part, directing the order of the performances and the manufacture of masks, costumes, etc., while the
oomailik or spokesman sinks back into insignificance for the time
being.

All these games, both private and public, take place in the
kashga. At the public performances the dancers and singers, men
and women, stand around the fire-hole ; and the men, to the time
of the drum and the singing, go through various contortions of the
body, shifting from one foot to the other without moving from the
spot, the skill of the dancer being displayed only in the endurance
and flexibility of his muscles. The women, on the other hand, with
their eyes cast down, motionless, with the exception of a spasmodic
twitching of the hands, stand around in a circle, forming, we may

---

* The Russian Imperial Government in 1841 ordered Governor Etholin,
of Sitka, to select a skilled engineer to make this exploration, and accordingly, on July 10, 1842, Zagoskin was started for St. Michael's. His expedition was the most extended of any white man ever made in Alaska prior to
American search.

say, a living frame to the animated picture within. The less motion a dancer displays the greater his skill. There is nothing indecent in the dances of our sea-board natives. The dancing dress of the men consists of short tight drawers made of white reindeer-skin and the summer boots of soft moose-hide, while the women on those occasions only add ornaments, such as rings and bracelets and bead-pendants, to their common dress, frequently weighting themselves down with ten or fifteen pounds of these baubles.

An entertainment of the women was described by Zagoskin as follows :

"We entered the kashga by the common passage and found the guests already assembled, but of the landladies nothing was to be seen. On three sides of the apartment stone lamps were lighted ; the fire-hole was covered with boards, one of them having a circular opening, through which the women were to make their appearance. Two other burning lamps were placed in front of the fire-hole. The guests then formed a chorus and began to sing to the sound of the drum, two men keeping them in order by beating time with sticks adorned with wolfs' tails and gulls' wings. Thus a good half-hour passed by. Of the song my interpreter told me that it consisted of pleasantry directed against the women ; that it was evident they had nothing to give, as they had not shown themselves for so long a time. Another song praised the housewifely accomplishments of some woman whose appearance was impatiently expected with a promised trencher of the mixed mess of reindeer-fat and berries. No sooner was this song finished than that woman appeared and was received with the greatest enthusiasm. The dish was set before the men, and she retreated amid vociferous compliments on her culinary skill. She was followed by another woman. The beating of drums increased in violence and the wording of the song was changed. Standing up in the centre of the circle this woman began to relate, in mimicry and gesture, how she obtained the fat, how she stored it in various receptacles, how she cleansed and melted it, and then, placing a kantag upon her head, she invited the spectators with gestures to approach. The song went on, while eagerness to partake of the promised luxury lighted up the faces of the crowd. At last the wooden spoons were distributed, one to each man, and nothing was heard for a time but the guzzling of the luscious fluid. Another woman advanced, followed still by another, and luxuries of all kinds were produced in quick succession and as quickly despatched, while the singers pointedly alluded to the praiseworthy Russian custom of distributing tobacco. When the desired article had been produced, a woman then represented with great skill all the various stages of stupefaction resulting from smoking and snuffing. The women dressed in men's parkas."

A man's entertainment witnessed by Zagoskin took place in the same village. The preparatory arrangements were similar; one of the women, a sorceress, lead the chorus. Her first song on that occasion praised a propensity of the Russian for making presents of tobacco, rings, and other trifles to women, who, in their turn, were always ready to oblige them. This, however, was only introductory, the real entertainment beginning with a chorus of men concealed in the fire-hole. The gist of their chant was that trapping, hunting, and trade were bad, that nothing could be made, and that they could only sing and dance to please their wives. To this the women answered that they had long been aware of the laziness of their husbands, who could do nothing but bathe and smoke, and that they did not expect to see any food produced, such as the women had placed before them, consequently it would be better to go to bed at once. The men answered that they would go and hunt for something, and shortly one of them appeared through the opening. This mimic, who was attired in female apparel, with bead-pendants in his nose, deep fringes of wolverine tails, bracelets, and rings, imitated in a most admirable and humorous manner the motions and gestures of the women in presenting their luxuries, and then gave imitations of the various female pursuits and labor, the guests chuckling with satisfaction. Suddenly the parka was thrown off, and the man began to represent how he hunted the mahklok, seated in his kayak, which performance ended with the production of a whole boiled mahklok, of which Zagoskin received the throat as his portion. Others represented a reindeer-hunt, the spearing of birds, the rendering of beluga-blubber, the preparation of seal-intestines for water-proof garments, the splitting of deer-tendons into thread, and so forth. One young orphan who, possessing nothing wherewith to treat the guests, brought on a kantag filled with water, which was drunk by the women amid much merriment. It sometimes happens on these occasions that lovers of fun sprinkle the women with oil, or with that fluid which they use in place of soap, squirted from small bladders concealed about their persons; and such jokes are never resented.

Another festival, in honor of the spirits of the sea (*ugiak*), is celebrated by the coast tribes during a whole month. The preparations for this gathering begin early in the autumn. Every hunter preserves during an entire year the bladders from all such animals as he kills with arrows; the mothers also save with the greatest care the bladders of all rats, mice, ground-squirrels, or other small animals killed by their children. At the beginning of December all these bladders are inflated, painted in various colors, and suspended in the kashga; and among them the men hang up a number of fantastically carved figures of birds and fish. Some of the figures of birds are quite ingeniously contrived, with movable eyes, heads, and legs, and are able to flap their wings. Before the fireplace there is a huge block wrapped up in dry grass. From morning until night these carved figures are kept in motion by means of strings, and during the whole time a chanting of songs continues, while dry grass and weeds are burned to smoke the suspended bladders. This fumigating process ends the day's performances, which are begun anew in the morning. In the evening of that culminating day of this festival those strings of bladders are taken down and carried by men upon painted sticks prepared for the occasion; the women, with torches in their hands, accompany them to the sea-shore. Arrived there, the bladders are tied to sticks and weighted with stones, and finally thrown into the water, where they are watched with the greatest interest to see how long they float upon the surface. From the time of sinking and the number of rings upon the water where a bladder has disappeared the shamans prophesy success or misfortune in hunting during the coming year.

A final memorial feast in honor of a distinguished ancestor is conducted as follows:

Eight old men clad in parkas enter the kashga, or council-house, each carrying a stone lamp, which they deposit around the fire-hole. They next produce three small mats and spread them upon the floor in three corners of the building, and from the spectators three men are selected who are willing to go to the grave. The three nearest relatives of the deceased then seat themselves on the mats and divest themselves of all their clothing, wash their bodies, and don new clothes, girding themselves with belts manufactured several generations back and preserved as heirlooms in the family. To each of these men a staff is given, and they advance together to

the centre of the kashga, when the oldest among the invited guests
sends them forth to call the dead. These messengers leave the
building, followed by the givers of this feast. After an absence of
ten minutes the former return, and through the underground passage the whole population of the village crowds in, from the old
and feeble down to children at the breast, and with them come the
masters of ceremonies, wearing long seal-skin gloves, and strings
of sea-parrot bills hanging about the breast and arms, with elaborate belts nearly a foot in width, consisting of white bellies of
unborn fawns trimmed with wolverine tails. All such ornaments
are carefully preserved and handed down from generation to generation, some of them being made of white sable—an exceedingly
rare skin—for which high prices are paid, as much as twenty or
thirty beavers or otters for one small skin. The women hold in
their hands one or two eagle-feathers, and tie around the head a
narrow strip of white sable. Each family, grouping itself behind
its own stone lamp, chants in turn in mournful measure a song composed for the occasion. These songs are almost indefinitely prolonged by inserting the names of all the relatives of the deceased,
living and dead. The singers stand motionless in their places, and
many of those present are weeping. When a "song of the dead"
is concluded the people seat themselves, and their usual feasting
and gorging ensues. The next morning, after a bath (indulged in
by all the males), the multitude again assembles in this kashga.
The chanting around the fire-hole is renewed in the same mournful
tone, until one old man seizes a bladder drum and takes the lead,
accompanied by a few singers, and followed in procession by all
participants in the feast. They walk slowly to every sepulchre in
succession, halting before each to chant a mourning song ; all visitors not belonging to the bereaved families in the meantime crowd
upon the sodded roofs of the houses and watch these proceedings.
In the evening all that remains of food in the village is set before
the people, and when every kantag is scraped of the last remnant
of its contents the feast is ended ; then those visitors at once depart
for their homes.

Occasionally the giver of such a feast, desiring to do special
honor to the object of it, passes three days sitting naked upon a
mat in a corner of his kashga, without food or drink, chanting a
song in praise of a dead relative. At the end of such a fast any
or all visitors present gifts to him ; the story of his achievement

THE SON OF AHGAAN

An Innuit boy, 6 or 7 years old

JEST OF AN INNUIT MOTHER

"Yes—me sell!—plenty tabak"

is carried abroad, and he is made famous for life among his fellows.

It has already been mentioned that many individuals give away all their property on such occasions. If it happens that during such a memorial feast a visitor arrives from a distant village who bears the same name with the subject of a celebration, he is at once overwhelmed with gifts, clothed anew from head to foot with the most expensive garments, and returns to his home a wealthy man.

The country in which the Innuit lives is one that taxes the utmost hardihood of man when it is traversed by land or by sea. It is not likely that it will ever be much frequented by white men—it will remain to us as it has been to the Russians, an immense area of desolate sameness, almost unknown to us, or to its savage occupants, for that matter. The general contour of the great Alaskan mainland interior is that of a vast undulating plain with high rounded granitic hills and ridges scattered in all lines of projection ; on the flanks of which, and by its countless lakes and water-courses, a growth, more or less abundant, of spruce, birch, willows, poplars, and a large number of hardy shrubs, will be encountered. Its summers are short, warm, and pleasant ; its winters are long, and bitterly cold and inclement.

The tundra, however, which fronts the whole of that extensive coast-line of Bering Sea and the Arctic Ocean, is indeed cheerless and repellant at any season. In the summer it is a great flat swale, full of bog-holes, shiny and decaying peat, innumerable sloughs, shallow and stagnant, and from which swarms of malignant mosquitoes rise to fairly torture and destroy a traveller unless he be clad in a coat of mail. In the winter and early spring fierce gales of wind at zero-temperature sweep over these steppes of Alaska in constant succession, making travel exceedingly dangerous, and as painful even as it is in the warmer months. During this period of the year all approach to the coast is barred in Bering Sea by a system of shoals and banks which extend so far seaward that a vessel drawing only ten feet of water will be hard aground, beyond the sight of land, sixty miles off the Yukon mouth.

At the head of the Bay of Bristol a small but deep and rapid river empties a flood of pure, clear water into an intricate series of sand and mud channels which belong there. The Kvichak is the name of this stream, and it rises less than forty miles away in the largest fresh-water lake known to Alaska—that inland sea of Ily-amna, over ninety miles in its greatest length, varying in width from fifteen to thirty. Those gusts and gales that sweep over its blue waters raise a heavy surf which beats sonorously upon its pebbly shores and under its cliffs, while the loud wailing cry of a great northern loon * echoes from one lonely shore to the other when disturbed by the unwonted passage of a native's canoe. Against the eastern horizon there springs from its bosom an abrupt and mighty wall of Alpine peaks, which stand as an eternal barrier between its pure sweet waters and the salt surges of the Pacific.

The ruins of an old Russian trading-post stand in the midst of a small native village at the outlet of, and on the slope of, a lovely grassy upland which rises from the lake. Its people are all living in log houses like those we noticed in Cook's Inlet ; but nevertheless they are true Innuits. The two other small hamlets on these Ily-amna shores are all that exist. Their inhabitants live in the greatest peace and solitary comfort that savages can understand. Two trails over the divide are travelled by these natives, who trade with the Cook's Inlet people, and who range over the mountain sides in pursuit of reindeer and of bears. A most noteworthy family of Russian Creoles lived here on the first portage. The father was a man of gigantic stature, and he reared four Anak-like sons, who are, as he was, mighty hunters, and of great physical power. This family lives all to itself in that beautiful wilderness of Ilyamna, a little way back from the lake on a hillside, where they command passes over to Cook's Inlet. They control the trade of this entire region and rule without a shadow of disputation.

A tragedy occurred in one of these small villages of Ilyamna, which has been fitly memorized by the Russian Church. In 1796 a priest of the Greek faith came over from Kadiak, and, enchanted by the scenery and pleased by a warm, kindly welcome received from the natives, he determined to tarry here with them and save their souls. He † was a man of the most handsome presence and the sweetest address, and for a moment prevailed. Then, as the

---

* *Colymbus arcticus.*

† The Archimandrite Jeromonakh Juvenal. The second of the priestly Russian service was Arch. Joassaf. He was drowned at sea in 1797. He was succeeded by Arch. Afanassy, who remained Bishop of Alaska until 1825, and he has been followed by many successors since.

heathenish rites and festivals were postponed at his bidding, surly shamans fomented seeds of hate and fear. Finally an hour
arrived when, at a preconcerted signal, the slumbering wrath of
the savages was aroused, and they fell upon and slew this unsuspecting missionary and destroyed every vestige of his existence
among them. The cause of Father Juvenal's death was his strong
opposition to polygamy. It is said that when he was attacked by
savages he neither fled nor did he defend himself, either of which
he might have successfully done ; but he delivered himself unre-
sistingly into the hands of his murderers, asking only for the safety
of his subordinates, which was granted. The natives say, in their
recitation of the event, that after the monk had been struck down
and left by the mob as dead, he "rose up once more, walked towards
them, and spoke." They fell upon him again, and again, and again,
for he repeated this miracle several times, until at last, in bewildered fury, they literally cut him into pieces.

Reindeer cross and recross the Kvichak River in large herds
during the month of September, as they range over to and from
the Peninsula of Alaska, feeding, and also to escape from mosquitoes. At the mouth of this stream is one of the broadest deer-roads
in the country. The natives run along the banks of the river
when reindeer are swimming across, easily and rapidly spearing
those unfortunate animals as they rise from the water, securing
in this way any number that fancy or want may dictate. At one
time a trader counted seven hundred deer-carcasses as they lay
here on the sands of the river's margin, untouched save by a removal of the hides ; not a pound of that meat out of the thousands
putrefying had been saved by these lazy Innuits ; who, improvident
wretches as they are, would be living, less than five months later, in
a state of starvation ! But all this misery of famine in March will
have been forgotten again next September, when the same surplus
of food is within their reach, for they will not store up against the
morrow—the labor is too great—the shiftless sentiment of a savage
forbids that exertion.

There is a curious distinction drawn by nature between the
Siberian and Alaskan reindeer. Everybody is familiar with the
fact that on the Asiatic side these animals are domesticated and
serve as a mainstay and support of large tribes, both savage and
civilized. But the spirit of the Alaskan deer is such that it will not
live under the control of man, or even within his presence. If con-

fined, it refuses food, and then perishes of self-imposed starvation.
The most patient and extended trials have been made at Nooshagak
by imported Kamschadales, who were raised to the life of deer-
driving over there ; yet, in no instance whatsoever were these
experts able to overcome the difficulty and accustom those timid
animals to the sight, sound, and smell of man.   The Alaskan species is much larger than its Asiatic cousin, but otherwise resembles
it closely, being, if anything, more uniformly gray in tint and less
spotted with white over the back and head.

Reindeer have a most extended range in Alaska, where an immense area of tundra and upland moors yield an abundance of
those mosses and lichens which they most affect.   Innumerable
sloughs and lakes afford these deer a harbor of refuge from cruel tor-
ments of mosquitoes, when the wind does not blow briskly in summer ; the wooded interior gives them shelter from the driving fury
of wintry snow-storms.   Big brown bears follow in the wake of
travelling herds, and feed fat upon all sickly or weaker members
and imprudent fawns of the drove ; so do wolves and wolverines ;
and the lop-eared lynx is not missing.

Nooshagak is a trading centre for that entire Bristol Bay district, which comprises the coast of Bering Sea from Cape Newenham, in the north, to the peninsular extremity at Oonimak, in the
south—an immense expanse in which some four thousand Innuits
abide, and live largely upon fish and deer-meat.   The Oogashik,
Igageek, Nakn_cek, Kvichak, Nooshagak, Igoosheek, and Togiak
Rivers all empty into this great shallow gulf.   Up their swollen
channels, after an opening of the ice during the last half of May,
salmon run from the sea in irregular but constant travel until the
end of August.   Inferior salmon run even as late as November,
while the various kinds of salmon-trout and white-fish exist under
the ice of deep streams and lakes all winter.   By the middle of
September hard frosts in the mountains congeal all sources of innumerable rivulets which have helped to swell the volume and
raise the level of a river's summer flood, and then these streams
which we have just named begin to fall rapidly in their channels.
If we chance to travel anywhere along their banks at this time, we
will find them covered with windrows and heaps of dead salmon
two and three feet in height.   The gravelly beaches of the lakes,
the bars and shoals of every stream, are then lined with decaying
and putrid bodies of these fish, while every overhanging bough

and projecting rock is festooned with their rotting forms—ah! the stench arising absolutely forbids the pangs of hunger, even though we have no provision. These are the salmon that have died from exhaustion and from bruises received in struggling with swift and impetuous currents, and the rocks and snags that beset their paths of annual reproduction.

North of the Togiak River are several small, rocky islets which, having a nucleus of solid granite, are the cause of a large series of sand and mud reefs. Upon those shoals the huge walrus of Bering Sea is wont to crawl and lazily sun himself in herds of thousands. He is practically secure here from attack, since the varying shifts of the tide and its furious rush in ebb and flood make a trip to the islets one of positive danger, even to a most hardy and well-acquainted hunter. Stragglers, however, are frequently surprised on the mainland shore opposite, and the southern coast of Hagenmeister Island toward Cape Newenham to the westward.

The muskrat catch of Alaska is secured almost wholly in the Nooshagak region—an immense number of these water-rodents are annually taken by Innuits here. Traders, however, do not prize them very highly, but to secure the natives' custom they are obliged to appear satisfied with all that these people bring in to the post. These skins are, however, not sold in this country; they are all shipped to France and Germany, where they meet with a ready sale, since the poor people there are not above wearing them. Also, most of the good Alaskan beaver peltries are from this district, where they have the best fur and are consequently prized above all other catches outside of that region. Land-otter is also in large quantity and fine quality, but the mink and martens and foxes are inferior. During summer seasons, on many lakes, flocks of big, white, trumpeting swans will be found frequenting nearly every one of those bodies of water. The natives hunt them at night, and capture unsuspecting birds as they sleep upon the water, by paddling noiselessly upon them. The traders encourage this industry for the sake of the swan's down which it produces. The most favored spot by swans is Lake Walker, which lies on the Nakneek portage over to Cook's Inlet. Perhaps its rare, unique beauty charms these giant natatores as it does ourselves, for, without question, it is incomparably the most lovely sheet of water, set in a frame of glorious mountains, which the fancy of an artist could possibly

400 OUR ARCTIC PROVINCE.

devise. It is an exceedingly fascinating spot, and language is utterly inadequate to portray its vistas, which alternate from absolute
grandeur to that of quiet loveliness, as you sail around its pebbly
shores and yellow sands.

The immediate banks of the Nakneek River, through which
Lake Walker empties its surplus water into Bristol Bay, are low and
flat, and covered with a luxuriant growth of bushes, grasses, and
amphibious plants, semi-tropical in their verdant vigor of life.
The timber on hill-slopes that rise from the plain is principally
clumps of birch and poplar, quickly passing to solid masses of
spruce as a higher ascent is made to the rolling uplands and
mountain sides. An old, deserted settlement—ruins of Paugwik,
marked by the decayed outlines of its cemetery, still is visible at
the debouchure of the Nakneck. With a strange disrespect for the
departed, those natives who live at an adjoining village come over
here to excavate salmon-holes in that ancient graveyard, wherein they
place their fish-heads, so that a process of moist rotting shall take
place prior to eating them! The Innuits of Kenigayat have no fear
of the " witching hour of night " in this burial site of their ancestors.

The seal and walrus hunters of the Nooshagak district are those
hardy Innuits who live at Kulluk and Ooallikh Bays, in plain sight
of these walrus islets and shoals which we were taking notice of a
short time ago. The large mahklok and a smaller, but quaintly
marked "saddle-backed" seal are taken by these people in large
numbers every year. The oil is their great stock-in-trade, for those
fur-bearing animals that belong to the land here are away below par
when brought to a trader. The coast between their villages and
the mouth of the Togiak River is one of a most remarkable series
of bluffy headlands, seven of them, being all of sandstone which
has weathered into queer, fantastic pinnacles and towers, and is
washed at the sea-level into hundreds of huge caverns wherein the
surf beats with a noise like the distant roar of artillery. Screaming flocks of water-fowl are breeding on their mural faces, and
troops of foxes lurk in the interstices, and roam incessantly for eggs
and unwary birds.

The Togiak River never was ascended by a white man until the
summer of 1880.* It is a very remarkable region with respect to

---

* Visited then by Ivan Petroff, who made an extended trip for the United
States Census.

Female

THE SADDLE-BACK, OR HARLEQUIN HAIR SEAL

[Histriophoca equestris]

Male

its people. Though the course of the river is only one hundred miles in length, yet we find upon it seven villages (one of them very large), having an aggregate population of 1,826 souls. No other one section of Alaska has so dense a population with reference to its inhabited area. The river is, however, a broad one, being a mile and a half in width, shoal and shallow, with deep pools and eddies here and there. Its banks are low, and the valley through which it runs is low and flat, with extensive bottom-lands that widen out at places to a distance of fifteen miles between the ridges and hills which direct its short course. Upon these flats grow most luxuriant and lofty grasses, high as the heads of natives— literally concealing, as it were, the dense human occupation of its extent.

The Togiaks are the Quakers of Alaska ; they are the simplest and the most unpretending of all her people ; they seem to live entirely to themselves, wholly indifferent as to what other folks have and they have not. They seldom ever view a white man, and then it is only when they go down to the river's mouth and visit a trader in his sloop or schooner. He never goes up to see them, for the best of reasons to him—they never have anything fit for barter save a few inferior mink and ground-squirrel skins to trade. .They have no chiefs ; each family is a law unto itself, and it comes and goes with a sort of free and easy abandon that must resemble the life and habit of primeval time. What little these people want and can-not get from each other, they do not go farther in search of, but do without, unless it be small supplies of tobacco which they pro-cure through other Innuits, second or third hand.

Entire families of them, during the summer, leave their winter huts and go out into the valley at such points as their fancy may indicate, where they pass two and three months with absolutely no shelter whatever erected during that entire lapse of time. When it rains hard they simply turn their skin boats bottom side up, stick their heads under, and consider themselves fully settled for protec-tion from tempestuous wind and sleet-storms, or any other climatic unpleasantness. How insensible to extremes of weather do these bodies of the Innuits become—their whole external form is as in-sensible to heat or cold as their stolid features are ! Were they living under Italian skies, they could not affect a greater disregard for the varying moods of that mild climate than they do for the chilly, boisterous weather of Alaska. The Togiakers never go far from the river upon which they build their rude winter villages, and never venture out from its mouth; hence they are not so happy in making the skin canoe or kayak, as their hardier brethren are: these boats on the Togiak are clumsy, broad of beam in proportion to length, and the hatch, or hole, so large that two persons can sit in it back to back. When a family concludes to go out for the summer camp, the man gets into his "kayak," takes the children who are under four or five years in with him, then pulls and paddles his way up against the current, or floats down, as the case may be; the women—wife, mother, and daughters—are turned ashore and obliged to find their way up or down through long grass and over quaking bogs—to toil in this manner from camp to camp, and as they plod along they shout and sing at the top of their voices to apprise any bear or bears, which may be in their path, of such coming, and thus stampede them; otherwise they would be in continual danger of silently stepping upon bruin as he lurked or slept in dense grassy jungles. When a bear first takes notice of the approach of a human being it invariably slinks away, rarely ever displaying, by the faintest sound, its departure; but that same animal, if surprised suddenly at close quarters, will turn and fight desperately, even unto death.

The bold, far-projected headland of Cape Newenham forms the southern pier of that remarkable funnel-like sea-opening to the Kuskokvim River—a river upon which the human ichthyophagi of the north do most congregate: three thousand savages are living here in a string of scattered hamlets that closely adjoin each other, and are nearly all located on the right-hand bank of the river as we ascend it. They are more like muskrat villages than human habitations—water, water all around and everywhere: situated on little patches, or narrow dikes, at the rim of the high tides, on the edge of the river proper, which is here, and for a long distance up, bordered by a strikingly desolate and forlorn country. A glance at our map will show to the reader that great funnel-fashioned mouth of the Kuskokvim, through which its strong and turbid, clay-white current is discharged into Bering Sea. The tides, in this enormous estuary, run with a rise and fall that simply beggars description—reaching an amazing vertical flow and ebb of fifty feet at the entrance! Such extraordinary change in tide-level is carried up, but much modified as it progresses, until lost at Mumtrekhlagamute; the entire physical aspect of that region, in which this sweeping

THE KUSKOKVIM RIVER AND TUNTUH MOUNTAINS

daily change in a level of the water prevails, is most repellant and discouraging.

From the high-tide bank-rims of the Kuskokvim, as we go up, across to the hills and to their rear in the east, extends a dreary expanse of swale and watered moors forty to sixty miles in width, flat and low as the surface of the sea itself. At high tide it appears to be nearly all submerged. It shimmers then like an inland ocean studded with myriads of small mossy islets. Again, when the tide in turn runs out, great far-expanded flats of mud and ooze supplant the waters everywhere, giving in this abrupt manner a striking shift of scenic effect. The eastern river bank is a queer, natural dike, formed by a rank and vigorous growth of coarse sedges, bul-rushes, and little sapling fringes of alders, willows, with birch and poplars interspersed. Upon this natural dike these native villages range in close continuity, each occupying all the dry land in its own immediate limits, and occupying it so thoroughly that a traveller cannot, without great difficulty, find bare land enough outside of their sites upon which to pitch his tent. Mud, mud everywhere—a whitish-clay silt, through which, at low tide, it is almost a phys-ical impossibility to walk from a stranded bidarka up to the vil-lages. Indeed, if you are unfortunate enough to reach a settlement here when coming down or going up the river as the tide is out, you are a wise man if you simply fold your arms, sit quietly in your cramped position until the rising, roaring flood returns and carries you forward and over to your destination. '. :.

On the Lower Kuskokvim the river width of itself is so great that the people living on its eastern banks never can see an oppo-site shore to the westward, for it is even more submerged there and swampy, if anything, than where they reside ; hence we find them located here on the east bank, to a practical exclusion of all settle-ment over on those occidental swales and bogs. The current of this singular stream flows quite rapidly. It discharges a great volume of water, which is colored a peculiar whitish tone by the contribution of a roiled tributary that heads in the Nooshagak divide. At its source and down to this muddy junction it is clear. It is a rapid stream in the narrows, and dull and sluggish in flow through wide openings.

The density of aboriginal population so remarkably manifested as we observe it on the Lower Kuskokvim does not, however, give all the testimony, inasmuch as during every summer two thousand or more natives from the Yukon delta come over here to fish with the Kuskokvims, making a sum-total of six or seven thousand fish-eaters, who catch, consume, and waste an astonishing quantity of salmon, which would, if properly handled, be sufficient to hand-somely feed the entire number of native inhabitants of Alaska, four times over, every year!

Snow lies deeply upon all this region, driven and packed in vast drifts and fields by the wrath of furious wintry gales, and the hunt-ing of land animals is thus made impossible. Then a native of the Kuskokvim Valley turns his attention to trapping white-fish * just as soon as the ice becomes firmly established, usually early in No-vember. The traps are made of willow and alder wicker-work, and nearly all in the same pattern as those employed for salmon, but of somewhat smaller dimensions, so as to be easier to handle, since they are not required to catch the huge "chowichic." Every morn-ing at dawn on the river the men of its many villages can be seen making their way out to these fish-traps, when it is not bitterly inclement, and even then, sometimes. They carry curiously shaped ice-picks, made or fashioned from walrus-teeth or deer-antlers, be-cause every night's freezing covers the trap anew with a solid cap of ice, which must be broken up and removed ere a savage can get at it, haul it out, and empty its "pot." Think of the physical hard-ihood required of a man who goes out from his hut to visit such a trap when the wind, away below zero, is blowing over an icy plain of the broad river at the rate of sixty miles an hour, whirling snowy spiculæ, like hot shot, into the faintest exposure which he dares to make of his face or eyes! He does not often go when a "poorga" prevails in this boisterous manner. Sometimes he feels as though he must, since a storm may have raged in wild, bitter fury for a week without sign of abatement. His children or his wife may be sick and half-starved; then, only then, does he vent-ure out to dare and endure the greatest hardship of savage life in Alaska.

It frequently happens after an unusually cold night that a trap, including its contents, is frozen solid. This is another dreaded accident, for it involves great labor, since the trap itself must be picked to pieces and built anew. In spite of all these difficulties, the natives get enough fresh fish during each winter by such method

---

* *Coregonus ssp.*

to eke out their scanty store of dried salmon and save themselves from starvation. On the lower river course, within the influence of that tremendous tidal action which has been described, a solid covering of ice never envelops the surface of the Kuskokvim. Here the natives hunt seals, the mahklok, and also the white whale or beluga, which furnishes them a full supply of oil * and blubber. A school of belugas puff and snort, like a fleet of tug-boats, as they push between and under tide-broken masses of ice in hot pursuit of fish that abound all over the broad estuary.

There is one particularly distressing and hideous feature that belongs to this entire area of the Alaskan coast tundra and marshy moors of the interior and its forests, its river-margins, and, in fact, to every place except those spots where the wind blows hard. It is the curse of mosquitoes—the incessant stinging of swarms of these blood-thirsty insects, which come out from their watery pupæ by May 1st (with the earliest growing of spring vegetation), and remain in perfect clouds until withered and destroyed by severe frosts in September and October. The Indians themselves do not dare to go into the woods at Kolmakovsky during the summer, and the very dogs themselves frequently die from effects of mere mosquito-biting about their eyes and paws only, for that thick woolly hair of these canines effectually shields all other portions of their bodies. Close-haired beasts, like cattle or horses, would perish here in a single fortnight at the longest, if not protected by man.

Universal agreement in Alaska credits the Kuskokvim mosquito

---

* The oil obtained from the beluga and the large seal (mahklok) is a very important article of trade between the lowland people and those of the mountains, the latter depending upon it entirely for lighting their semi-subterranean dwellings during the winter, and to supplement their scanty stores of food. It is manufactured by a very simple process. Huge drift-logs are fashioned into troughs much in the same manner as the Thlinket tribes make their wooden canoes. Into these troughs filled with water the blubber is thrown in lumps of from two to five pounds in weight. Then a large number of smooth cobble-stones are thrown into a fire until they are thoroughly heated, when they are picked up with sticks fashioned for the purpose and deposited in the water, which boils up at once. After a few minutes these stones must be removed and replaced by fresh ones, this laborious process being continued until all oil has been boiled out of the blubber and floats on the surface, when it is removed with flat pieces of bone or roughly fashioned ladles, and decanted into bladders or whole seal-skins, then cached on pole-frames until sold or used by the makers.

as being the worst. They do not appear elsewhere in the same
number or ferocity, but they are quite unendurable at the best
and most-favored stations. Breeding here, as they do, in these
vast extents of tundra sloughs and woodland swamps, they are able
to rally around and embarrass an explorer beyond all reasonable
description. Language is simply inadequate to portray that misery
and annoyance which the Alaskan mosquito-swarms inflict upon
us in the summer, whenever we venture out from the shelter of
trading-posts, where mosquito-bars envelop our couches and cross
the doors and windows to our living-room. Naturally, it will be
asked, What do the natives do? They, too, are annoyed and suffer ;
but it must be remembered that their bodies are daily anointed with
rancid oil, and certain ammoniacal vapors constantly arise from their
garments which even the mosquito, venomous and cruel as it is,
can scarcely withstand the repellant power of. When the natives
travel in this season, they gladly avail themselves, however, of any
small piece of mosquito-netting that they can secure, no matter how
small. Usually they have to wrap cloths and skins about their
heads, and they always wear mittens in midsummer. The traveller
who exposes his bare face at this time of the year on the Kuskokvim
tundra or woodlands will speedily lose his natural appearance ; his
eyelids swell up and close ; his neck expands in fiery pimples, so
that no collar that he ever wore before can now be fastened around
it, while his hands simply become as two carbuncled balls. Bear
and deer are driven into the water by these mosquitoes. They are
a scourge and the greatest curse of Alaska.

Two hundred miles up from the Kuskokvim mouth is a focal
centre of the trade in this district. It is Kolmakovsky, established
by the Russians in 1839. It consists of seven large, roughly built
frame dwellings and log warehouses, and a chapel, which stand on
a flat, timbered mesa well above the river, on its right or southern
shore. Here the current of the stream has narrowed, and flows
between high banks over a gravelly bed. These terraces, which
rise from the water, are flat-topped, and covered with a tall growth
of spruce. Mossy tundras and grassy meadows roll in between
forest patches. The timber is much larger here than it is anywhere else in the great Alaskan interior, and that scenery along
this river is far wilder and more agreeable than any which is so
monotonous and characteristic of the Upper Yukon. The desolate flatness and muddy wastes of the Lower Kuskokvim are now

KOLMAKOVSKY, ON THE KUSKOKVIM

Old Russian trading-post, established in 1832, two hundred miles up the River: these houses were once surrounded with a stockade, but such a defence has long been needless. This view is taken from the opposite bank of the River, looking over to the high hills of the Nooshagak divide, and Mount Tamahloopat in the distance

replaced by this pleasing change, which we have just mentioned, a short distance below Kolmakovsky.

Back of that post, and clearly defined against the horizon, are the snowy-capped summits of those mountains that form a Nooshagak divide. One of them rises in an oval-pointed crest to a very considerable elevation * above all the rest, and is the landmark of every traveller who comes over the Yukon divide to Kolmakov. The river here, as it brawls swiftly in its course, is about seven hundred feet in width, with bends above and below where it expands to fully twice that distance.

While the Kuskokvim is the only considerable rival of the Yukon in this whole Alaskan country, yet when seriously contrasted with the great Kvichpak † itself, then the Kuskokvim bears about the same resemblance to it that the Ohio River does to the Mississippi.

Kolmakovsky marks the limit of inland migration allotted to the Innuit race on its banks, who are not permitted by those Tinneh tribes of the interior to advance farther up the river. It is also removed from that disagreeable influence of Bering Sea, where the prevalence of rain and of furious protracted gales of wind make life a burden to a white man on the Lower Kuskokvim. Its environing forests break the force of these storms, and there is also less fog, so that the sun usually shines out clear and hot, especially in July and August.

In the winter season, when frost has locked up miry swales and swamps, and snow lies in deep, limitless drifts, a white hunter at Kolmakov can join the Kuskokvamoots in trailing and shooting giant moose which come down from the mountains of the Nooshagak divide. This animal is quickly apprehended by the native dogs, so that whenever winter weather will permit, a native Innuit spends most of his time, not employed by ice-fishing on the Kuskokvim, in this sport.

The fur-trade at Kolmakovsky is quite active, but it is almost exclusively transacted with a few Indians up the river, and not with the numerous Innuits below. The latter are, commercially speaking, very poor, having not much of anything but little stores of

---

* Mount Tamahloopat : two thousand eight hundred feet.

† The Russians and natives always called the Yukon River by this name. Our change was first made by those Hudson Bay traders who came over to it from the Mackenzie, and was subsequently universally adopted.

"mahklok" seal-oil. These big phocaceans are almost as great fishermen as the Innuits are themselves, and find the mouth of the Kuskokvim as attractive as it is to their human foes. In this frame of mind the mahklok ventures on to those tidal banks of the estuary below, and this rash habit enables the natives to capture a great many of them there every year. Those Innuits below Kolmakovsky have no land-furs whatever, save a few inferior mink-skins ; but they trade their surplus seal-oil with the Indians above and on the Yukon for that ground-squirrel parka and tanned moose-skin shirt which they universally wear. There is an exceeding rankness to an odor of rancid fish-oil, but the aroma from a bag of putrescent seal-oil is simply abominable and stifling to a Caucasian nose—an acrid funk, which pervades everything, and hangs to it for an indefinite length of time afterward in spite of every effort made to disinfect.

The Indians of the Upper Kuskokvim were once said to be a very numerous tribe ; but the severity of successive cold winters has so destroyed them, as a people, that to-day they exist there as a feeble remnant only of what they once were. An intelligent trader, Sipari, who has traversed their entire country, in 1872–76, declares that "forty tents," or one hundred souls is an ample enumeration of their number.

The Innuits of the Lower Kuskokvim are much better physical specimens of humanity than are those of their race living on the Lower Yukon. These latter are called by all traders the most clumsy and degraded of Alaskan savages. The portage from Kolmakovsky to the Kvichpak is only three days' journey in winter, or five days by water in canoes, during summer. It is a trip made by large numbers of the natives of both streams, in the progress of their natural barter and moose-hunting.

The forests of the Kuskokvim and the Nooshagak mountains and uplands are frequently swept by terrible conflagrations, which utterly destroy whole areas of timber as far as the eye can reach. This ruin of fire, of course, absolutely extinguishes all trapping for any fur-bearing animal hitherto found in those brulé tracts, and entails much privation upon the natives who have been accustomed to gain their best livelihood largely by hunting in those sections. A burnt district presents a desolate front for years after ; the fire does not, in its swift passage, do more, at first, than burn the foliage and smaller limbs of trees in a dense spruce forest ; but it roasts the bark and kills a trunk, so that all sap-circulation is forever at an end in it. As the years roll by, these trunks gradually bleach out to almost a grayish white, the charred, blackened bark is all weathered off, and gradually such trees fall, as they decay at the stump, in every conceivable direction upon the ground, across one another, like so many jack-straws, making a perfectly impassable barricade to human travel without tedious labor. A brisk growth of small poplars, birch, and willow springs up in place of the original spruce forest, but none of these trees and shrubs ever grow to any great size. At rare intervals a young evergreen is seen to rise in sharp relief, towering over all deciduous shrubbery, and in the lapse of long years it will succeed in supplanting every growing thing around with its own kind again.

" Brulé " Desolation ; Alaskan Interior.
[*A view on the Stickeen Divide : bears, in search of larvæ, ripping open decayed logs.*]

The traders at Kolmakovsky make up their furs into snug bales and descend the river in wooden and skin boats, every June, to a point below, about one hundred and fifty miles, where they meet their respective schooners, or go still lower to an anchorage of larger vessels, and renew their annual supplies. These river-boats are then poled and rope-walked up the river back to the post. The principal trade here is beaver, red foxes, mink, marten, land-otter, and brown and black bears.

The traders say it is exceedingly seldom that a white man ever comes in contact with the natives of the Lower Kuskokvim, and that there is nothing to call them there ; also, that the labors of the Russian missionaries of the Yukon never extended to this region, though their registers and reports show quite a number of Christians on the Kuskokvim River. The only trace of Christianity among this tribe, outside of the immediate vicinity of a trading-station with its chapel, consists of a few scattered crosses in burial-places adjoining the settlement. At the village of Kaltkhagamute, within three days' travel of the Russian mission on the Yukon, a graveyard there contains a remarkable collection of grotesquely carved monuments and memorial posts, indicating very clearly the predominance of old pagan traditions over such faint ideas of Christianity as may have been introduced for these people. Among monuments in this place the most remarkable is that of a female figure with four arms and hands, resembling closely a Hindoo goddess, even to its almond eyes and a general cast of features. Natural hair is attached to its head, falling over the shoulders. The legs of this figure are crossed in true oriental style, and two of the hands, the lower pair, hold rusty tin plates, upon which offerings of tobacco and scraps of cotton prints have been deposited. The whole is protected by a small roof set upon posts.

Other burial posts are scarcely less remarkable in variety of feature and coloring, and the whole collection would afford a rich harvest of specimens to any museum. Nearly all these figures are human effigies, though grotesque and misshapen, and drawn out of proportion. No images of animals or birds, which would have indicated the existence of totems and clans in the tribe, were to be seen ; but here and there, over apparently neglected graves, a stick, surmounted by a very rude carving of a fish, a deer, or a beluga, indicative of the calling of the deceased hunter, could be discovered.

Petroff, who has made the only hand-to-hand examination ever conducted, by a white man, of the people of the Lower Kuskokvim, says that they resemble in outward appearance their Eskimo neighbors in the north and west, but their complexion is perhaps a little darker. The men are distinguished from those of other Innuit tribes by having more hair on the faces ; mustaches being quite common, even with youths of from twenty to twenty-five, while in other tribes this hirsute appendage does not make its appearance until the age of thirty-five or forty. Their hands and feet are small, but both sexes are muscular and well developed, inclined rather to embonpoint. In their garments they differ but little from their neighbors hitherto described, with an exception of the male upper garment, or parka, which reaches down to the feet, even dragging a little upon the ground, making it necessary to gird it up for purposes of walking. The female parkas are a little shorter. Both garments are made of the skins of ground-squirrels, orna-

AN INNUIT TOMB

mented with pieces of red cloth and bits of tails of that rodent. The women wear no head-covering except in the depth of winter, when they pull the hoods of reindeer parkas over their heads. The men wear caps, made of the skin of an Arctic marmot, resembling in shape those famous Scotch "bonnets," so commonly worn by Canadians. Many young men wear a small band of fur around the head, into which they insert eagle and hawk-feathers on festive occasions. A former custom of this tribe, of inserting thin strips of bone or the quills of porcupines through an aperture cut in the septum, seems to have become obsolete, though the nasal slit can still be seen on all grown male individuals. Their ears are also universally pierced for an insertion of pendants, but these seem at present to be worn by children only, who discard them as they grow up. In fact, all ornamentation in the shape of beads, shells, etc., appears to be lavished upon their little ones, who toddle about with pendants rattling from ears, nose, and lower lip, and attired in frocks stiff with embroidery of beads or porcupine-quills, while the older girls and boys run almost naked, and the parents themselves are imperfectly protected against cold and weather by a single fur garment.

The use of the true Eskimo kayak is universal among the Kuskokvagmute, but in timbered regions of the upper river, in the vicinity of Kolmakovsky, the birch-bark canoe also is quite common. The latter, however, is not used for extended voyages or for hunting, but is reserved chiefly for attending to fish-traps, for the use of women in their berrying and fishing expeditions, and for crossing rivers and streams.

The only indigenous fruit which this large population of the Lower Kuskokvim can enjoy is that of the pretty little "moroshkie," or red raspberry,* which grows in great abundance on its short, tiny stalks throughout all swales and over rolling tundra. These berries are saturated in rancid oil, however, before they are eaten to any great extent, being air-dried first and pressed into thin cakes; then, as wanted, they are pounded up in mortars and boiled, or simply thrown into a wooden basin (or kantag) of oil. Then the fingers, or rude horn spoons, are dipped in by happy feeders, who apparently relish this ill-savored combination just as keenly as one of our Gothamitic gourmands appreciates the flavor of a Chesapeake terrapin stewed in champagne.

---

* *Rubus chamœmorus.*

CHAPTER XIII.

LONELY NORTHERN WASTES.

> Lo! to the wintry winds the pilot yields,
> His bark careering o'er unfathomed fields,
> Now far he sweeps, where scarce a summer smiles,
> On Behring's rocks, or Greenland's naked isles.
> —CAMPBELL.

Is it not a little singular that the lonely and monotonous course of
the Yukon River, reaching as it does to the very limits of the path-
less interior of a vast, unexplored region on either side, should be
that one section of all others in Alaska the best known to us ? An
almost uninterrupted annual march has been made up and down
its dreary banks since 1865, by men * well qualified to describe its

---

* The first white man to enter the Yukon and behold its immense volume
was Glazoonov, a Russian post-trader of the old Company, who, with a small

varying moods and endless shoals—every turn in its flood, every
shelving bank of alluvium or rocky bluff that lines the margin of
its turbid current, has been minutely examined, named and renamed
to suit the occasion and character of a traveller.

The Yukon River is not reached by traders as any other stream
of size is in Alaska, by sailing into its mouth. No ocean-going
craft can get within sixty miles of its deltoid entrance. Were a
sailor foolhardy enough to attempt such a thing, he would be hard
aground, in soft silt or mud, a hundred miles from land in a
direct line from the point of his destination. Therefore it is the
habit of mariners to sail up as far north as Norton's Sound, and
then turn a little to the southward and anchor their schooners or
steamers under a lee of Stuart's and St. Michael's Islands, where
the old post of Michaelovsky is established on the latter.

The "Redoute Saint Michael" was founded here in 1835 by
Lieutenant Tebenkov, and has been ever since, and is to-day, the
most important post in the Alaskan North. This post is a shipping point for the accumulated furs gathered by all traders from
the Lower and Upper Yukon, and the Tannanah, the annual yield
from such points being the largest and the most valuable catch of
land-furs taken in Alaska. A vessel coming into St. Michael's at any
time during the summer will find, encamped around its warehouses many bands of Innuits and Indians who have come in there,
over long distances of hundreds of miles, from the north, east, and
south. They are there as traders and middlemen. The fur-trading on the Yukon is very irregular as to its annual time and place —
the traders constantly moving from settlement to settlement, because this year they may get only a thousand skins where they got

band of promishlyniks, managed to overcome the hostility of the natives sufficiently to get up as far as the present site of Nulato. This was in 1833.
Lieutenant Zagoskin, of the Russian Navy, made a thorough engineering examination of the river up as far as the "Ramparts," between the years 1842-
45, inclusive, locating its positions and courses by astronomical and magnetic
observations. After him, named in regular order of their priority in visiting
the river, came the following Americans, the first in 1865, the last in 1885 :—
Kennicott, Pease, Adams, Ketchum, Dall, Whymper, Mercier, Raymond, Hill
and Shaw (two miners, from its very source), Nelson, Petroff, then Schwatka
and Everett (also from its source). All of these men have given to the world
more or less elaborate accounts of the Yukon through the medium of published works, letters, and lectures. The literature on the single subject of the
Kvichpak is decidedly voluminous.

five thousand last season, and *vice versa*. It is impossible to locate
the best single spots for trade ; the catch in different sections will
vary every winter according to the depth of snow, the severity of
climate, the prevalence of forest fires, or starvation of whole villages, owing to unwonted absence of fish, and so on.

In midsummer the Yukon is reached by small, light-draft, stern-
wheel steamers, which, watching their opportunity, run down from
St. Michael's and enter its mouth, towing behind them a string of
five or six large wooden boats which are each laden with several
tons of merchandise. The scream of their whistles and puffing of
these little trading-steamers as they slowly drag such tows against

Trader's Steamer towing Bateaux laden with Goods up the Yukon.
[*The Kvichpak just below Mercier's Station.*]

a rapid current, is the only enlivenment which the immense lonely
solitudes of the Yukon are subjected to by our people. That area
of watery waste is so wide and long, and the boats are so small and
few in number, that even this innovation must be watched for every
year with a hawk's eye, or it will pass unobserved.

The waters of the Kvichpak are discharged into Bering Sea
through a labyrinth of blind, misleading channels, sloughs, and
swamps, which extend for more than one hundred miles up until
they unite near Chatinak with the main channel of that great river.
This enormous deltoid mouth of the Yukon is a most mournful and
depressing prospect. The country itself is scarcely above the level
of tides, and covered with a monotonous cloak of scrubby willows
and rank sedges. It is water, water—here, there, and everywhere

—a vast inland sea filled with thousands upon thousands of swale islets scarcely peeping above its surface. Broader and narrower spaces between low delta lands are where the whirl of its current is strongly marked by a rippling rush and the drift-logs that it carries upon its muddy bosom. These are the channels, the paths through the maze that leads from the sea up to the river proper; and where they unite, at a point above Andrievsky and Chatinak, the Yukon has a breadth of twenty miles; and again, at many places, away on and up this impressive stream as far as seven or eight hundred miles beyond, this same great width will be observed, but the depth is very much decreased.

Myriads of breeding geese, ducks, and wading water-fowl resort to this desolation of the deltoid mouth of the Yukon, where, in countless pools and the thick covers of tall grass and sedge, they are provided with a most lavish abundance of food and afforded the happiest shelter from enemies; but the stolid Innuit does not affect the place. The howling wintry gales and frightful curse of mosquitoes in the summer are too much even for him. His people live in only six or seven small wretched hamlets below Andrievsky and Chatinak—less than five hundred souls in all, including the entire population found right on the coast of the delta, between Pastolik in the north and Cape Romiantzov on the south. Above Anvik on the main river the Innuit does not like to go. He has no love for those Indians who claim that region all to themselves and resent his appearance on the scene. Whenever he does, however, he is always in company with the traders, and he never gets out of their sight and protection, even when making that overland portage from St. Michael's by the Oonalakleet trail.

As we emerge from those dreary, low and watery wastes of the delta at Chatinak, the bluffs there, though desolate enough themselves, with their rusty barren slopes, yet they give us cheerful assurance of the fact that all Alaska is not under water, and that the borders of its big river are at last defined on both sides. High rolling hills come down boldly on the left bank as we ascend; but the right shore is still low and but little removed from the flatness of a swale. The channel of the river now zigzags from side to side (in the usual way of running bodies of water which wash out and undermine), building up bars and islets, and sweeping in its resistless flood an immense aggregate of soil and timber far into Bering Sea. The alluvial banks, wherever they are lifted above this surging current, which runs at an average rate of eight miles an hour, are continually caving down, undermined, and washed away. So sudden and precipitate are these landslides, sometimes, that they have almost destroyed whole trading expeditions of the Russians and natives, who barely had time to escape with their lives as the earthy avalanches rolled down upon the river's edge and into its resistless current.

Above the delta large spruce and fir-trees, aspens, poplars, and plats of alders and willows grow abundantly on the banks; but they do not extend far back from the river on either side into any portions of the country, which is low and marshy, and which embraces so large a proportion of the entire landscape. Small larch-trees are also interspersed. The river is filled with a multitude of long, narrow islands, all timbered as the banks are, and which are connected one with the other by sand and gravel bars, that are always dry and fully exposed at low-water stages. Immense piles of bleached and splintered drift-logs are raised on the upper ends of these islands, having lodged there at intervals when high water was booming down.

Between Anvik and Paimoot are many lofty clay cliffs, entirely made up of clean, pure, earth of different bright colors—red, yellow, straw-colored, and white, with many intermediate shades. The Yukon runs down from its remote sources at the Stickeen divide in British Columbia, down through a wild, semi-wooded country, a succession of lakes and lakelets, through a region almost devoid of human life. That extensive area, wherein we find such scant or utter absence of population, is, south of the Yukon, very densely timbered with spruce-trees on the mountains, and with poplars, birch, willow, along the courses of the stream and margins of the lakes. Its immediate recesses only are occasionally penetrated by roving parties of Indian hunters, who now and then leave the great river and the Tannanah for that purpose. It is a silent, gloomy wilderness. To the northward of the Yukon this variety in timber still continues; indeed, it reaches as far into the Arctic Circle and toward the ocean there as the seaward slopes of those low and rolling mountains extend, which rise in irregular ridges trending northeast and southwest. These hills are between one hundred and one hundred and fifty miles from the banks of the Kvichpak. Beyond this divide and water-shed of the northern tributaries of the Yukon a forest seldom appears in any case whatsoever, except where a low, straggling spur of hills stretches itself down to the shores of an icy sea ; but it is stunted and scant in its hyperborean distribution thereon.

It is not necessary to enter into a description of the appearance and disposition of these Yukon Indians who live on this great river above Anvik, since they resemble those savages which we are so familiar with in the British American interior, Oregon, and Dakota. The Russians, in regarding them, at once took notice of their marked difference from the more stolid Innuits, so that they were styled, jocularly, by Slavonian pioneers, "Frenchmen of the North," and "Gens de Butte." The Innuits called them "Ingaleeks," and that is their general designation on the river to-day. They differ from our Plain Indians in this respect only : they are all dog-drivers. They rely upon the river and its tributaries largely for food, using birch-bark canoes—no skin-boats whatever. They have an overflowing abundance of natural food-supply of flesh, and fowl also, and when they suffer, as they often do, from starvation, it is due entirely to their own startling improvidence during seasons of plenty, which occur every year. A decided infusion of Innuit blood will be observed in the faces of the Indians who live at Anvik, and some distance up the river from that point of landed demarcation between Innuit and Ingaleek. In olden times the latter were wont to raid upon the settlements of the former, and carried off Innuit women into captivity whenever they could do so, treating the Eskimo just as the Romans raped the Sabines.

An Innuit is not thrifty at all, but when brought into comparison with the Indian he is a bright and shining light in this respect. Among the Ingaleeks of the Yukon a spring famine regularly prevails every year during the months of April and May, or until the ice breaks up and the salmon run. One would naturally think that the bitter memories of gnawing hunger endured for weeks before an arrival of abundant food, would stimulate that savage to glad exertion when it did arrive so as to lay by of such abundance enough to insure him and his family against recurring starvation next year. Strange to say, it does not. The fish come ; the famished natives gorge themselves, and thus engorged, loaf and idle that time away which should be employed in drying and preserving at least sufficient to keep them in stock when the fish have left the stream. Often we will actually see them lazily going to their slender store which they have newly prepared, and eat thereof, while salmon are still running in the river at their feet ! Such im-

27

providence and reckless disregard of the need of the morrow is hard indeed for us to realize. Many of the beasts of the field and forest with which the savage is well acquainted set him annually, but in vain, a better example.

White traders during the last twenty years have so thoroughly traversed the course of the Yukon, and, since our control of Alaska, little stern-wheel steamers annually make trips from the sea, accompanied with retinues of white men—these incidents have thoroughly familiarized the Indians here with ourselves. But the wilder Ingaleeks of the Tannanah, only six or seven hundred souls in number, however, are as yet comparatively unknown to us. With an exception of a white trader's visit to their country in 1875,[*] and the recent descent of the Tannanah by a plucky young officer of the United States Army,[†] these Koltchanes have been unknown at home and wholly undisturbed by us. There are less than sixteen hundred Indians living over the entire Yukon region—a fact which speaks eloquently for an exceeding scantiness of the population of that vast landed expanse of this interior of the Alaskan mainland—a great arctic moor north of the Kvichpak, which is a mere surface of slightly thawed swale, swampy tundra, lakes and pools, sloughs and sluggish rivers, in the summer solstice, while the wildest storms of frigid winds, laden with snow and sleet, career in unchecked fury over them during winter. Such an extreme climate is the full secret of its marked paucity of human life. But that desolation of winter does not prevent an immense migration of animal life to this repellant section every summer from the south. Myriads of water-fowl, such as geese, ducks, and the smaller forms, breed and moult here then in all security, and free from molestation, while great herds of reindeer troop over the lichen-bearing ridges. The musk-ox, however, has never been known to range here or anywhere in Alaska within the memory of man. Its fossil remains have been disinterred from the banks of the Yukon, at several places (just as those of the mammoth have), but that, with a few bleached skulls, is the only record of this animal we can find which we would most naturally anticipate meeting with on such ground, apparently so well adapted for it.

---

[*] François Mercier, according to whom " Ingaleek " signifies " incomprehensible."

[†] Lieutenant H. T. Allen, Engineer Corps, United States Army.

There is some difference of opinion as to whether the Yukon or the Mississippi is the larger river, with respect to the volume of their currents. The variation in this regard can hardly be very great, either one way or the other. The Tannanah is the Missouri of the Kvichpak, and swells the flood of that river very perceptibly below its junction.

Michaelovsky has been, and will continue to be, the chief rendezvous of a small white residency of the Alaskan North. It is an irregularly built *omnium* of old Russian dwellings, warehouses, and a few of our own structure. The stockade which once encircled it has long ago been dispensed with, though the antique bas-

Michaeiovsky.
[*Extreme northern settlement of white Americans.*]

tions and old brass cannon still stand at one or two corners as they stood in early times, well placed to overawe and intimidate a bold and hostile savage people then surrounding them. The buildings are clustered together on a small peninsula of an island, about twenty-five or thirty feet above high-water mark; littered all around them are the small outbuildings and the summer tents of Innuit and Indian tourists who are loitering about for the double purpose of gratifying a little curiosity, and of trading. An abundance of drift-wood from the Yukon lies stranded on the beaches, and a large pile of picked, straight logs have been hauled from the water and stacked upon one side of a slope. The whole country, hill and plain, in every direction from this post is a flat and alternately rolling moor-land, or tundra, the covering of which is composed principally of

mosses and lichens, and a sphagnous combination which produces
in the short growing season a yellowish-green carpet, with patches
of pale lavender gray where the lichens are most abundant. At
sparse and irregular intervals bunches of coarse sedge grasses rise,
and the entire surface of moor is crossed at various angles with
lines of dwarf birches and an occasional clump of alders and
stunted willows. The most attractive feature in such an arctic landscape, when summer has draped it as we now behold it, is the nodding seed-plumes of the equisetum grasses—they are tufts of a pure,
fleecy white that, ruffled in the breeze, light up the sombre russet swales with an almost electrical beauty. Everywhere here, in
less than eighteen inches or two feet beneath this blossoming flora,
will be found a solid foundation of perpetual frost and ice—it never
thaws lower. The flowers of that tundra embrace a list of over
forty beautiful species, chief among them being phloxes, a pale-
blue iris, white and yellow poppies, several varieties of the red-
flowered saxifrages, the broad-leaved archangelica, and many deli-
cately fronded ferns.

Twittering, darting flocks of barn-swallows hover and glide over
the old faded roofs and walls of Michaelovsky, and the bells of a
red-painted church, just beyond, come jangling sweetly across the
water, mingled with that homelike chattering of these swallows.
But a pious mission here is a practical failure in so far as any effect
upon the Innuit mind is concerned. During summer-time, in the
Upper Yukon country, thunder-showers are very common; down
here, on the coast, they are never experienced. The glory, however, of an auroral display is divided equally between them, when
from September until March luminous waves and radii of pulsating
rose, purple, green, and blue flames light up and dance about the
heavens—gorgeous arches of yellow bands and pencil-points of
crimson fire are hung and glitter in the zenith. These exhibitions
beggar description; they are weirdly and surpassingly beautiful,
far beyond all comparison with anything else of a spectacular nature
on earth.

In the autumn and in the early days of December, a low declination of the sun tints up the clouds at sunrise and sunset into
beautiful masses of colors that rapidly come and go in their orig-
ination and fading. Twilight is a lovely interval of the day in this
latitude, and is even enjoyed by the hard-headed traders themselves.
Winter is a weary drag here—about seven months—lasting from

October until well into May; but, in spite of its intense cold, there are many long periods of its endurance characterized by clear, lovely weather, while the warmer summer is rendered disagreeable by a large number of cold misty days, rain, and gloomy palls of overhanging clouds which shut down upon everything like a leaden cover.

We are accustomed to associate an occurrence of a real mirage with dry, arid, desert countries, where the thirsty and sun-burned traveller is mocked by illusions of clear lakes and a green oasis just ahead. In truth, the mirage of an Alaskan tundra in midwinter is fully as remarkable, and quite as tantalizing. When the trader starts out with his dog-team, on an intensely still, cold day, the vibrations of the air are so energetic that those blades of grass which stick out from the snow, just ahead, seem to him like thickets of willow- and birch-trees, around which he must make a painful detour. Then, again, the ravines and valleys are transformed into vast lakes, with the loftiest and most precipitous shores. On the coast here, during cool, clear days in March, hills, which are thirty or seventy-five miles away from the windows of Michaelovsky, are lifted up and transported to the very beach of the island itself, contorted and fantastic changes constantly taking place in the picture, until suddenly a slight something, or a change perhaps in an observer's position, causes the singular delusion to vanish.

St. Michael's is all by itself to-day; yet it, at one time, was not the only settlement on the island; for, close by the fort, there were two Mahlemoöt villages, Tahcik and Agahliak, whose inhabitants were first to cordially invite the Russians to locate here in 1835. But in 1842 the ravages of small-pox absolutely depopulated these native towns, and a few survivors fled in dismay from the place— they never came back, nor have their descendants returned. For some reason or other the Russians made the most persistent and energetic attempt to develop a successful vegetable garden in this region and to keep cattle. But, beyond a small exhibit of eatable cabbages, good radishes and turnips, and a few inferior potatoes, grown in the warm sand-dunes of Oonalakleet, nothing more, substantially, ever resulted from it.

Generally the snow falls, at Michaelovsky, as the beginning of its hyemal season, about October 1st, and by October 20th ice has formed, and has firmly locked up the Yukon by November 1st to 5th. These icy fetters break away by June 5th, and in a week or

ten days the great river is entirely clear. The sea is usually covered by sludgy floes as early as the middle or end of every October, which remain opening and closing irregularly until next June.
The months of July and August are the warmest, ranging from
48°·to 54° Fahr. during daytime.*

From St. Michael's to the westward a low basaltic chain of hills
borders the coast, and, parallel to it some thirty miles inland, a
few peaks attain an elevation of one thousand to fifteen hundred
feet. Jutting out at a sharp angle from this volcanic range stands
that low peninsula, tipped with the granitic headland named (by
Cook more than a century ago) Cape Denbigh. This point forms
the southern wall for that snug, tightly enclosed Bay of Norton,
thus partitioned off from a sound of the same title. The Oonalakleet River empties into Norton's Sound, at a point about midway between Michaelovsky and Denbigh. The debouchure of this
stream is marked by the richest vegetation to be found anywhere
in all of this entire region north of Bristol Bay. It is due to the
warm sand-dune flats which are located here ; and here is one of the
liveliest Mahlemoöt villages of that north. That river is an exclusive gateway to the Yukon during the winter season, from and to
Michaelovsky, and these Innuits are the chief commission merchants of Alaska. In a village, now called Kegohtowik, near by,
Zagoskin received his first initiation into the wild life which he led
up here as an explorer, since it was the first camp † he ever made
among the Innuits after he had started out from Michaelovsky.
This young Russian was kindly received by the wondering natives,
who unharnessed his dogs and hung up his sleds on the cache scaffolds as a token of their hospitality. Into their kashga he was
taken with every demonstration of regard and curiosity. He happened to have arrived just as these people were preparing for and
celebrating a great festival of homage to an Eskimo sea-god who
rules the icy waters of Bering and the Arctic Ocean. He quaintly
records their proceeding in this language :

---

* An average temperature prevails in this region for the year as follows :

January,	—.5°	April, 22.1°	July,	53.1°	October,	28.0°
February,	—.6°	May, 32.8°	August,	52.1°	November,	18.3°
March,	9.5°	June, 45.2°	September, 43.3°		December,	8.9°

† December 5, 1842. The refreshing honesty and frankness of this explorer's thorough work on the Yukon and Kuskokvim deserve to be better
known.

"I had an opportunity of observing the natives preparing for a great festival called by them 'drowning little bladders in the sea.' In the front part of the kashga, on a strip of moose or other skin, there were suspended about a hundred bladders taken from animals killed by arrows only. On these bladders are painted various fantastic figures. At one end of the trap hangs an owl with a man's head and a gull carved from wood; at the other end are two partridges. By means of threads running to the crop-beam these images are made to move in imitation of life. Below the bladders is placed a stick six feet in height, bound about with straw. After dancing in front of the bladders a native takes from the stick a small wisp of straw, and lighting it, passes it under the bladders and birds so that the smoke rises around them. He then takes the stick and straw outside. This custom of 'drowning little bladders in the sea' is in honor of the sea-spirit called 'Ug-iak;' but I cannot discover," says Lieutenant Zagoskin, "how the custom originated, or why they use bladders from animals killed by arrows in preference to those killed by other means. To all questions upon the subject the natives answered : 'It is a custom which we took from our fathers and our grandfathers.' It seems to be of great antiquity, as the natives can give no information as to its origin or the reasons for its adoption. * Before these bladders they dance all day in their holiday dress, which consists of light parka, warm boots, and short under-dress for the men ; and parkas, reindeer-trousers, colored in Innuit style, for women, and ornamented with glass beads and rings."

And again, in this connection, the pleasures of a dog-sled journey overland to the Yukon are graphically narrated by the same traveller, who resumed his trip, after spending the night as above related, on snow-shoes and dog-sleds laden with his provisions and instruments. On the morning of December 9, 1842, he struck the Oonalakleet River and started up its frozen channel. He says :

"The weather was at first favorable, but it soon changed, and a driving snow-storm set in, blinding our eyes so that we could not distinguish the path. A blade of grass seventy feet distant had the appearance of a shrub, and sloping valleys looked like lakes with high banks, the illusion vanishing upon nearer approach.

---

* It is the same reply that is honestly given to any query made as to the reason of almost every one of these Innuit mummeries. Too many attempts have been made to attach serious meaning to such idle ceremonies.

On December 9th, at midnight, a terrible snow-storm began, and in the short space of ten minutes covered men, dogs, and sledges, forming a perfect hill above them. We sat at the foot of a hill, with the wind from the opposite side, and our feet drawn under us to prevent them from freezing, and covered with our parkas. When we were covered by the snow, we made holes with sticks through to the open air. In a short time the warmth of the breath and perspiration melted the snow so that a man-like cave was formed about each individual. In these circumstances our travellers passed five hours, calling to one another at intervals to keep awake, it being certain death to sleep in that intense cold. If we had been on the other side of the hill, exposed to the full fury of the wind, we would have been buried in the snow and suffocated."

Such are the experiences of all travelling traders on the Yukon, who encounter these wintry "poorgas" in the pursuit of their calling every year of their lives spent in that great Alaskan moorland. Familiarity with this subject never breeds contempt for it in the minds of those hardy men—that pain and privation to which these characteristic storms subject all human beings who are caught and chained on a tundra, or in the mountains, by their wild rushing and bitterly cold breath, is never forgotten.

On the shores of Norton's Sound are many low clayey bluffs, which, as they are annually undermined by the surf and chiselled by frost, fall in heavy crumbled masses upon the beach. This exposes their long-concealed deposit of the tusks and bones of those preglacial elephants, the mammoth and the mastodon. Such fossil ivory has been used by all Innuits from time immemorial in making their sleds and in tipping their spears, lances, and arrows.

A party of Americans spent the summer of 1881 exploring the country at the head of that deep indentation in the north shore of Norton's Sound called Golovin Bay. They were miners, and engaged in locating the sources from which the Innuits had been bringing large masses of lead-ore with a micaceous sparkle. The hope of a silver-mine had allured these hardy prospectors, who had not reckoned, however, on what they would have to face during the long winter, on the ice that was always left in the soil. Still, in the summer this bay of Golovin is an attractive anchorage—the most agreeable landscape presented anywhere on our Arctic coast. Several rivers empty into it, and on the slopes of the uplands of the northwest side is a growth of white pines that reach a height of fifteen or twenty feet. These small rounded conifers, scattered in clumps over the green and russet tundra, an absence of underbrush, and the dark-green lines of stunted willows and birches that fill the ravines on the sloping sides of gently rising hills, suggest the parking of an old-country place where the orchards are separated by hedges.

The beaches everywhere are profusely littered with drift-logs from the Yukon, twenty to forty feet in length, thickly strewn. They are pushed high above tides by the ice-floes in winter. What the result would be of failure to gain that abundant supply of fuel, now so easy of attainment, upon the natives of this entire region, is not difficult to determine. As they live to-day they are steadily, rapidly diminishing in number. The whalemen have substantially exterminated their chief sources of life—the whale and the walrus. Seals are not as abundant as on the Greenland coasts, and if, in addition to their extra labor of securing food-supply, they were obliged to do without wood, a practical depopulation of the Alaskan coast of Bering Straits and the Arctic Ocean would be effected soon.

As the trader shapes his course from St. Michael's for Port Clarence and Kotzebue Sound, his little vessel skirts the low north shore of Norton's Sound very closely. He may stop for an hour or two, if the weather permits, at Sledge Islet, standing "off and on" while the Innuits come out to the schooner in their skin "oomiaks" or bidarrahs. This barren rock was so named by Captain Cook, who, when he landed on it, found nothing but a native's hand-sled. Its inhabitants were all sojourning on the mainland, berrying. It is only about a mile in its greatest length, less than half a mile wide, and raised almost perpendicularly from the sea to a height of five or six hundred feet. When the modicum of walrus-oil and ivory which these natives have to barter has been hoisted on board, the schooner shapes her course for another islet—the curious "Ookivok," or King's Island—which stands, a mere rock as it were, in the flood that sweeps through Bering Straits. It is rugged, and strewn with immense quantities of basaltic fragments, scoriæ, and rises so precipitately from the sea that no place for a beach-landing can be found.

Here on the south side, clinging like nests of barn-swallows, are the summer houses of the Ookivok walrus-hunters. They are from fifty to one hundred feet above the brawling surf that breaks in-

cessantly beneath them, and secured to the perpendicular cliffs by
lashings and guys of walrus-thongs. The wooden poles thus fastened to the rocks are covered with walrus-hides. On these unique
brackets those hardy Innuits spend the warmer weather. Their
winter residences are mere holes excavated in the interstices and
fissures of the same bluff to which their flimsy summer dwellings are attached, the entrances to most of them being directly under the frail platforms upon which these Mahlemoöt families are
perched with all of their rude household belongings. The nakedness of the island is so great as to forbid life to even a spear of
grass or moss—nothing but close, leathery lichens, that grow so
tightly to its weathered rocks that they appear to be part and par-

Ookivok.

cel of the splintered basaltic cubes or olivine bluffs themselves. A
more uninviting spot for human habitation could not be found in
all the savage solitudes of the north. But the Innuit is here, not
for the pleasure of location ; he is here for that command which
this station gives him over all walrus-herds floating up and down
on the ice-floes of Bering Sea at the sport of varying moods of wind
and current.

From the rugged crests of King's Island the natives can appre-
hend drifting sea-horses as they sleep heavily on broad ice-cakes,
and make ample preparation for their capture. The violence of the
wind is so great that the small, flat summit of this islet cannot be
utilized as a place of residence—the winds that howl over and
around its rock-strewn head would hurl the Innuits, bag and baggage, into those angry waves which thunder incessantly below.

Long experience at plunging through surf with their handsomely made kayaks, and returning to land on these perilous shores of King's Island, has made the Ookivok people the boldest and the best watermen in the north. Their little skin canoes are of the finest construction, and their surplus time is largely passed in carving walrus-ivory into all fashions of rude design for barter in the summer, when the ice shall disappear and the sails of whaling-ships and fur-trading schooners challenge their attention in the offing.

What a winter these people must witness! What a succession of furious storms and snow-laden gales! When their summer comes it brings but little sunlight to their rocky retreat; for, standing, as it does, in the full sweep of that warmer flood which flows up from the Japanese coast into the Arctic, cold, chilly fogs and obstinate clouds envelope them most of the time. But sympathy is utterly wasted; were they to be transported to California, and surrounded with all the needs of a creature existence, they would soon entreat, beg, implore us to return them to the inhospitable rock from which they were taken. The whalers have, at various intervals during the last twenty years, carried Innuits down to spend the winter with them at the Sandwich Islands, under an idea that these people would be delighted with the soft, warm climate there, and such fruits and flowers, and be grateful for the trip. But in no instance did an individual of this hyperborean race fail to sigh for his home in Bering Sea, or the Arctic Ocean, soon after landing at Hawaii. Those Innuits who were without kith or kin became just as homesick and forlorn as any natives did who had relatives behind awaiting their return.

A few hours' sailing, with a free wind, to the north from King's Island, brings you into full view of a bold headland at the entrance to Port Clarence. Cape York is a noted landmark in this well-travelled highway to the Arctic Ocean—well travelled by the whaling fleets of the whole world until recently; now, an elimination of cetacean life from these waters has caused their substantial abandonment by those vessels, and no others come, save a trading-schooner ever and anon at wide intervals. A roomy harbor, sheltered from the south by a long pier of alluvium, is Port Clarence. Leading beyond it is an immense inner basin, walled in all about by steep slate precipices: this is Grantley Harbor. High hopes and great expectations were centred here in 1865-66, by the location of that short cable-end which, underrunning Bering Straits, was to

unite an overland telegraph wire from St. Petersburg with that one
we were to build, in the same fashion, from Portland, Ore., thus
to span the Old and New Worlds by this short submarine link.
Naturally, then, it made this point of its beginning a most interesting locality. In obedience to an order of a few wealthy, energetic
capitalists, who did not then believe in the practicability of the Atlantic cable, many stately ships, freighted with men and goods, left
San Francisco in the summer of 1865, and, again, in the succeeding
season of 1866, for divers points in Alaska and Siberia. These men
were to build the line overland. They were landed at St. Michael's
and at Port Clarence, and at several harbors on the Asiatic coast.
They had fairly got to work, when, late in 1866, the success of the
submarine cable between Newfoundland and Ireland was assured.
That success compelled an abandonment of the Collins Overland
Telegraph, and these men were consequently recalled, and sailed
back to California in those handsome vessels of the telegraph fleet.
How the Innuits of Port Clarence marvelled when these smart,
richly dressed men disembarked, and put up houses in which to
store their treasures of food and telegraph materials, as well as to
actually live in—to stay there with them, in their own rude country,
where no such thing had ever been even dreamed of before. After
the ships had squared their yards and filled away, without calling
the Americans on board, then the Mahlemoöt heart was filled with
unknown and strange emotions of joy and curiosity—both of these
passions were fully satisfied ere the white men left Grantley Harbor.

With Cape York just astern, you pass under the lee of those
sheer and lofty walls of that shoulder to our continent, Cape Prince
of Wales. Its bold front stands in full but silent recognition of an
Asiatic coast westward, just thirty-six miles away, over the shallow
flood of Bering Straits. What changes in a great northland and
seas would have been wrought had a tithe of such volcanic energy
which raised up the Aleutian archipelago been only exerted here in
throwing a basaltic dike across from continent to continent! Had
the upheaval and power that elevated the large island of Oonimak
alone been focused here, we, should have no division of the Old
World and the New. That ocean-river which flows steadily into
the icy wastes of a known and unknown polar basin above Alaska
and Siberia would not now give that life which it so freely grants
both animals and vegetables in the wide reach of the North Pacific.
A dam of adamantine rock or basalt across the Straits of Bering

would cause a startling revision of all the natural order of life in Bering Sea and our Arctic Ocean.

Cape Prince of Wales, which forms the extreme narrowing of Bering Straits, is a high, rugged promontory, with walls on the south side that are abrupt precipices of a full thousand feet, while the uplands rise, culminating in a snowy crown that is twenty-five hundred feet above the level of the sea. Deep gulches seam these vertical walls, and are the paths of numerous tiny rivulets that trickle and run in cascades down from the spongy moorlands above. When, however, you stand in to the straits, homeward bound from the Arctic Ocean, this cape on that side presents a wholly different outline. It slopes up gradually from the beaches, and presents the appearance of a tundra gently rising to a small ridge-like summit. This lowland on the north side is projected under the sea for a distance of over eight miles in a northerly direction, making an exceedingly dangerous shoal, and justly dreaded by the mariner.

The Siberian side and opposite headland is the bold and lofty East Cape, and is connected with the mainland by a low neck of rolling tundra, which is characteristic of Cape Prince of Wales also. Both of these outposts of two mighty continents present, at a small distance, the resemblance of islands.

On June 20th, two hundred and thirty-eight years ago (1648), Simeon Deschnev, a Cossack chief trader, sailed from the mouth of the Siberian river Kolyma, standing to the eastward, where he intended to cruise until the country of those Chookchie natives, who had ivory for trade, should be reached. His party sailed in three small "kotches," which were rude wooden shallops, decked over, about thirty feet long and twelve in beam, drawing but little water. They pushed on and on in that region to the eastward, from which direction the nomadic natives of the Kolyma had always returned laden with walrus-ivory. Fields of ice retarded them ; no populous trading-villages rewarded their scrutiny of the rugged coast as they advanced. The known waters behind them closed up with floes, so returning was impossible ; while the unknown waters ahead were open and invited exploration. In this manner, hugging the coast, Deschnev and his companions sailed through the straits, landing once there in September. He called it an "isthmus," and described the appearance of the Diomede Islands, which he plainly saw from the shore. Although no mention is made by any one of this party of having seen the American continent, yet it must

have been observed by them, for the bold headland of Cape Prince
of Wales can be easily descried on any clear day from the Asiatic
side. Deschnev's vogage had been quite forgotten until Müller, in
hunting over old records in 1764, found the narrative then, and at
once published it in the "Morskoi Sbornik."

A long interregnum elapsed between the hardy voyage of Simeon Deschnev and the next or second passage of the straits by the
keel of a white man's vessel. Not until August, 1728, did Bering
sail through here. He went only a short distance above, into the
Arctic Ocean, and returned without giving any sign thereafter of
the importance of the pass or its nature, believing, most likely,
that what land he saw on the eastern side was a mere island and
not a great American continent. But that intrepid navigator, Cap-

The Diomedes.
"Fairway Rock."   " Ignalook" (America).   "Noornabook " (Asia).
[*Viewed from the Arctic Ocean ; looking S.S.W. 7 m.*]

tain Cook, who comes third in this early initiation of our race,
made no mistake : he fully realized that the division of two hemi-
spheres was here effected, and so declared the fact, and then gave
to these straits, in a most chivalric manner, the name of Bering,
August, 1778.

Midway, stepping-stones as it were, across those straits are the
Diomedes, two barren, rocky islets and a sheer rock. The largest and
the most western is about three miles long and one in width ; it is
seven or eight hundred feet in abrupt elevation from the water, and
the line of division between the Siberian possessions and our own
just takes it in. The sister island is somewhat smaller, less than
half as large, but it is as bold and sheer in its rocky elevation,
leaving a channel-width of two miles only in between. The first
is named Ratmanov, or *Noornabook ;* the second, Kroozenstern, or

Ignalook ; while that high isolated hay-cock mass, about seven miles south of Kroozenstern, is called Fairway Rock. Bering Straits has an average depth of only twenty-six fathoms, with a hard, regular bottom of sand, gravel, and silt.

This gateway to the Arctic Ocean is closed by ice-floes usually by the middle or end of October every year, and opened again in the following season by May 25th or June 1st, but the ice-fields do not allow much room for navigation north until the middle or end of June, sometimes not until the month of July has been well passed.

On that low, northern tundra slope of Cape Prince of Wales is the largest Innuit village in the Alaskan northland. Four hundred souls live there in a settlement which they style Kingigahmoot, and they bear unmistakable evidence of the vicious and degrading influence which evil whalers and rum-traders have exerted. We are struck by their saucy flippancy, their restless, meddlesome, and impertinent bearing. It is because these people have been for a great many years thoroughly familiarized with and degraded by all the tricks and petty treacheries of dishonest and disreputable white men. They do not draw a line in favor of any decency in our race to-day, and hence their disagreeable manner. Otherwise, beyond shaving the crowns of their heads, they do not differ from the Innuits whom we have met heretofore. They are seamen in the full sense of the word—hardy, reckless navigators who boldly launch themselves into stormy waters and cross from land to land in tempest and in fogs, depending solely upon the frail support of their walrus-skin baidars, or oomiaks. These are very neatly made, however, the covering of seal- and walrus-hides being stretched and sewed tightly over wooden frames that are lashed at the joints with sinew and whalebone-thongs. They hoist a square sail of deer-skins or cotton drilling, and run before the wind in heavy gales ; or they employ paddles and oars, and urge their craft against head-winds and perverse currents. Their poverty is the only redemption which they have had from absolute destruction ; for were they possessed of furs that would encourage the regular visits of traders, they would, with their disposition to debauchery, have been utterly exterminated long before this time, But they are poor, very poor, having nothing to tempt the cupidity of white traders—nothing but small stores of walrus-oil and teeth, and a few red and white foxes, perhaps. Therefore our people never stop long near them, just laying the vessel's sails aback for a few minutes, or an hour, while the dusky paddling crews of the oomiaks surrounding the schooner exhibit their slim stocks of oil and ivory.

These northern Innuits are not known anywhere to have a village located far back from the sea save at three places, where, on the Selawik, the Killiamoot, and the Kooak Rivers, are settlements of a few people who are at least fifty and one or two hundred miles inland ; but they are the exceptions only to their rule of living. Some thirty-five villages of these hyperborean Innuits of Alaska are scattered along the coast between St. Michael's and Point Barrow.; they possess an aggregate (estimated) inhabitation of three thousand men, women, and children. The Diomede and Prince of Wales natives are the most active middlemen or commission merchants among their people ; they conduct all the trade between the Asiatic Chookchie savages and the American Innuits, chiefly with those of Kotzebue Sound. Before a wholesale destruction by our people, in 1849–57, of the whales that once were so abundant in these waters, the life of those natives was a comparatively easy struggle for existence, and they were far more numerous then than they are to-day ; but a fleet of four and five hundred whaling-ships, manned by the hardiest men of all nations, literally swept that cetacean life from the North Pacific and Bering Sea, and drove it so far into the Arctic Ocean that its remnant, which is still there, is practically safe and beyond human reach.

As you leave the Straits of Bering behind, your little vessel cuts the cold, green waves of the Arctic Ocean rapidly, especially if under the pressure of a warm southwester which funnels up stiffly through the pass. You find nothing to catch your eye in all that long reach from Cape Prince of Wales to the entrance of Kotzebue Sound, which is an objective point of all the traders who come into the Arctic. Here is the last safe Alaskan harbor for a sea-going vessel as we go north. It is a big one ; and it is a famous place for a geologist and Innuits alike. To the latter it is of especial significance, since the small rivers which empty there mark an extreme northern limit of salmon-running in America.

The shores which bound this large gulf rise as perpendicular bluffs, either directly from the water or from a shelving beach. In some places the land is remarkably low (as it always is when bordering the coast), and only so much raised above tide-level as to render the idea probable that it is of an alluvial formation, the result of accumulated mud and sand, brought down in former times by the melting and running of large glacial rivers, and then thrown up later by recent ice-floes of the Arctic Sea. The cliffs are, in part, abrupt and rocky ; others are made up of falling masses of mud, sand, and ice. ' The rocky cliffs are dominant on the western and southern shores, while the diluvial bluffs and flats complete that remaining east and northeast circuit of the sound. Lowlands border a major portion of the Bay of Good Hope, and form the land of Cape Espenberg and contiguous country.

A most striking natural feature of this final rendezvous of the salmon-loving Innuits is the Peninsula of Choris, which divides the inner waters of the Bay of Escholtz from those of Good Hope. It is a narrow, variously indented tongue ; its northern end is separated from the southern, and connected by a slender neck of very low land. This lower point assumes the shape of a round and somewhat conical eminence, surmounted by a flat, hut-like peak, the sides of which rise a few feet perpendicularly above a surrounding surface, as though raised artificially by masonry. The whole height is about six hundred feet above sea-level. Both sides of that quaint headland terminate in rocky cliffs which, toward the west, are one hundred and fifty or two hundred feet high, stratified, unbroken, and dipping to the west at an angle of thirty degrees. They are composed of micaceous slate, with no included minerals. This slate is of a greenish hue, with a very considerable predominance of mica. In it are garnets, veins of feldspar enclosing crystals of schorl, and fissures filled with quartz. At one point, nearly midway between the southern end of this peninsula and its low neck, is a singular bed of pure milk-white quartz, that marks its locality from a long distance by the masses of large white blocks which have fallen down by natural processes of cleavage and frost-chiselling, and these remain unaltered in their snowy color in spite of the corroding action of time and weather. Again, still nearer the neck, a narrow bed of limestone forms a distinct protrusion above some mica-schist, about thirty .feet in length and five in depth. It reappears in such strength, however, at the southern end of the peninsula, that it forms most of the rock exposed, and produces four perpendicular and contiguous promontories, separated from each other by small, receding bays, that present curious walls striped a white and blue tint in beautifully blended stratification, most unique and attractive to the eye. The upper part of this limestone contains iron pyrites,

28

and has cavities filled with chlorite. The lower strata are more abundantly mixed with micaceous schistus, containing compact actynolite, and flat prisms of a glassy shade of it, crystals of tourmaline, and those various concretions of iron pyrites. The quartz is, in some places, colored a real topaz tint. Such, in brief, is a faint description of those geological attractions which the Arctic rocks of Kotzebue Sound present to a student.

The country everywhere, that borders the Arctic Ocean and this sound, is low. The land rises by faint and gradual slopes; it is covered with clay soils and the characteristic vegetation of a tundra. The many low, projecting points of Kotzebue Sound are thickly strewn with large and smaller masses of vesicular and of compact lava, containing olivine. Some of these blocks extend into the sea; others are embedded in the sandy soil of the beach; but many are insulated and awash above the surf. They are honeycombed with empty cavities. The sands of this Arctic Ocean beach partake of the black and volcanic nature of those blocks. These large and numerous erratic blocks of basalt, collected chiefly on such jutting points, must have been conveyed there by ice-sheets from a very considerable distance, for no volcanic formation is to be seen in their vicinity.

A suggestive wreck lies half buried in the sand and drift of the north shore of Choris Peninsula—it is the scant and weathered remnants of a large whaling-bark, which was run ashore here and burned. Its own crew did so to prevent its capture by the *Shenandoah*—that cruiser which, during our civil war, swooped down upon our Asio-Alaskan whaling-fleet, as a fish-hawk drops upon a flock of startled gulls. Again, on the south side of Good Hope Bay, in this same remarkable sound of Kotzebue, is a bluff of solid blue clay, from the face of which the frost-king annually strikes large masses. The weathered débris of these fallen sections reveal many fine specimens of well-preserved remains of huge pachyderms—mammoths—and their finding has given a fit name of "Elephant Point" to the place.

Across that peninsula, which Choris Point and its comical little tender of Chamisso Islet project from, lies the long and narrow estuary of Hotham Inlet, where all Innuits, from Icy Cape to the far north and Bering Straits in the south, annually repair for salmon-fishing in August. Into the mouths of a half-dozen small streams which empty there, and that large one, of Kooak River, the humpbacked salmon runs, for a brief period, in great numbers: then the harvest of the Eskimo is at hand. Nowhere else above this point can a salmon ever be taken, and as it is the last chance of these natives, they improve it. Flocks of fat ducks and geese hover over and rest upon the smooth, shallow waters of this inlet, alternately feeding. there and then alighting upon the tundra where crowberries and insects abound. Our whalers have taught these Innuits how to make and use gill-nets, with which they now catch their fish almost exclusively; and not unwisely have those natives made the change, for they have not got any slender willow brush and alder-saplings which their brethren use so effectually in making rude traps on the Yukon, Kuskokvim, and Nooshagak Rivers. They also stretch these gill-nets over certain narrow places, from shore to shore, of lagoons and lakes, where flocks of water-fowl are wont to fly (in early morning and late in the evening), and succeed in capt- . uring a great many luckless birds by this simple method.

CHAPTER XIV.

MORSE AND MAHLEMOÖT.

An Innuit village is in plain sight on the low shores of Cape Kroozenstern, which forms a northern pier-head of Kotzebue Sound, and
its inhabitants greet your vessel as it passes out and up the coast with
the usual dress-parade—climbing upon the summits of their winter
houses, and by running in light-hearted mirth along the beach.
A most dreary expanse of low moorland borders the coast as the little schooner reviews it, swiftly heeling on her course to the north.
Not until the bluffs of Cape Thompson are in sight does a noteworthy landmark occur. This is an abrupt headland capped by
carboniferous limestone full of fossils, shells, corals and the like,
which are peculiar to that age. It is also traversed by veins of a
blackish chert varying in thickness from six inches to three feet or
more, causing a decided network tracery to appear very plainly on
its gray-white face. Half-way down from the top, the limestone is
succeeded by blue, black, and gray argillaceous shades, the colors
of which alternate in layers of horizontal strata, six or eight feet in

thickness, nearly down to the base; it is then composed of black carboniferous shales alone, which abound in organic remains and are occasionally interstratified by limestone much deflected. This contortion is so great as to form two regularly banded arches. Several tiny snow-water cascades tumble down its ravines and boldly plunge over the bluffs, which are about four hundred feet high in their greatest elevation.

This chert is that which the Eskimo of the entire Alaskan arctic region (before the coming of white men) used for tipping their lance- and arrow-heads when ivory was not employed. They, aided with a small piece of bone, were able to " flake " it off in slices that were easily reduced to the desired forms. They still work a little of it up every year, in a desultory or perfunctory manner, however, more for amusement than anything else, since they have a profusion of iron and steel now in their possession. The fashion in which they chip it gives ample evidence of their full understanding of a flat conchoidal fracture peculiar to flint, and of which they take advantage.

To the northwest of Cape Thompson the coast runs out abruptly as a low spit, projected into the Arctic Ocean for a distance of twenty miles. This is Port Hope. The beach everywhere is principally formed of dark basaltic gravel. To the north of a considerable stream not far from this point, and on a low and diluvial shore, is a large hamlet of Innuits, who have covered the turfy thatches on their winter houses with heavy blocks of angular clink-stones picked up from the sea-beach. The whole surface of the interior country here is raised several hundred feet above tide-level, and is diversified with saddle-backed hills of gray and bronzed tints, separated by wide valleys in which a rich green summer verdancy is characteristic. Here and there conical eminences and perpendicular shelving cliffs arise from a general evenness of the whole landscape. These cliffs seem to be composed of limestones, while their acclivities are of slate and shale.

As we near Cape Lisburne a jutting range of bluffs, stratified in bands of grayish-brown and black, receive the full wash of the sea, and are called Cape Dyer ; but Cape Lisburne is the striking landmark, and a most important one for the navigator to recognize. It is composed of two remarkable promontories : the southwestern one rises abruptly from the surf, is covered with loose gray stones, divested of the smallest traces of vegetation. The northeastern one rises gradually, and, although but thinly clad with verdure, it forms a pleasing and marked contrast with the gray head of the other. The first is elevated from the sea in distinct strata, with a south-western dip, and consists of layers of impure chert in its central and most prominent projections, and of a soft, friable slate and shale in its worn and more retiring sides. The front of the second is rugged and shelving, with very indistinct bandings ; it is partly covered with tundra vegetable-growths, and with fallen masses of gray flint. Both points to this double-headed cape of Lisburne are easily accessible ; they are about one thousand feet in height from the shore of the ocean, and both stretch their ridges away inland far to the southeast.

The highly elevated country here ceases at once to the northeast of Cape Lisburne, where the entire coast-line, away on and off to Icy Cape, and beyond again, forms a deep and extensive bay skirted by a dark, low beach. A gravel-flat fronts this again, filled with shallow estuaries and lagoons. The land of the interior rises from that beach in a series of low, earthy cliffs and in gradual acclivities.

The coal-veins, which Beechey visited in 1826, are about fifty miles to the eastward of Lisburne, embedded in a ridge some three hundred feet high where it juts into the ocean. This point is known as Cape Beaufort. A narrow vein of pure carboniferous coal is exposed there, about a quarter of a mile from the beach. " It was slaty, but burned with a bright, clear flame and rapid consumption." Again, at a point about midway between Beaufort and Lisburne, directly at the surf-margin, the officers of the United States Revenue Marine cutter *Corwin* mined a few tons of this same coal in 1880–81. But no harbor for a coaling ship is near by ; the steady north and westerly winds of summer, which blow right on shore almost all of that short time in which a vessel can navigate the Arctic, make it very doubtful whether these remote mines of Alaskan "black diamonds" will ever be of real economic value.

That sand- and shingle-spit ahead of us, which the whalers have named Icy Cape with perfect fitness, is in itself almost invisible, since it is a mere continuation of the outer rim to a remarkable lagoon which borders this coast from Cape Beaufort to Wainright Inlet, over one hundred miles in length, and varying in width from five to ten miles, with an average depth of two fathoms. It is spanned by occasional sand-bars, some of them entirely dry, so that it is not navigable except for those small boats and oomiaks of the natives, who haul these craft across as they journey, thus safe and snug, up and down a desolate coast. This lagoon of the Arctic Ocean has several openings to the sea itself. Small schooners can run in and escape from ice-pack "jams," if they draw less than eight or ten feet of water. The coast-line of the mainland at Icy Cape is a series of low mud-cliffs, varying from ten to fifty feet in height above a shingly beach, which is everywhere composed of fine, minutely comminuted, pebbly bases of granite, of chert, of sienite, and of indurated clay, the last being a predominant form.

From this point clear around to the boundary of our Alaskan Arctic coast at Point Demarcation that country presents the same appearance which we note here. It is low and slightly rolling, and falls in small cliffs of mud or sandstone at the sea-shore. During

Innuit Whaling-camp at Icy Cape.

the midsummer season it wears a hue of gray and brown, with little patches of bright green where the snow has melted early in sunny, sheltered spots. The lines of many streams, as they course in carrying off melting snows, are plainly marked over a dreary tundra by the dark fringes of dwarfed willows, birches, and alders which only grow upon their banks.

All along this cheerless northern sea-shore are small and widely scattered settlements of our Innuits, who burrow in their turfy underground winter huts, and who tent outside in summer-time upon these shingly gravels and clink-stones of the Arctic coast. They then live upon the walrus and kill an occasional calf-whale. For the better apprehension of these animals they erect lookouts on the beach by setting up drift-wood scaffolds, and climbing as lookouts to an elevated platform thus made. In the winter, when the weather permits, they net a ringed seal (*Phoca fœtida*) under the ice, make short inland trips, where they camp for weeks at a time in rude snow-houses, hunting reindeer, which are shy though abundant, and they trap a few wolves and foxes. Every July and August they expect the visit of a few whaling-vessels at least, and they are seldom disappointed, for such craft are compelled by ice-floes to hug this shore very closely, in order to get as far to the eastward as the whales are found; sometimes, in spite of all the wariness and skill of our own hardy whalemen, great floe-booms, of icy make, suddenly shut down on that land so quickly from the north as to catch and crush the staunchest ships like egg-shells under foot. Then, indeed, is the sadness and the distress of the white men sharply contrasted with that great joy and happy anticipation of an Innuit who feasts his eyes and gloats in fancy over the abandoned vessels as they lie riven by ice upon those shallow strands of Icy Cape or Point Barrow.

It is more than sixty years now since Captain Beechey * camped upon and located Point Barrow, our extreme limit of northern landed possession, and in that time few changes, other than depopulation of the natives, have taken place on this coast. That same village of Noowuk, which he graphically described, still stands there on the tip of a low gravel-spit which extends out from the mainland twelve miles into the chill flood of an Arctic Ocean. All the land at its extremity not inundated by the sea in storms is now, as it was then, occupied by the winter houses of the natives. Blooming here in the short summer of July, on those desolate moors adjacent to Point Barrow, is the same dandelion and buttercup which filled the Englishman then, as it does us now, with thoughts of meadows at home, and some bright little poppies still nod their yellow heads again to us, as they did to him, on this low north end of Alaska. A tiny golden butterfly flits from flower to flower, and as they fade, it, too, disappears over frost-bitten swales.

Big ice-fields seldom ever fail to threaten the coast here, even

---

* Captain F. W. Beechey H. M. S. *Blossom*, voyage 1825-28, inclusive. The seasons of 1826 and 1827 were passed in these waters. Murdoch, who passed the winters of 1881-83, inclusive, here, has given an interesting *résumé* of the natural history, etc., of the spot. Beechey's account of the people and country are confirmed by him.

when they relax their grasp in July. In a few short weeks, however, they return to stay for the rest of the year and best part of
the next. Such brief intervals for navigation in the Arctic Ocean
during every July and August are those which lure whaling-ships,
and the dark lanes of open water in white ice-floes are the last
refuge of many hard-hunted whales, unless they dive, and rise to
breathe again in that conjectured clear yet frigid flood of a polar
sea, far away under the north star.

There is nothing more to see, or noteworthy to learn, at or beyond Point Barrow, even were you to live and drag out a wretched
year's existence in looking for it, so you gladden the heart of your
skipper and his hardy crew by telling them to shape a course home-

The Ringed Seal (Phoca foetida).
[*The common Hair-Seal of the Arctic Ocean.*]

ward. Back through the Straits of Bering, wrapped in a chill
thick fog, the little schooner heels, with a singing northwester on
her quarter that holds her canvas just as taut as if made of tough
wood. She fairly scrapes by the Diomedes—the walls of Noornabook loom up high in a cold, gray fog-light, as though its bold,
gray cliffs were right over her spars—but the crew know at the time
that they are more than two miles away from that surf which noisily
thunders on the dark rocks of these islets. That same chill wind,
and gloomy fog-surrounding, follows them into Bering Sea—not a
glimpse of all the land and mountain, which they so plainly discerned going up, have they caught going down.

What trifles often determine our success or failure in life ! Had
it not been for a sudden sunburst from the gloom of a leaden fog

which shrouded all about it in its misty darkness, and thus lighted
up a lofty russet head of the East Cape of St. Lawrence Island, a
little vessel bearing the author would have been piled up and
thrown into foaming breakers which beat upon a low, rocky reef
that reaches out from its feet.   This gleam of light reflected from
that headland warned a startled man on the lookout just in time to
have her wheel put hard up, and thus luff our light trim craft in
season to shave safely by.

St. Lawrence is the largest island in Bering Sea.   It is directly
south of Bering Straits, one hundred and eighty miles distant
from the Diomedes ; it is eighty to eighty-five miles in length, with
an average width of fifteen or twenty.   The sea has built onto it
quite extensively, in very much the same manner as it·has filled
out and extended the coast of St. Paul, of the Pribylov group.   At
Kagallegak, on the east shore, the island is made up of coarse feldspathic, red granitic flats and hills, with extensive lagoons and
lakelets.   The skeleton of this island seems to have been originally
one of low hills and ranges of granite, with volcanic outbursts everywhere manifested at their summits, especially on the north shore.
Between them stretch long, low plains, or gently rolling uplands,
and perfectly smooth reaches of sand and gravelly beaches that
border the sea everywhere not so marked by bluffs.

At Kagallegak your eye sweeps over extensive level plains to the
northward, upon which a green-stalked and white-plumed tundra
grass (*Eriophorum*) principally grows everywhere on the wet and
boggy surface, while, on those sand-beach margins, the "wild wheat"
(*Elymus*) springs up most abundantly, short and stunted, however.
These extended low areas of moorland so peculiar to this island
are made up of fine granitic drift and clays, lined at their sea-bor-
ders with a low, broad sand-belt.   The hills and· hill ranges of St.
Lawrence are rich in color, with dark blue-black patches interspersed which indicate a location of trap-protrusions.   No shrubbery whatever grows upon these wind-swept tundra and hills save
dwarfed and creeping willows ; yet, a series of characteristic rock-
lichens color such bare summits in their bright relief which we
have just noted.   The rocks themselves are reddish, coarse-grained,
shining granites, with abundant trap-protrusions, that weather out
and fall down upon the flanks of the peaks and ridges in dusky
patches and streaks, so as to contrast, from a slight distance, very
sharply with the main ground of pinkish rock, which is moss- and

VILLAGE AND ISLET OF POONOOK

Mahlemoot Winter Houses on the Poonook Islets, 6 miles East of St. Lawrence

lichen-grown, and colored here and there with areas of that peculiar and characteristic greenish-russet tinge of sphagnous origin. This dark marking of those trap-dikes appears like the presence of low-growing shrubbery from the vessel, as an observer sails by. Snow and ice lie all the year around in small bodies within the gullies and on the hill-sides.

The lower plains have a richer, warmer, yellowish-green tone than that cold tint of the uplands, while the sand of the sea-shore is a bright light-brown. Small streams flow down from these hills, and twist and turn sluggishly through the tundra as they lead to lakes or empty directly into the sea—a few parr, or young-salmon, being the only fish in them that can be found ; most of the fresh-water lakes and lagoons are, however, fairly stocked with familiar-looking mullets (*Catastomus*), but nothing else.

The entire expanse of these lowlands of St. Lawrence are precisely like all of those vast reaches of Alaskan tundra—they are great saturated, earthy sponges, filled and overrunning with water in midsummer—the chief and happiest vegetation upon them being that same beautiful tufted or plumed grass which we noticed at Michaelovsky, since the white and silken tassels of its feathery inflorescence never fail to charm even tired and travel-worn eyes. This grass, in conjunction with several rank-growing mosses, the trailing runners of the crowberry-vines, and little patches of the humble arctic raspberry (*Rubus chamœmorus*) make up that conventional tundra color of russet-green (flecked with grayish-blue spots on the slopes of stern northern exposures) which mark these great marshy tracts of Alaska, and under which eternal frost is found, even in midsummer, a foot or two only from their surfaces. Small white shells of a land-mollusk (*succinea*) are scattered thickly over these moorlands.

On the flats of the east shore of St. Lawrence a great abundance of drift-wood was piled in much confusion. Here the natives had a wood-cutting camp, hewing and carving ; its chips were scattered all along the beach-levels for miles. There are places, here, where the ice in some unusual seasons has carried large logs and pieces of drift-wood far back, full half a mile from the sea, and a vigorous growth of tundra vegetation now flourishes in between ; and there they lie to-day deeply embedded in the swale, settling down in decay—that slow, hungering eremacausis of the Arctic.

The Innuits, living here as they do, some three or four hundred in number, are great walrus-hunters. They enjoy a location that enables them to secure these animals at all seasons of a year. In winter the sea-horse floats on big ice-fields ; but during summer-time the " aibwook" hauls up to sun and rest his heavy body in and on the inviting peace of those beaches of St. Lawrence. A famous spot for this landing of the walrus is on the rocky and pebbly shores of Poonook (three small rocky islets), just five miles east of the summer tents of Kagallegak. These tiny, detached fragments of St. Lawrence stand in the full sweep of those air- and water-currents which keep broad ice-floes in constant motion, and thus bring walrus-herds into range of Mahlemoöt hunters, who have a winter village dug deep into sandy flats of "Poonookah."

The Walrus-hunter.

[A St. Lawrence Mahlemoöt—in winter parka with the hood removed. August 16, 1874.]

Naturally enough we regard the walrus with more than passing interest, for it plays so large and so vital a part in sustaining the life of human beings who reside in these arctic and subarctic regions of Alaska. Perhaps the only place in all this extended area in which these clumsy brutes are found, where the creature itself can be closely observed and studied, is that unique islet, six miles east of St. Paul (Pribylov group) and about four hundred miles south of St. Lawrence.

Here the morse rests upon some rocky, surf-washed tables characteristic of this place without being disturbed ; hence the locality afforded me a particularly pleasant and advantageous opportunity of minutely observing these animals. My observations, perhaps, would not have passed over a few moments of general notice, had I found a picture presented by them such as I had drawn in my mind from previous descriptions ; the contrary, however, stamping itself so suddenly and decidedly upon my eye, set me to work with

pen and brush in noting and portraying such extraordinary brutes,
as they lay grunting and bellowing, unconscious of my presence,
and not ten feet away from the ledge upon which I sat.*

Sitting as I did to the leeward of them, with a strong wind blowing in at the time from seaward, which, ever and anon, fairly covered
many of them with foaming surf-spray, therefore they took no notice
of me during the three or more hours of my study. I was first astonished at observing the raw, naked appearance of the hide : it was
a skin covered with multitudes of pustular-looking warts and large
boils or pimples, without hair or fur, save scattered and almost invisible hairs ; it was wrinkled in deep, flabby seam-folds, and marked by
dark-red venous lines, which showed out in strong contrast through
the thicker and thinner yellowish-brown cuticle, that in turn seemed
to be scaling off in places as if with leprosy ; indeed, a fair expression of this walrus-hide complexion if I may use the term, can be
understood by the inspection of those human countenances in the
streets and on the highways of our cities which are designated as
the faces of "bloats." The forms of *Rosmarus* struck my eye at
first in a most unpleasant manner, and the longer I looked at them
the more heightened was my disgust ; for they resembled distorted,
mortified, shapeless masses of flesh ; those clusters of big, swollen,
watery pimples, which were of a yellow, parboiled flesh-color, and

---

* These favored basaltic tables are also commented upon in similar connection by an old writer in 1775, Shuldham, who calls them "echouries ;" he is
describing the Atlantic walrus as it appears at the Magdalen Islands: "The
echouries are formed principally by nature, being a gradual slope of soft rock,
with which the Magdalen Islands abound, about eighty to one hundred yards
wide at the water-side, and spreading so as to contain, near the summit, a very
considerable number." The tables at Walrus Island and those at Southwest
Point are very much less in area than those described by Shuldham, and are
a small series of low, saw-tooth jetties of the harder basalt, washed in relief,
from a tufa matrix ; there is no room to the landward of them for many walruses to lie upon. The *Odobœnus* does not like to haul up on loose or shingly
shores, because it has the greatest difficulty in getting a solid hold for its fore
flippers with which to pry up and move ahead its huge, clumsy body. When it
hauls on a sand-beach, it never attempts to crawl out to the dry region back of
the surf, but lies just awash, at high water. In this fashion they used to rest
all along the sand-reaches of St. Paul prior to the Russian advent in 1786-87 ;
and when Shuldham was inditing his letters on the habits of *Rosmarus*, *Odobœnus* was then lying out in full force and great physical peace on the Pribylov Islands.

principally located over the shoulders and around the necks, painfully suggested unwholesomeness.

On examining the herd individually, and looking upon perhaps
one hundred and fifty specimens directly beneath and within the
sweep of my observation, I noticed that there were no females
among them ; they were all males, and some of the younger ones
had considerable hair, or enough of that close, short, brown coat to
give a hirsute tone to their bodies—hence I believe that it was only
the old, wholly matured males which offered to my eyes such bare
and loathsome nakedness.

I noticed, as they swam around, and before they landed, that
they were clumsy in the water, not being able to swim at all like

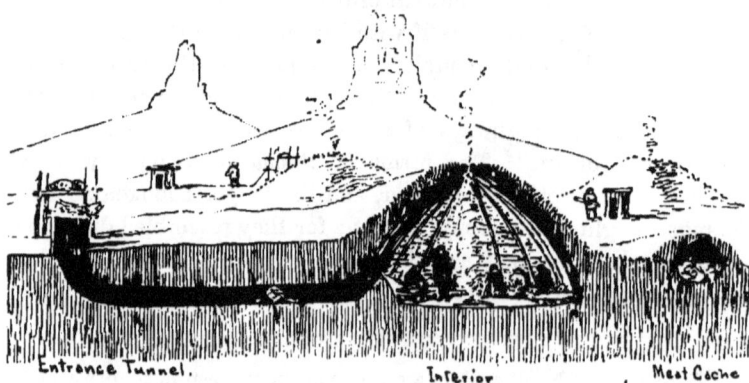

Section showing Construction of Mahlemoöt Winter Houses at Poonook.

the *Phocidæ* and the *Otariidæ* ; yet their progress in the sea was
wonderfully alert when brought into comparison with that terrestrial action of theirs ; the immense bulk and weight of this walrus,
contrasted with the size and strength of its limbs, renders it simply impotent when hauled out of the water on those low, rocky
beaches or shelves upon which it rests. Like the seals, however, it
swims entirely under water when travelling, but it does not rise, in
my opinion, so frequently to take breath ; when it does, it blows or
snorts not unlike a whale. Often have I heard this puffing snort
of those animals (since the date of these observations on Walrus
Islet), when standing on the bluffs near the village of St. Paul and
looking seaward ; on one cool, quiet morning in May I followed
with my eye and ear a herd of walrus, tracing its progress some
distance off and up along the east coast of the island by those tiny

jets of moisture or vapor from its confined breath which the animals blew off as they rose to respire.

Mariners, while coasting in the Arctic, have often been put on timely footing by a walrus fog-horn snorting and blowing as the ship dangerously sails silently through dense fog toward land or ice-floes, upon which those animals may be resting ; indeed, these uncouth monitors to this indistinct danger rise and bob under and around a vessel like so many gnomes or demons of fairy romance, and sailors may well be pardoned for much of that strange yarning which they have given to the reading world respecting the sea-horse during the last three centuries.

When a walrus-herd comes ashore, after short preliminary surveys of the intended spot of landing, an old veteran usually takes the lead of a band which is so disposed.

Finally the first one makes a landing, and no sooner gets composed upon the rocks for sleep than a second one comes along, prodding and poking with its blunted tusks, demanding room also, thus causing the first to change its position to another location still farther off and up from the water, a few feet beyond ; then the second is in turn treated in the same way by a third, and so on until hundreds will be slowly packed together on the shore as thickly as they can lie—never far back from the surf, however—pillowing their heads upon the bodies of one another : and, they do not act at all quarrelsome toward each other. Occasionally, in their lazy, phlegmatic adjusting and crowding, the posteriors of some old bull will be lifted up, and remain elevated in the air, while the passive owner continues to sleep, with its head, perhaps, beneath the pudgy form of its neighbor.

These pinnipeds are, perhaps, of all animals, the most difficult subjects that an artist can find to reproduce from life. There are no angles or elbows to seize hold of. The lines of body and limbs are all rounded, free and flowing ; yet, the very fleshiest examples never have that bloated, wind-distended look which most of the published figures give them. One must first become familiarized with the restless, varying attitudes of these creatures by extended personal contact and observation ere he can satisfy himself with the result of his drawings, no matter how expert he may be in rapid and artistic delineation. Life-studies by artists of the young of the Atlantic walrus have been made in several instances ; but of the mature animal, until my drawing, there was nothing extant of that character.

As the walrus came ashore they made no use of their tusks in assistance ; but such effort was all done by their fore flippers and the " boosting " of exceptionally heavy surf which rolled in at wide intervals, and for which marine assistance the walrus themselves seemed to patiently wait. When moving on land they do not seem to have any real power in the hinder limbs. These are usually pulled and twitched up behind, or feebly flattened out at right angles to its body. Terrestrial progression is slowly and tediously made by a dragging succession of short steps forward on the forefeet; but if an alarm is given, it is astonishing to note the contrast which they present in their method of getting back to sea : they fairly roll and hustle themselves over and into the waves within an exceedingly short lapse of time.

When sleeping on drifting ice-floes of the Arctic Ocean, or on rocks at St. Matthew's or Walrus Island, they resort to a very singular method of keeping guard, if I may so term it. In this herd of three or four hundred male walrus that were beneath my vision, though nearly all were sleeping, yet the movement of one would disturb the other, which would raise its head in a stupid manner for a few moments, grunt once or twice, and before lying down to sleep again it would strike the slumbering form of its nearest companion with its tusks, causing that animal to rouse up in turn for a few moments also, grunt, and pass the blow on to the next, lying down in the same manner. Thus the word was transferred, as it were, constantly and unceasingly around, always keeping some one or two aroused, which consequently were more alert than the rest.

On Walrus Island a particularly large individual walrus was selected and shot, out of a herd of more than two hundred. This was done at the author's instance, who made the following memoranda : It measured twelve feet seven inches from its bluff nostrils to the tip of its excessively abbreviated tail, which was not more than two and one-half or three inches long ; it had the surprising girth of fourteen feet. An immense mass of blubber on the shoulders and around the neck made the head look strangely small in proportion, and the posteriors decidedly attenuated ; indeed, the whole weight of the animal was bound up in its girth anteriorly. It was a physical impossibility for me to weigh this brute, and I therefore can do nothing but make a guess, having this fact to guide me— that the head, cut directly off at the junction with the spine, or the occipital or atlas joint, weighed eighty pounds ; that the skin, which

I carefully removed with the aid of these natives, with the head, weighed five hundred and seventy pounds. Deducting the head and excluding the flippers, I think it is safe to say that the skin itself would not weigh less than three hundred and fifty pounds, and the animal could not weigh much less than a ton, from two thousand to two thousand two hundred pounds.

The head had a decidedly flattened appearance, for the nostrils, eyes, and ear-spots seem to be placed nearly on top of the cranium. The nasal apertures are literally so, opening directly over the muzzle. They are oval, and closed parallel with the longitudinal axis of the skull, and when dilated are about an inch in their greatest diameter.

The eyes are small, but prominent; placed nearly on top of the head, and, protruding from their sockets, they bulge like those of a lobster. The iris and pupil of this eye is less than one-fourth of its exposed surface; the sclerotic coat swells out from under the lids when they are opened, and is of a dirty, mottled coffee-yellow and brown, with an occasional admixture of white; the iris itself is light-brown, with dark-brown rays and spots. I noticed that whenever the animal roused itself, instead of turning its head, it only rolled its eyes, seldom moving the cranium more than to elevate it. The eyes seem to move, rotating in every direction when the creature is startled, giving the face of this monster a very extraordinary attraction, especially when studied by an artist. The expression is just indescribable. The range of sight enjoyed by the walrus out of water, I can testify, is not well developed, for, after throwing small chips of rock down upon the walruses near me, several of them not being ten feet distant, and causing them only to stupidly stare and give vent to low grunts of astonishment, I then rose gently and silently to my feet, standing boldly up before them; but then, even, I was not noticed, though their eyes rolled all over from above to under me. Had I, however, made a little noise, or had I been standing as far as one thousand yards away from them to the windward, they would have taken the alarm instantly, and tumbled off into the sea like so many hustled wool-sacks, for their sense of smell is of the keen, keenest.

The ears of the walrus, or rather the auricles to the ears, are on the same lateral line at the top of the head with the nostrils and eyes, the latter being just midway between. The pavilion, or auricle, is a mere fleshy wrinkle or fold, not at all raised or devel-

29

oped ; and from what I could see of the *meatus externus* it was very narrow and small; still, the natives assured me that the *Otariidæ* had no better organs of hearing than *Rosmarus.*

The head of the male walrus, to which I have alluded, and from which I afterward removed the skin, was eighteen inches long between the nostrils and the post-occipital region ; and, although its enormous tusks seemed to be firmly planted in their osseous sockets, judge of my astonishment when one of the younger natives flippantly struck a tusk with a wooden club quite smartly, and then easily jerked the tooth forth. I had frequently observed that it was difficult to keep such teeth from rattling out of their alveoli in any of the best skulls I had gathered of the fur seals and sea-lions, especially difficult in the case of the latter.

Its tusks, or canines, are set firmly under the nostril-apertures in deep, massive, bony pockets, giving that strange, broad, square-cut front of the muzzle so characteristic of its physiognomy.

The upper lips of this walrus of Bering Sea are exceedingly thick and gristly, and its bluff, square muzzle is studded, in regular rows and intervals, with a hundred or so, short, stubby, gray-white bristles, varying in length from one-half to three inches. There are a few very short and much softer bristles set, also, on the fairly hidden chin of its lower jaw, which closes up under a projecting snout and muzzle, and is nearly concealed by the enormous tushes, when laterally viewed.

The thickness of the skin of the walrus is a marked and most anomalous feature. I remember well how surprised I was, when I followed the incision of a broad-axe used in beheading the specimen shot for my benefit, to find that the skin over its shoulders and around the throat and chest was three inches thick—a puffy, spongy epidermis, outwardly hateful to the sight, and inwardly resting upon a slightly acrid fat or blubber so peculiar to this animal. Nowhere was that hide, upon the thinnest point of measurement, less than half an inch thick. It feeds exclusively upon shell-fish (*Lamellibranchiata*), or clams principally, and also upon the bulbous roots and tender stalks of certain marine plants and grasses which grow in great abundance over the bottoms of broad, shallow lagoons and bays of the main Alaskan coast. I took from the paunch of the walrus above mentioned more than a bushel of crushed clams in their shells, all of which that animal had evidently just swallowed, for digestion had scarcely commenced. Many of those clams in that stomach, large as my clinched hands, were not even broken ; and it is in digging this shell-fish food that the services rendered by its enormous tusks become apparent.*

I am not in accord with some singular tales told, on the Atlantic side, about the uses of these gleaming ivory teeth, so famous and conspicuous : I believe that the Alaskan walrus employs them solely in his labor of digging clams and rooting bulbs from those muddy oozes and sand-bars in the estuary waters peculiar to his geographical distribution. Certainly, it is difficult for me to reconcile my idea of such uncouth, timid brutes, as were those spread before me on Walrus Islet, with any of the strange chapters written as to the ferocity and devilish courage of a Greenland "morse." These animals were exceedingly cowardly, abjectly so. It is with the greatest difficulty that the natives, when a herd of walruses are surprised, can get a second shot at them. So far from clustering in attack around their boats, it is the very reverse, and a hunter's only solicitude is which way to travel in order that he may come up with the fleeing animals as they rise to breathe.

On questioning the natives, as we returned, they told me that the walrus of Bering Sea was monogamous, and that the difference between the sexes in size, color, and shape is inconsiderable ; or, in other words, that until the males are old the young males and the females of all ages are not remarkably distinct, and would not be at all if it were not for their teeth. They said that the female brings

---

* It is, and always will be, a source of sincere regret to me and my friends that I did not bodily preserve this huge paunch and its contents. It would have filled a half-barrel very snugly, and then its mass of freshly swallowed clams (*Mya truncata*), filmy streaks of macerated kelp, and fragments of crustaceans, could have been carefully examined during a week of leisure at the Smithsonian Institution. It was, however, ripped open so quickly by one of the Aleutes, who kicked the contents out, that I hardly knew what had been done ere the strong-smelling subject was directly under my nose. The natives then were anxious that I should hurry through with my sketches, measurements, etc., so that they might the sooner push off their egg-laden bidarrah and cross back to the main island before the fogs would settle over our homeward track, or the rapidly rising wind shift to the northward and imperil our passage. Weighty reasons these, which so fully impressed me, that this unique stomach of a *carnivora* was overlooked and left behind ; hence, with the exception of curiously turning over the clams (especially those uncrushed specimens), which formed the great bulk of its contents, I have no memoranda or even distinct recollection of the other materials that were incorporated.

forth her young, a single calf, in June usually, on the ice-floes in
the Arctic Ocean, above Bering Straits, between Point Barrow and
Cape Scartze Kammin ; that this calf resembles the parent in general proportions and color when it is hardly over six weeks old, but
that the tusks (which give it its most distinguishing expression) are
not visible until the second year of its life ; that the walrus mother
is strongly attached to her offspring, and nurses it later through
the season in the sea ; that the walrus sleeps profoundly in the
water, floating almost vertically, with barely more than the nostrils
above water, and can be easily approached, if care is taken as to the
wind, so as to spear it or thrust a lance into its bowels ; that the
bulls do not fight as savagely as the fur-seal or the sea-lion ; that
the blunted tusks of these combatants seldom do more than bruise
their thick hides ; that they can remain under water nearly an hour,
or about twice as long as the seals, and that they sink like so many
stones immediately after being shot at sea.

I personally made no experiments touching the peculiarity of
sinking immediately after being shot. Of course, on reflection, it
will appear to any mind that all seals, no matter how fat or how lean,
would sink instantly out of sight, if not killed, at the shock of a
bullet ; even if mortally wounded, the great involuntary impulse
of brain and muscle would be to dive and speed away, for all swimming is submarine when pinnipeds desire to travel.

Touching this mooted question, I had an opportunity when in
Port Townsend, during 1874, to ask a man who had served as a
partner in a fur-sealing schooner off the Straits of Fuca. He told
me that unless a seal was instantly killed by the passage of his
rifle-bullet through its brain, it was never secured, and would sink
before they could reach the bubbling wake of its disappearance.
If, however, the aim of a marksman had been correct, then its body
was invariably taken within five to ten minutes after the rifle discharge. Only one man does the shooting ; the rest of such a
crew, ten to twelve white men and Indians, man canoes and boats
which are promptly despatched from the schooner, after each report, in the direction of a victim. How long one of the bodies
of these "clean" killed seals would float he did not know ; the
practice always was to get it as quickly as possible, fearing that
the bearings of its position, when shot from a schooner, might be
confused or lost. He also affirmed that, in his opinion, there were
not a dozen men on the whole northwest coast who were good

enough with a rifle, and expert at distance calculation, to shoot fur-seals successfully from the deck of a vessel on the ocean. The Indians of Cape Flattery do most of their pelagic fur-sealing by cautiously approaching from the leeward when these animals are asleep, and then throw line-darts or harpoons into them before they awaken.

The finest bidarrah skin-boats of transportation that I have seen in this country were those of the St. Lawrence natives. These were made out of dressed walrus-hides, shaved and pared down by them to the requisite thickness, so that when they were sewed with sinews to the wooden whalebone-lashed frames of such boats they dried into a pale greenish-white prior to oiling, and were even then almost translucent, tough and strong.

When I stepped, for the first time, into the baidar of St. Paul Island, and went ashore, from the *Alexander*, over a heavy sea, safely to the lower bight of Lukannon Bay, my sensations were of emphatic distrust ; the partially water-softened skin-covering would puff up between the wooden ribs, and then draw back, as the waves rose and fell, so much like an unstable support above the cold, green water below, that I frankly expressed my surprise at such an outlandish craft. My thoughts quickly turned to a higher appreciation of those hardy navigators who used these vessels in circumpolar seas years ago, and of the Russians who, more recently, employed bidarrahs chiefly to explore Alaskan and Kamchatkan *terræ incognita*. There is an old poem in Avitus, written by a Roman as early as 445 A.D. ; it describes the ravages of Saxon pirates along the southern coasts of Britain, who used just such vessels as this bidarrah of St. Paul.

"Quin et armoricus piratim Saxona tractus
Spirabat, cui pelle falum fulcare Britannum
Ludus, et assuto glaucum mare findere lembo."

These boats were probably covered with either horse's or bulls' hides. When used in England they were known as *coracles ;* in Ireland they were styled *curachs*. Pliny tells us that Cæsar moved his army in Britain over lakes and rivers in such boats. Even the Greeks used them, terming them *karabia ;* and the Russian word of *korabl'*, or "ship," is derived from it. King Alfred, in 870-872, tells us that the Finns made sad havoc among many Swedish settlements on the numerous "meres" (lakes) in the moors of that country, by "carrying their ships (baidars) overland to the meres whence they make depredations on the Northmen ; their ships are small and very light."

Until I saw these bidarrahs of the St. Lawrence natives, in 1874, I was more or less inclined to believe that the tough, thick, and spongy hide of a walrus would be too refractory in dressing for use in covering such light frames, especially those of the bidarka ; but the manifest excellence and seaworthiness of those Eskimo boats satisfied me that I was mistaken. I saw, however, abundant evidence of a much greater labor required to tan or pare down this thick cuticle to that thin, dense transparency so marked on their bidarrahs ; for the pelt of a hair-seal, or sea-lion, does not need any more attention, when applied to this service, than that of simply unhairing it. This is done by first sweating the "loughtak" in piles, then rudely, but rapidly, scraping with blunt knives or stone flensers the hair off in large patches at every stroke ; the skin is then air-dried, being stretched on a stout frame, where, in the lapse of a few weeks, it becomes as rigid as a board. Whenever wanted for use thereafter, it is soaked in water until soft or "green" again ; then it is sewed with sinews, while in this fresh condition, tightly over the slight wooden skeleton of the bidarka or the heavier frame of a bidarrah. In this manner all boats and lighters at the islands are covered. Then they are air-dried thoroughly before oiling, which is done when the skin has become well indurated, so as to bind the ribs and keel as with an iron plating. The thick, unrefined seal-oil keeps the water out for twelve to twenty hours, according to the character of the hides. When, however, the skin-covering begins to "bag in" between the ribs of its frame, then it is necessary to haul the bidarrah out and air-dry it again, and then re-oil. If attended to thoroughly and constantly, those skin-covered boats are the best species of lighter which can be used in these waters, for they will stand more thumping and pounding on the rocks and alongside ship than all wooden, or even corrugated-iron, lighters could endure and remain seaworthy.

The flesh of the walrus is not, to our palate, at all toothsome ; it is positively uninviting. That flavor of the raw, rank mollusca, upon which it feeds, seems to permeate every fibre of its flesh, making it very offensive to the civilized palate ; but the Eskimo, who do not have any of our squeamishness, regard it as highly and feed upon it as steadily, as we do on our own best corn-fed beef. Indeed, the walrus to an Eskimo answers just as the cocoa-palm does to a South Sea islander : it feeds him, it clothes him, it heats and illuminates his " igloo, " and it arms him for the chase, while he builds a summer shelter and rides upon the sea by virtue of its hide. The morse, however, is not of much account to the seal-hunters on the Pribylov Islands. They still find, by stirring up the sand-

Newack's Brother, with a Sealskin full of Walrus-oil.
[*Mahlemoöt boy—fourteen or fifteen years of age.*]

dunes and digging about them at Northeast Point, all the ivory that they require for their domestic use on the islands, nothing else belonging to a walrus being of the slightest economic value to them. Some authorities have spoken well of walrus-meat as an article of diet. Either they had that sauce for it born of inordinate hunger, or else the cooks deceived them. Starving explorers in the arctic regions could relish it—*they* would thankfully and gladly eat anything that was juicy, and sustained life, with zest and gastronomic fervor. The Eskimo naturally like it; it is a necessity to their existence, and thus a relish for it is acquired. I can readily understand, by personal experience, how a great many, perhaps a majority of our own people, could speak well, were they north, of seal-meat, of whale " rind," and of polar-bear steaks ; but I know that a mouthful of fresh or " cured " walrus-flesh would make their "gorges rise." The St. Paul natives refuse to touch it as an article of diet in any shape or manner. I saw them removing the enormous testicles of an old morse which was shot, for my purposes, on Walrus Island. They told me they did so in obedience to the wishes of a widow doctress at the village, Maria Seedova, who desired a pair for her incantations.

Curiosity, mingled with a desire to really understand, alone tempted me to taste some walrus-meat which was placed before me at Poonook, on St. Lawrence Island ; and candor compels me to say that it was worse than the old beaver's tail which I had been victimized with in British Columbia, worse than the tough brown-bear steak of Bristol Bay—in fact, it is the worst of all fresh flesh of which I know. It had a strong flavor of an indefinite acrid nature, which turned my palate and my stomach instantaneously and simultaneously, while the surprised natives stared in bewildered silence at their astonished and disgusted guest. They, however, greedily put chunks, two inches square and even larger, of this flesh and blubber into their mouths as rapidly as the storage room there would permit ; and with what grimy gusto ! as the corners of their large lips dripped with the fatness of their feeding. How little they thought, then, that in a few short seasons they would die of starvation, sitting in these same igloos—their caches empty and · nothing but endless fields of barren ice where a life-giving sea should be. The winter of 1879-80 was one of exceptional rigor in the Arctic, although in the United States it was unusually mild and open. The ice closed in solid around St. Lawrence Island—so firm and unshaken by the giant leverage of wind and tide that all walrus were driven far to the southward and eastward beyond the reach of those unhappy inhabitants of that island, who, thus unexpectedly deprived of their mainstay and support, seemed to have miserably starved to death then, with an exception of one small village on the north shore : thus, the residents of Poonook, Poogovellyak, and Kagallegak settlements perished, to a soul, from hunger ; nearly three hundred men, women, and children. I recall that visit which I made to these alert Innuits, August, 1874, with sadness, in this unfortunate connection, because they impressed me with their manifest superiority over the savages of our northwest coast. They seemed, then, to be living, during nine months of the year, almost

"Newack" and "Oogack."

[*St. Lawrence Mahlemoöts: pen portraits made at Poogovellyak, August, 1874.*]

wholly upon the flesh and oil of the morse. Clean-limbed, bright-eyed, and jovial, they profoundly impressed me with their happy reliance and subsistence upon the walrus-herds of Bering Sea. I could not help remarking then, that these people had never been subjected to the temptations and subsequent sorrow of putting their trust in princes ; hence their independence and good heart. But now it appears that it will not do to put your trust in *Rosmarus* either.

I know that it is said by Parry, by Hall, and lately by others, that the flesh of the Atlantic walrus is palatable ; perhaps the nature of its food-supply is the cause. We all recognize a wide difference in pork from hogs fed on corn and those fed on beech- mast and oak-acorns, and those which have lived upon the offal of the slaughtering houses, or have gathered the decayed castings of the sea-shore; the sea-horse of Bering Sea lives upon that which does not give a pleasant flavor to its flesh.

The range of our Alaskan walrus now appears to be restricted in the Arctic Ocean to an extreme westward at Cape Chelagskoi, on the Siberian coast, and an extreme eastward between Point Barrow and the region of Point Beechey, on the Alaskan shore. It is, however, substantially confined between Koliutchin Bay, Siberia, and Point Barrow, Alaska. As far as its distribution in polar waters is concerned, and how far to the north it travels from these coasts of two continents, I am unable to present any well-authenticated data illustrative of the subject ; the shores of Wrangel Island were found in possession of walrus-herds during the season of 1881.

This walrus has, however, a very wide range of distribution in Alaska, though not near so great as in prehistoric times. They abound to the eastward and southeastward of St. Paul, over in Bristol Bay, where great numbers congregate on the sand-bars and flats, now flooded, now bared by the rising and ebbing of the tide ; they are hunted here to a considerable extent for their ivory. No morse are found south of the Aleutian Islands ; still, not more than forty-five or fifty years ago, small gatherings of these animals were killed here and there on some islands between Kadiak and Oonimak Pass ; the greatest aggregate of them, south of Bering Straits, will always be found in the estuaries of Bristol Bay and on the north side of the peninsula of Alaska.

I have been frequently questioned whether, in my opinion, more than a short space of time would elapse ere the walrus was exterminated, or not, since our whalers had begun to hunt them in Bering Sea and the Arctic Ocean. To this I frankly make answer that I do not know enough of the subject to give a correct judgment. The walrus spend most of their time in waters that are within reach of these skilful and hardy navigators ; and if they (the walrus) are of

sufficient value to a whaler, he can and undoubtedly will make a
business of killing them, and work the same sad result that he has
brought about with the mighty schools of cetacea which once
whistled and bared their backs, throughout the now deserted waters
of Bering Sea, in perfect peace and seclusion prior to 1842. The
returns of the old Russian America Company show that an annual
average of ten thousand walrus have been slain by the Eskimo since
1799 to 1867. There are a great many left yet ; but, unless the oil
of *Rosmarus* becomes very precious commercially, I think the shoal
waters of Bristol Bay and Kuskokvim mouth, together with the ec-

The Death-stroke.
[*Mahlemoöts Morse-hunting in the summer.*]

centric tides thereof, will preserve the species indefinitely. Forty
years ago, when the North Pacific was a rendezvous of the greatest
whaling-fleet that ever floated, those vessels could not, nor can they
now, approach nearer than sixty or even eighty miles of many'
muddy shoals, sands, and bars upon which the walrus rest in Bristol Bay, scattered in herds of a dozen or so to bodies of thousands,
living in lethargic peace and almost unmolested, except in several
small districts which are carefully hunted over by the natives of
Oogashik for oil and ivory. I have been credibly informed that
they also breed in Bristol Bay, and along its coast as far north as
Cape Avinova, during seasons of exceptional rigor in the Arctic.

The Innuits of St. Lawrence, and all of their race living above them, hunt the walrus without any excitement other than that of securing such quarry. They never speak of real danger. When they do not shoot them as these beasts drift in sleepy herds on ice-floes, then they surprise them on the beaches or reefs and destroy a herd by spearing and lancing. When harpooned or speared, a head of the weapon is so made as to detach itself from its shank, and by thus sticking in the carcass a line of walrus-hide is made fast to the plethoric body of *Rosmarus*. When this brute has expended its surplus vitality by towing the natives a few miles in a mad, frenzied burst of swimming, their bidarrah is quietly drawn up to its puffing form close enough to permit of a *coup* by an ivory-headed lance ; it is then towed to a beach at high water. When the ebb is well out, the huge carcass is skinned by its dusky butchers, who cut it up into large square chunks of flesh and blubber, which are deposited in queer little "Dutch-oven" caches of each family that are made especially for its reception.

Dressing walrus-hides is the only serious hard labor which the Alaskan Innuit subjects himself to. He cannot lay it entirely upon the women, as the Sioux do when they spread buffalo bodies all over the plains. It is too much for female strength alone, and so the men bear a hand right lustily in this business. It takes from four to six stout natives, when a green walrus-hide is removed, to carry it to a sweating-hole, where it is speedily unhaired. Then, stretched alternately upon air-frames and pinned over the earth, it is gradually scraped down to a requisite thinness for use in covering the bidarrah skeletons, etc.

There are probably six or seven thousand human beings in Alaska who live largely by virtue of the existence of *Rosmarus*, and every year, when the season opens, they gather together by settlements, as they are contiguous, and discuss the walrus chances for a coming year as earnestly and as wisely as our farmers who confer over their prospects for corn and potatoes. But an Eskimo hunter is a sadly improvident mortal, though he is not wasteful of morse life, while we are provident, and yet wasteful of our resources.

If the North Pole is ever reached by our people, they will do so only when they can eat walrus-meat and get plenty of it—at least that is my belief—and, knowing now what the diet is, I think the journey to that hyperborean *ultima* is a long one, though there is plenty of meat and many men who want to try it.

· Unless we spend a winter in the Arctic Ocean above Bering
Straits we will not be able to see a polar bear ; but there is one place,
and one place only, in Alaska where in midsummer we can land,
and there behold on its swelling, green, and flowery uplands hundreds of these huge ursine brutes. That place is the island of St.
Matthew, and it is right in our path as we leave St. Lawrence and
head for Oonimak Pass and home.

St. Matthew Island is an odd, jagged, straggling reach of bluffs
and headlands, connected by bars and lowland spits. The former,
seen at a little distance out at sea, resemble half a dozen distinct isl-

Mahlemoöts Landing a Walrus.
[*An Innuit " double purchase." St. Lawrence Island.*]

ands. The extreme length is twenty-two miles, and it is exceedingly
narrow in proportion. Hall Island is a small one that lies west from
it, separated from it by a strait (Sarichev) less than three miles in
width, while the only other outlying land is a sharp, jagged pinnacle-rock, rearing itself over a thousand feet abruptly from the sea,
standing five miles south of Sugar-loaf Cone on the main island.
From a cleft and blackened fissure, near the summit of this serrated pinnacle-rock, volcanic fire and puffs of black smoke have been
recorded as issuing when first discovered, and they have issued ever
since.

Our first landing, early in the morning of August 5th, was at the
spot under Cub Hill, near Cape Upright, the easternmost point of

the island. The air came out from the northwest cold and chilly, and snow and ice were on the hill-sides and in the gullies. The sloping sides and summits of these hills were of a grayish, russet tinge, with deep-green swale flats running down into the lowlands, which are there more intensely green and warmer in tone. A pebble-bar formed by the sea between Cape Upright and Waterfall Head is covered with a deep stratum of glacial drift, carried down from the flanks of Polar and Cub Hills, and extending over two miles of this water-front to the westward, where it is met by a similar washing from that quarter. Back, and in the centre of this neck, are several small lakes and lagoons without fish; but emptying into them are a number of clear, lively brooks, in which were salmon-parr of fine quality. The little lakes undoubtedly receive them; hence they were land-locked salmon. A luxuriant growth of thick moss and grass, interspersed, existed almost everywhere on the lowest ground; and occasionally strange dome-like piles of peat were lifted four or five feet above marshy swales, and appeared so remarkably like abandoned barraboras that we repeatedly turned from our course to satisfy ourselves personally to the contrary.

As these lowlands ascend to the tops of higher hills, all vegetation changes rapidly to a simple coat of cryptogamic gray and light russet, with a slippery slide for the foot wherever a steep flight or climbing was made. Water oozes and trickles everywhere under foot, since an exhalation of frost is in progress all the time. Sometimes these swales rise and cross hill-summits to the valleys again without any interruption in their wet, swampy character. The action of ice in rounding down and grinding hills, chipping bluffs, and chiselling everywhere, carrying the soil and *débris* into depressions and valleys, is most beautifully exhibited on St. Matthew. The hills at the foot of Sugar-loaf Cone are bare and literally polished by ice-sheets and slides of melting snow. Rocks and soil from these summits and slopes are carried down and "dumped," as it were, in numberless little heaps beneath, so that the foot of every hill and out on the plain around strongly put us in mind of those refuse-piles which are dropped over the commons or dumping-grounds of a city. Nowhere can the work of ice be seen to finer advantage than here, aided and abetted, as it undoubtedly is, by the power of wind, especially with regard to that chiselling action of frost on the faces of ringing metallic porphyry cliffs.

The flora here is as extensive as on the Seal Islands, two hundred miles to the southward; but the species of *gramma* are not near so varied. Indeed, there is very little grass around about. Wherever there is soil it seems to be converted by the abundant moisture into a swale or swamp, over which we travelled as on a quaking water-bed; but on the rounded hill-tops and ridge-summits wind-driven and frost-splintered shingle makes good walking. Both of these climatic agencies evidently have a permanent iron grip on this island.

The west end of St. Matthew differs materially from the east. A fantastic weathering of the rocks at Cathedral Point, Hall Island, will strike the eye of a most casual observer as his ship enters the straits going south. This eastern wall of that point looms up from the water like a row of immense cedar-tree trunks. The scaling off of basaltic porphyry and a growth of yellowish-green and red mossy lichens made the effect most real, while a vast bank of fog lying just overhead seemed to shut out from our vision the foliage and branches that should be above. This north cape of Hall Island changes when approached, with every mile's distance, to a new and altogether different profile.

Our visit at the west end of the island of St. Matthew was, geologically speaking, the most interesting experience I have ever had in Alaska. A geologist who may desire to study the greatest variety of igneous forms *in situ*, within a short and easy radius, can do no better than make his survey here. These rocks are not only varied by mineral colors, together with a fantastic arrangement of basalt and porphyry, but are rich and elegant in their tinting by the profuse growth of lichens—brown, yellow, green, and bronze.

An old Russian record prepared us, in landing, to find bears here, but it did not cause us to be equal to the sight we saw, for we met bears—yea, hundreds of them. I was going to say that I saw bears here as I had seen seals to the south, but that, of course, will not do, unless as a mere figure of speech. During the nine days that we were busy in surveying this island, we never were one moment, while on land, out of sight of a bear or bears; their white forms in the distance always answered to our search, though they ran from our immediate presence with a wild celerity, travelling in a swift, shambling gallop, or trotting off like elephants. Whether due to the fact that they were gorged with food, or that the warmer weather of summer subdued their temper, we never could coax one of these animals to show fight. Its first impulse and its last one, while within our influence, was flight—males, females, and cubs—all, when surprised by us, rushing with one accord right, left, and in every direction, over the hills and far away.

After shooting half a dozen, we destroyed no more, for we speedily found that we had made their acquaintance at the height of their shedding-season, and their snowy and highly prized winter-dress was a very different article. from the dingy, saffron-colored, grayish fur that was flying like downy feathers in the wind, when ever rubbed or pulled by our hands. They never growled, or uttered any sound whatever, even when shot or wounded.

Here, on the highest points, where no moss ever grows, and nothing but a fine porphyritic shingle slides and rattles beneath our tread, are bear-roads leading from nest to nest, or stony lairs, which they have scooped out of frost-splintered *débris* on the hill-sides, and where old she-bears undoubtedly bring forth their young : but it was not plain, because we saw them only sleeping, at this season of the year, on the lower ground ; they seemed to delight in stretching themselves upon, and rolling over, the rankest vegetation.

They sleep soundly, but fitfully, rolling their heavy arms and legs about as they doze. For naps they seem to prefer little grassy depressions on the sunny hill-sides and along the numerous water-courses, and their paths were broad and well beaten all over the island. We could not have observed less than two hundred and fifty or three hundred of these animals while we were there ; at one landing on Hall Island there were sixteen in full sight at one sweep of our eyes, scampering up and off from the approach of the ship's boat.

Provided with more walrus-meat than he knows what to do with, the polar bear, in my opinion, has never cared much for the Seal Islands ; the natives have seen them, however, on St. Paul, and its old men have their bear stories, which they tell to a rising generation. The last "medvait" killed on St. Paul Island was shot at Bogaslov in 1848 ; none have ever come down since, and very few were there before, but those few evidently originated at and made St. Matthew Island their point of departure. Hence I desire to notice this hitherto unexplored spot, standing, as it does, two hundred miles to the northward of St. Paul, and which, until Lieutenant Maynard and myself, in 1874, surveyed and walked over its entire coast-line, had not been trodden by white men, or by natives, since that dismal record made by a party of five Russians and seven Aleutes who passed the winter of 1810–11 on it, and who were so stricken down with scurvy as to cause the death of all the Russians save one, while the rest barely recovered and left early the following year. We found the ruins of those huts which had been occupied by this unfortunate and discomfited party of fur-hunters ; they were landed there to secure polar bears in the depth of winter, when such shaggy coats should be the finest.

As we complete our review of St. Matthew and its ursine occupation, the circuit of Alaska has been made—its impression we have recorded, and the path from here home again is a bee-line to the Golden Gate over

> " Nothing, nothing but the sea—
> Vast in its immensity ? "

INDEX.

www.ingramcontent.com/pod-product-compliance
Lightning Source LLC
Chambersburg PA
CBHW020854210326
41598CB00018B/1662